Circular Strategies for Waste Management

This book encapsulates the concept of circular economy and its application to waste management. It provides an overview of the various technical approaches that can be used to implement circular strategies for managing waste. The book discusses the benefits of circular waste management, including reduced environmental impact, increased resource efficiency, and economic advantages. It also offers guidance on implementing circular waste management strategies within communities or organizations.

Key features:

- Provides a comprehensive fundamental overview of the circular economy, circular economy strategies, and waste management
- Discusses circular strategies for waste management across different sectors
- Explores the use of Industry 4.0 tools and techniques for waste management
- Reviews the relevance of social elements in the circular economy
- Includes a range of use cases that illustrate circular economy principles in practice for waste management

This book is intended for researchers and graduate students in the fields of waste management, circular economy, and sustainability.

Industrial Engineering, Systems, and Management

Series Editors: Kapil Gupta and Jayakrishna Kandasamy

Industrial engineering, systems, and management is a broad term that includes theories, principles, concepts, tools, and techniques. This series will cover industrial engineering practices from raw material extraction to customer feedback, business processes, and information exchange. It will showcase state-of-the-art technological advancements for enhancement of product quality, process productivity, and sustainability. The book series will include industry 4.0 and sustainability interventions in industrial engineering, systems, and management area. The series is aimed at senior undergraduate students, graduate students, academic researchers, and professionals in diverse areas including mechanical engineering, industrial engineering, human factors, ergonomics, and manufacturing engineering.

Supply Chain Management: Strategic Implementation in Manufacturing
Edited by Bharti, Shikha Singh, Anand Pandey and Amit Sachan

Industry 4.0 for Manufacturing Systems: Concepts, Technologies, and Applications
Edited by Vishal Ashok Wankhede, Vimal K.E.K and Pankaj Sahlot

Circular Strategies for Waste Management
Edited by Jayakrishna Kandasamy and Rajyalakshmi G

For more information about this series, please visit: www.routledge.com/Industrial-Engineering-Systems-and-Management/book-series/CRCIESM

Circular Strategies for Waste Management

Edited by
Jayakrishna Kandasamy and Rajyalakshmi G

CRC Press
Taylor & Francis Group
Boca Raton London New York

CRC Press is an imprint of the
Taylor & Francis Group, an **informa** business

Designed cover image: Jayakrishna Kandasamy and Rajyalakshmi G

First edition published 2027
by CRC Press
2385 NW Executive Center Drive, Suite 320, Boca Raton FL 33431

and by CRC Press
4 Park Square, Milton Park, Abingdon, Oxon, OX14 4RN

CRC Press is an imprint of Taylor & Francis Group, LLC

ISBN: 9781032814254 (hbk)
ISBN: 9781041068921 (pbk)
ISBN: 9781003637677 (ebk)

DOI: 10.1201/9781003637677

Typeset in Times
by Newgen Publishing UK

Dedication

To the unsung heroes of waste collection, recycling, and disposal, whose tireless efforts sustain our communities and protect our planet for future generations.

Thank you.

Contents

Foreword

In an age defined by escalating environmental concerns, resource scarcity, and mounting waste, the shift from a linear "take-make-dispose" model to a circular, regenerative system is not just an option—it is an imperative. *Circular Strategies for Waste Management* emerges as a timely and essential contribution to this global transition, offering a comprehensive and interdisciplinary roadmap to rethink how waste is managed, minimized, and transformed into value.

This book stands out for its breadth and depth, covering a wide array of topics that extend beyond conventional waste management. It presents a holistic framework encompassing sustainable development principles, circular economy models, and sector-specific strategies for industries such as agriculture, healthcare, manufacturing, and construction. Each chapter is thoughtfully structured to include theoretical foundations, practical approaches, and illustrative case studies that bridge academic rigor with real-world relevance.

What makes this work particularly compelling is its integration of global perspectives with local solutions. It explores the role of Sustainable Development Goals (SDGs), government policies, and community-driven initiatives, while also addressing region-specific challenges and opportunities. Furthermore, the inclusion of advanced technologies—such as Industry 4.0 tools, machine learning for waste segregation, and micro-recycling techniques—highlights the role of innovation in achieving circularity.

Social dimensions, often overlooked in technical literature, are also given due emphasis, recognizing the critical role of human behavior, collaborative consumption, and social equity in building circular systems. The book not only informs policymakers and researchers but also empowers educators, students, and practitioners with actionable insights and frameworks.

Above all, this volume serves as a tribute to the countless individuals whose daily work in waste collection, recycling, and recovery sustains our communities. As the world moves toward a circular future, this book offers the knowledge, strategies, and inspiration to lead that change.

Otávio José de Oliveira
Professor, Production Department
Faculty of Engineering and Sciences
Sao Paulo State University, UNESP, Brazil

Preface

The world is facing a growing waste crisis. Every year, we produce billions of tons of waste, much of which ends up in landfills or incinerators. This waste has a significant environmental impact, contributing to climate change, air pollution, and water pollution. The circular economy offers a more sustainable alternative to the traditional linear economy, which is based on the principle of "take, make, waste." In a circular economy, waste is seen as a resource, and products are designed to be reused, recycled, or composted. This book provides a comprehensive overview of the circular economy and its applications in waste management. The book covers a wide range of topics, from the principles of circular economy to specific case studies of circular waste management practices. The book discusses the principles and practices of circular economy in the context of waste management. It covers a wide range of topics, including sustainable development, circular bioeconomy, circular strategies for waste management in manufacturing, agro-waste, hazardous solid waste, biomedical waste, tannery waste, grey water, e-waste, plastic recycling, building waste, social factors in circular economy, and the use of I4.0 tools and techniques in upcycling waste. This book is intended for a wide audience, including students, researchers, policymakers, and practitioners. This book is further intended for businesses and individuals who are interested in learning more about the circular economy and its potential to reduce waste and protect the environment.

About the Editors

Jayakrishna Kandasamy is a professor in the School of Mechanical Engineering at Vellore Institute of Technology (VIT), Vellore, India. His research focuses on the design and management of manufacturing systems and supply chains to improve efficiency, productivity, and sustainability. He works extensively on sustainable manufacturing, circular economy strategies, industrial symbiosis, and hybrid biocomposites. He has mentored seven PhD, three MTech, and 38 BTech theses, resulting in 114 journal publications, 48 book chapters, 94 conference papers, four authored books, and 12 edited volumes. Dr Jayakrishna teaches undergraduate and graduate courses in manufacturing and industrial systems, and his innovative teaching practices have earned national recognition. He is ranked among the Top 2% most influential researchers globally by Stanford University (2023 and 2024) and has received multiple awards, including the Global Engineering Education Award (2021), Outstanding Professor in Supply Chain Management (2024), and the Institution of Engineers (India) Young Engineer Award (2019). He serves as Coordinator of the Circular Economy Club at VIT and holds editorial roles in several journals, including *Operations Management Research*, *Circular Economy – Frontiers in Sustainability*, as well as CRC Press's book series *Industrial Engineering, Systems, and Management.*

Rajyalakshmi G is a professor in the Department of Manufacturing Engineering at Vellore Institute of Technology (VIT), Vellore, India. She received her PhD degree in Production Engineering in 2015 and her MTech degree in Industrial Engineering in 2003 from Sri Venkateswara University, Tirupati, India. Her research focuses on sustainable manufacturing, waste management, and optimisation, with more than 19 years of teaching and research experience in Mechanical Engineering. She has mentored three PhD, 15 MTech, and 50 BTech theses. She has published two books with international publishers (as author and co-author), 15 book chapters, and contributed over 60 articles to journals and conferences. Dr Rajyalakshmi serves as a reviewer for more than ten prestigious Web of Science journals and contributes actively to advancing sustainable manufacturing practices through her research and academic leadership.

Contributors

Deepa Aakepati
School of Mechanical Engineering
Vellore Institute of Technology
Vellore, India

Sahik Naseem Ahmed
School of Mechanical Engineering
Vellore Institute of Technology
Vellore, India

Kamal Aryan
School of Mechanical Engineering
Vellore Institute of Technology
Vellore, India

Suresh Babu C
School of Mechanical Engineering
Vellore Institute of Technology
Vellore, India

Jishnu Priyan G
School of Mechanical Engineering
Vellore Institute of Technology
Vellore, India

Rajyalakshmi G
School of Mechanical Engineering
Vellore Institute of Technology
Vellore, India

Akhil Gokulan
School of Mechanical Engineering
Vellore Institute of Technology
Vellore, India

Jonathan Jacob
School of Mechanical Engineering
Vellore Institute of Technology
Vellore, India

Jayakarthikeyan
School of Mechanical Engineering
Vellore Institute of Technology
Vellore, India

Jayakrishna Kandasamy
School of Mechanical Engineering
Vellore Institute of Technology
Vellore, India

Koushik Manchala
School of Mechanical Engineering
Vellore Institute of Technology
Vellore, India

Raajasekar Mathiyalagan
School of Mechanical Engineering
Vellore Institute of Technology
Vellore, India

Mayank Mishra
School of Mechanical Engineering
Vellore Institute of Technology
Vellore, India

Neelesh Mishra
School of Mechanical Engineering
Vellore Institute of Technology
Vellore, India

Kartik Mittal
School of Mechanical Engineering
Vellore Institute of Technology
Vellore, India

Rahul Prasad Nandhakumar
School of Mechanical Engineering
Vellore Institute of Technology
Vellore, India

Sri Charan P
School of Mechanical Engineering
Vellore Institute of Technology
Vellore, India

Pratyush Pandey
School of Mechanical Engineering
Vellore Institute of Technology
Vellore, India

Abhishek Patel
School of Mechanical Engineering
Vellore Institute of Technology
Vellore, India

J C Nikhil Rao
School of Mechanical Engineering
Vellore Institute of Technology
Vellore, India

Sharvesh Ramesh
School of Mechanical Engineering
Vellore Institute of Technology
Vellore, India

Mahajan Gargi Ravindra
School of Mechanical Engineering
Vellore Institute of Technology
Vellore, India

Surriya S
School of Mechanical Engineering
Vellore Institute of Technology
Vellore, India

Joyston Jacob S
School of Mechanical Engineering
Vellore Institute of Technology
Vellore, India

Burada Durga Sai
School of Mechanical Engineering
Vellore Institute of Technology
Vellore, India

Aravind Raj Sakthivel
School of Mechanical Engineering
Vellore Institute of Technology
Vellore, India

Shiven Shanmuganathan
School of Mechanical Engineering
Vellore Institute of Technology
Vellore, India

Adhith Krishna Erath Shine
School of Mechanical Engineering
Vellore Institute of Technology
Vellore, India

Nishkarsh Singhal
School of Mechanical Engineering
Vellore Institute of Technology
Vellore, India

Abhyudaya Sinha
School of Mechanical Engineering
Vellore Institute of Technology
Vellore, India

Kota Suraj
School of Mechanical Engineering
Vellore Institute of Technology
Vellore, India

Aniket Yadav
School of Mechanical Engineering
Vellore Institute of Technology
Vellore, India

Acknowledgements

The editors would like to express their heartfelt appreciation to all those who contributed to the successful completion of this book. This work is the result of the collective efforts of many individuals, and we are deeply grateful for their dedication and support.

We extend our sincere thanks to all the contributing authors for their insightful chapters, commitment to excellence, and timely submissions. Their expertise and scholarly contributions have been instrumental in shaping the depth and diversity of this volume.

We are equally grateful to the reviewers for their valuable time and constructive feedback. Their meticulous evaluations and suggestions greatly enhanced the quality, coherence, and clarity of the chapters, ensuring that the content meets high academic and professional standards.

Special thanks are also due to our management for providing the academic environment and resources that made this work possible.

With sincere gratitude,

Jayakrishna Kandasamy
Professor, School of Mechanical Engineering
Vellore Institute of Technology
Vellore, Tamil Nadu, India

Rajyalakshmi G
Professor, School of Mechanical Engineering
Vellore Institute of Technology
Vellore, Tamil Nadu, India

1 Sustainable Development

Abhishek Patel, Aniket Yadav, Kartik Mittal, and Mayank Mishra

1.1 INTRODUCTION TO SUSTAINABILITY

Sustainability is no longer just a concept—it is a necessity. With growing global challenges such as climate change, resource depletion, pollution, and social inequalities, the need for sustainable solutions has never been more urgent. Sustainability is about finding a balance between economic growth, environmental responsibility, and social well-being, ensuring that development today does not come at the cost of future generations.

The idea of sustainability gained momentum with the 1987 Brundtland Commission Report, *Our Common Future* (Kuhlman and Farrington, 2010) which defined sustainable development as development that "meets the needs of the present without compromising the ability of future generations to meet their own needs." Since then, the concept has expanded, influencing global policies, shaping industrial practices, and transforming everyday decision-making processes. Today, sustainability is not just about conservation but about innovation—developing new technologies, systems, and business models that align with the principles of long-term environmental and social stability.

One of the key approaches to achieving sustainability is through circular economy principles, which focus on reducing waste, improving resource efficiency, and designing systems that allow materials to be continuously reused and repurposed. Unlike the traditional linear economy—where resources are extracted, used, and discarded—a circular economy aims to minimize waste and environmental impact while maximizing economic value. The integration of sustainability engineering into industries plays a crucial role in achieving this balance, enabling innovations in renewable energy, green manufacturing, sustainable infrastructure, and waste management (Ajirotutu, Adeyemi, Ifechukwu, et al., 2024).

As the world transitions toward a more sustainable future, engineering and technology must evolve to support this shift. Emerging trends in sustainability engineering highlight the role of advanced materials, energy-efficient systems, and innovative design strategies in creating sustainable solutions. From renewable energy development to eco-friendly product design, modern engineering approaches are transforming industries and setting the foundation for a greener and more resilient world.

DOI: 10.1201/9781003637677-1

1.1.1 Definition and Conceptual Framework

Sustainable development is often defined as an integrative process that seeks to balance economic progress, social inclusion, and environmental stewardship. This multidimensional framework challenges traditional models that have historically prioritized short-term economic gains, often at the expense of ecological health and social welfare. Instead, sustainability promotes a long-term vision in which economic activities are harmonized with the carrying capacity of natural systems and the equitable distribution of resources among all members of society.

The conceptual framework underlying sustainability recognizes that economic, social, and environmental systems are deeply interdependent. For instance, rapid industrialization and economic expansion can lead to the depletion of natural resources, increased waste and pollution, and the degradation of ecosystems—effects that, in turn, can undermine human health and social stability. Conversely, a focus on environmental protection without consideration of economic realities can lead to policies that stifle innovation and limit opportunities for growth. Thus, sustainable development is not about sacrificing one dimension for another; rather, it is about finding synergies among these dimensions that foster resilient, adaptive systems capable of thriving in the long run (David, Aigbavboa, Nula, et al., 2024).

This holistic view compels us to reconsider the metrics by which we judge progress. Traditional economic indicators, such as gross domestic product (GDP), no longer capture the full spectrum of human well-being or the health of our environment. Instead, sustainability calls for new indicators that integrate social and environmental dimensions—such as measures of ecological footprint, social equity, and quality of life—into our assessments of progress. In this way, the conceptual framework of sustainability serves not only as a guide for policy but also as a lens through which we can evaluate the true cost and benefit of our development choices.

1.1.2 Historical Context

The term "sustainable development" gained prominence during a period of intense environmental and social change. In the latter half of the 20th century, rapid industrialization and the ensuing environmental degradation prompted widespread concern about the long-term impacts of human activity on the planet. The 1987 report *Our Common Future* crystallized these concerns into a single, powerful definition that has since become a cornerstone of environmental policy and academic discourse.

Following the publication of this seminal report, the concept quickly gained traction in international policy circles. The 1992 United Nations Conference on Environment and Development (UNCED), held in Rio de Janeiro, saw the adoption of Agenda 21—a comprehensive action plan aimed at integrating environmental and development policies at the local, national, and global levels. Agenda 21 was instrumental in operationalizing the ideas of sustainable development by providing practical guidelines and strategies for governments and communities to follow. It laid the groundwork for later initiatives and underscored the importance of collaboration among nations, institutions, and local stakeholders in addressing global environmental challenges (Schaltegger, Beckmann, Hockerts et al., 2018).

The historical evolution of sustainable development reflects a growing recognition that the problems we face—be they environmental, social, or economic—cannot be solved in isolation. The emergence of sustainability as a guiding principle marked a significant shift in thinking, moving away from siloed approaches toward more integrated, cross-sectoral strategies. This historical context not only underscores the urgency of adopting sustainable practices but also sets the stage for understanding the principles and policies that continue to shape sustainable development today.

1.1.3 Key Principles of Sustainable Development

At the heart of sustainable development lie several key principles that provide a foundation for both theory and practice. One of the foremost principles is that of *normativity*, which asserts that sustainability is inherently value-laden. It reflects the collective aspirations and ethical commitments of society, compelling us to ask what kind of world we wish to leave for future generations. This normative dimension is essential because it places sustainability beyond mere technical or economic considerations, embedding it within the realm of moral responsibility.

Another central principle is **equity**. Sustainable development demands fairness in both the distribution of resources and the opportunities available to individuals and communities. This principle of equity encompasses not only the fair allocation of material resources but also procedural justice—ensuring that all stakeholders have a voice in the decisions that affect their lives. By addressing both distributive and procedural aspects of justice, sustainability aims to foster social cohesion and reduce disparities that can lead to social unrest and environmental degradation (Eizenberg and Jabareen 2017).

The **integration** principle is equally critical. It underscores the need for a coordinated approach in which environmental, economic, and social policies are designed to complement rather than conflict with one another. Effective integration means that actions taken in one sphere (for example, environmental regulation) are aligned with broader economic and social goals. This integrated approach is fundamental to achieving outcomes that are resilient and mutually reinforcing—a key consideration as nations work to meet the challenges of climate change, resource scarcity, and social inequality.

A further principle is **dynamism**. Recognizing that both natural ecosystems and human societies are in a state of constant flux, sustainability must be viewed as an adaptive process. This dynamism requires flexibility and the capacity to respond to emerging challenges, whether they arise from technological innovations, economic shifts, or environmental crises. The adaptive nature of sustainable development is reflected in the continuous refinement of policies and practices, ensuring that they remain relevant in a rapidly changing world.

Together, these principles form the backbone of sustainable development, offering a framework that is both aspirational and practical. They provide the guiding values for policymakers and practitioners alike, setting the stage for strategies that promote long-term well-being across all facets of society. As the chapter progresses into subsequent sections—where the roles of social factors, key initiatives, and major components of sustainability are explored—the interplay among these principles will become increasingly evident (Lang, Wiek, Bergmann, et al., 2012).

1.1.4 Evolution of the Concept

Over the decades since its formal introduction, the concept of sustainable development has undergone significant evolution. Initially, the focus was predominantly on environmental conservation and resource management, as societies grappled with the visible impacts of industrial pollution, deforestation, and biodiversity loss. Early efforts in sustainability were largely reactive, aiming to mitigate the damage caused by decades of unchecked development.

However, as our understanding of environmental systems deepened, so too did our appreciation for the complex interactions between nature and human society. Scholars and policymakers began to recognize that economic development and environmental protection are not mutually exclusive goals; rather, they can be mutually reinforcing if approached with foresight and integrated planning. This realization paved the way for a more holistic understanding of sustainable development—one that goes beyond mere environmentalism to incorporate social and economic dimensions.

Modern interpretations of sustainable development now encompass a broad range of factors, including cultural, psychological, and even spiritual dimensions. The evolution of the concept reflects a shift from a narrow focus on mitigating harm to a more proactive approach that seeks to enhance human well-being in all its forms. In this expanded view, development is not solely measured by the growth of industrial output or GDP but also by improvements in quality of life, community resilience, and cultural vibrancy. This evolution has been instrumental in shaping international agreements and national policies that aim to foster an environment where economic prosperity, social justice, and ecological health are seen as interdependent objectives (Adanma and Ogunbiyi, 2024).

The dynamic nature of sustainable development means that its definition and scope will likely continue to evolve as new challenges emerge. For instance, the increasing impacts of climate change, the rapid pace of technological innovation, and shifts in global economic power all necessitate continual adaptation of sustainable practices. This evolving character is critical not only for addressing current challenges but also for ensuring that the framework remains relevant for future generations.

1.1.5 Global Agenda

The global agenda for sustainable development reached a new milestone with the adoption of the 2030 Agenda by the United Nations in 2015. This landmark agreement, which is encapsulated in the 17 Sustainable Development Goals (SDGs), represents the most ambitious attempt yet to align global efforts across economic, social, and environmental domains. The SDGs provide a comprehensive blueprint for tackling some of the world's most pressing issues—ranging from poverty and inequality to climate change and environmental degradation (Hassan and Umar, 2024).

What sets the 2030 Agenda apart is its emphasis on universality and integration. Unlike previous frameworks that might have focused on individual sectors, the SDGs are designed to be holistic and indivisible, reflecting the interconnected nature of the challenges they aim to address. Each goal is accompanied by a set of targets and indicators that are intended to guide policy, measure progress, and

FIGURE 1.1 Evolution of sustainability.

promote accountability at all levels of governance. This integrated approach not only reinforces the core principles of sustainable development but also ensures that the global community works in concert toward common objectives.

The emphasis on international cooperation and partnerships is particularly notable. In an era marked by globalization, no country can achieve sustainable development in isolation. The SDGs call for strengthened multilateral cooperation, knowledge sharing, and the mobilization of resources on an unprecedented scale. This global commitment creates a framework within which local, national, and international efforts can be harmonized, setting the stage for future sections that delve into the roles of social factors, key initiatives, and the various dimensions that underpin sustainable development (Adanma and Ogunbiyi, 2024).

1.2 ROLE OF SOCIAL FACTORS IN SUSTAINABLE DEVELOPMENT

The social dimension of sustainable development is as critical as its economic and environmental counterparts. Social factors shape the way communities engage with development policies, influence the distribution of resources, and determine the overall well-being and resilience of societies. In this section, we explore how social equity and justice, community cohesion, participatory governance, health and quality of life, and cultural dimensions together form the backbone of sustainable development efforts. By understanding and integrating these elements, policymakers and

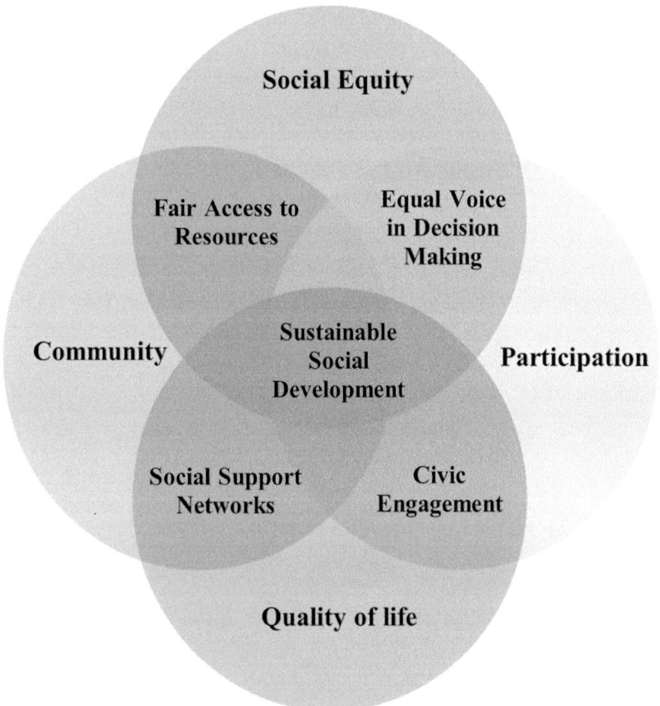

FIGURE 1.2 Social Factors of sustainability.

practitioners can create development strategies that are inclusive, resilient, and truly sustainable (Fitzgibbons and Mitchell, 2021).

1.2.1 Social Equity and Justice

At the heart of sustainable development lies the imperative of social equity and justice. Social equity refers to the fair distribution of resources and opportunities among all members of society. It encompasses both distributive justice—ensuring that economic gains and resources are shared equitably—and procedural justice, which involves inclusive and transparent decision-making processes.

Economic disparities, if left unchecked, can exacerbate social tensions and undermine long-term development. For example, when wealth and resources are concentrated in the hands of a few, large segments of the population may be excluded from the benefits of progress. Such exclusion not only breeds discontent but also hampers the overall potential for growth and innovation. Scholars have long argued that addressing these disparities is essential for building a just society. In this light, sustainable development becomes a vehicle for social transformation, promoting policies that reduce inequality and expand opportunities for marginalized communities (Hariram, 2023).

The normative basis of social equity is rooted in ethical and moral considerations. It compels societies to consider what is fair—not only in terms of material wealth but also in terms of access to education, healthcare, and social services. For instance, initiatives aimed at improving access to quality education or affordable housing contribute significantly to social equity by enabling individuals to participate fully in economic and civic life. The notion of social justice, therefore, extends beyond mere economic transactions to include the rights and dignity of every individual.

Moreover, social equity is closely linked to environmental justice. Communities that are socially and economically marginalized often bear the brunt of environmental degradation. Polluted neighbourhoods, inadequate public infrastructure, and exposure to industrial hazards are not randomly distributed; they are frequently concentrated among disadvantaged populations. By integrating environmental considerations with social policies, sustainable development seeks to redress these imbalances. This integrated approach ensures that policies aimed at protecting the environment do not inadvertently perpetuate social inequities but rather work in tandem to create healthier, more just communities.

Ultimately, the pursuit of social equity and justice within sustainable development requires a commitment to long-term systemic change. It demands that both policymakers and community leaders adopt a holistic perspective—one that recognizes the interconnectedness of social well-being, economic opportunity, and environmental stewardship. Only by addressing these interdependencies can we build a future that is truly inclusive and sustainable (Hosseini, 2023).

1.2.2 Community and Social Cohesion

Strong, cohesive communities are indispensable for the realization of sustainable development. Community cohesion refers to the strength of relationships and the sense of solidarity among members of a community. It is a critical component of social resilience, enabling communities to collectively respond to challenges, share resources, and support one another in times of need.

Social cohesion is fostered by the quality of interpersonal relationships, mutual trust, and shared values. When community members feel a sense of belonging and connection, they are more likely to collaborate on local projects, participate in decision-making processes, and engage in collective problem-solving. This collaborative spirit is essential for the success of sustainable initiatives, whether at the local or national level. For example, community-driven environmental projects—such as local recycling programs, community gardens, or neighbourhood clean-up efforts—often yield better results because they leverage local knowledge and social capital.

The role of community cohesion in sustainable development is also evident in the way it can buffer against external shocks. Communities that are tightly knit tend to be more resilient when facing economic downturns, natural disasters, or social upheavals. The collective strength derived from a well-integrated community helps distribute risks and provides a support network that can mitigate the adverse effects of such events. As a result, fostering social cohesion is not just about improving quality of life; it is a strategic imperative for enhancing the overall resilience of development systems (Chiroli, 2023).

Several factors contribute to strong community cohesion. Access to shared public spaces, effective local governance, and inclusive community programmes all play pivotal roles. Urban design, for instance, can facilitate social interaction by creating parks, community centres, and pedestrian-friendly neighbourhoods. These environments encourage residents to interact and build bonds, thereby enhancing the social fabric of the community. Likewise, local governments that prioritize participatory planning and transparent communication can build trust among citizens, further strengthening community ties.

Education and cultural activities also contribute significantly to social cohesion. Schools, libraries, and community theatres are more than just institutions for learning and entertainment; they are vital platforms for fostering dialogue, understanding, and mutual respect among diverse groups. When individuals from various backgrounds come together to share their experiences and perspectives, they build bridges that can lead to more inclusive and harmonious communities. This social integration is a cornerstone of sustainable development, ensuring that all voices are heard and that development benefits are equitably distributed (Selvakumar, 2025).

In summary, community and social cohesion are central to sustainable development because they create the conditions for collective action and resilience. A society that values strong interpersonal connections and mutual support is better equipped to tackle the multifaceted challenges of modern development.

1.2.3 Participation and Local Democracy

Participation in decision-making processes is a defining feature of sustainable development. Local democracy, which empowers communities to have a direct say in the policies that affect their lives, is fundamental to ensuring that development strategies are responsive, inclusive, and effective.

When citizens are actively involved in governance, they are more likely to support and sustain development initiatives. Participatory approaches to decision-making provide a platform for diverse perspectives, allowing for the identification of local needs and the co-creation of solutions that are context-specific. This bottom-up approach not only enhances the legitimacy of policy measures but also fosters a sense of ownership and responsibility among community members.

One of the critical benefits of participatory governance is the enhancement of accountability. When local communities are involved in setting priorities and monitoring implementation, public officials are more likely to be transparent and responsive. This increased accountability can lead to more effective and efficient use of resources, as policies are closely aligned with the needs and aspirations of the people they serve. Moreover, participatory governance encourages the sharing of knowledge and expertise, which can drive innovation and improve the overall quality of public services (Florini and Pauli, 2018).

The concept of local democracy is particularly important in the context of sustainable development because it recognizes that solutions must be tailored to the unique challenges and opportunities of each community. Decentralized decision-making allows local governments to be more agile and responsive, adapting policies to the specific environmental, social, and economic conditions of their areas. For instance, while one community might prioritize renewable energy projects to address local pollution, another might focus on improving public transportation to reduce urban congestion. Both approaches, though different in focus, contribute to the overarching goal of sustainability by addressing localized issues through participatory means.

However, ensuring effective participation is not without its challenges. Marginalized groups—such as ethnic minorities, women, and the economically disadvantaged—often face barriers to full participation in decision-making processes. These barriers can include limited access to information, social stigmas, or institutional biases that favour certain groups over others. Addressing these challenges requires deliberate efforts to democratize access and create inclusive platforms where all voices can be heard. Policies that support capacity-building, education, and targeted outreach are essential to overcoming these obstacles and ensuring that the benefits of development are shared equitably.

Successful examples of participatory governance can be found around the world. In many regions, local councils and community boards serve as models of how inclusive governance can lead to sustainable outcomes. These bodies often work in tandem with nongovernmental organizations, academic institutions, and community groups to develop and implement policies that are responsive to local conditions. By fostering a culture of active participation, these initiatives help create a dynamic and engaged citizenry that is committed to long-term, sustainable development (Vargas and Cooper, 2024).

In conclusion, local democracy and citizen participation are not only democratic ideals but also practical necessities for sustainable development. When communities are empowered to shape their own futures, they are more likely to develop innovative solutions that address both current challenges and future uncertainties.

1.2.4 Health, Quality of Life, and Well-Being

The quality of life and well-being of individuals are fundamental indicators of sustainable development. A healthy society is not merely one with robust economic growth; it is one in which citizens enjoy physical, mental, and social well-being. This holistic perspective on health goes beyond the absence of disease to include the overall quality of life, influenced by factors such as access to education, housing, clean environments, and recreational opportunities.

Social determinants of health—such as income, education, and social support—play a crucial role in shaping the well-being of individuals and communities. Disparities in these determinants often result in significant differences in health outcomes among different social groups. For example, communities with lower socio-economic status frequently face higher rates of chronic illness, limited access to healthcare services, and greater exposure to environmental hazards. Addressing these disparities is a key aspect of sustainable development, as it promotes social equity and contributes to the overall resilience of communities (Williams, Mohammed, Leavell et al., 2010).

Investments in public health are essential for sustainable development. Governments and local authorities that prioritize health initiatives—such as improving access to healthcare, promoting preventive care, and ensuring clean air and water—can create a foundation for long-term well-being. Moreover, public health initiatives often yield significant economic benefits by reducing healthcare costs and increasing productivity. A healthy population is more capable of contributing to economic growth and social progress, creating a virtuous cycle that reinforces the principles of sustainability.

Mental health and social well-being are equally important components of a sustainable society. The stresses of modern life—exacerbated by economic uncertainties, environmental degradation, and social isolation—can have profound impacts on mental health. Community programmes that foster social interaction, provide counselling services, and create safe spaces for dialogue can help mitigate these challenges. By addressing mental health as an integral part of well-being, sustainable development initiatives can ensure that individuals are equipped not only to survive but also to thrive in the face of adversity.

Furthermore, quality of life is enhanced by access to education and opportunities for lifelong learning. Education empowers individuals to improve their socio-economic status, engage in informed decision-making, and contribute meaningfully to community development. When education is accessible and inclusive, it becomes a catalyst for social change, enabling communities to overcome barriers and realize their full potential. In this way, investments in education are intimately linked to both social equity and sustainable development (Binagwaho, 2022)

In urban settings, quality of life is often associated with the availability of green spaces, recreational facilities, and efficient public transportation. Urban design that prioritizes pedestrian-friendly streets, parks, and community centres contributes to both physical health and social cohesion. These public spaces serve as venues for social interaction and cultural expression, reinforcing the social fabric of the community while promoting environmental sustainability.

Taken together, the pursuit of health, quality of life, and well-being is not only a moral imperative but also a practical necessity for sustainable development. When individuals are healthy and empowered, communities become more resilient, economies more vibrant, and societies more just. The integration of health and well-being into sustainable development strategies ensures that the benefits of growth are distributed equitably, setting the stage for long-term social progress.

1.2.5 Cultural Dimensions

Cultural diversity and heritage represent another critical pillar of sustainable development. Culture shapes the way communities perceive the world, informs social norms, and provides a sense of identity and belonging. Recognizing and respecting cultural diversity is therefore essential to developing sustainable policies that resonate with local values and traditions.

Cultural dimensions influence sustainable development in several ways. First, cultural practices and traditions often embody sustainable principles that have evolved over centuries. Indigenous knowledge systems, for example, include practices of land stewardship, resource management, and community governance that are inherently sustainable. By integrating traditional ecological knowledge with modern science, communities can develop innovative strategies to manage natural resources and mitigate environmental degradation (Al-Mansoori and Hamdan, 2023).

Furthermore, cultural diversity fosters creativity and innovation. Societies that celebrate multiple perspectives and value diverse forms of expression are better equipped to generate novel solutions to complex challenges. Artistic expression, storytelling, and communal rituals can all play a role in raising awareness about environmental issues and mobilizing collective action. By embedding sustainability within cultural narratives, communities can strengthen their commitment to preserving both their heritage and their natural environment.

Cultural preservation is also integral to social cohesion. When communities maintain and celebrate their cultural heritage, they reinforce a shared sense of identity and belonging. This shared identity is a powerful motivator for collective action and social solidarity. Cultural festivals, traditional ceremonies, and local crafts are more than just expressions of heritage—they are avenues for fostering mutual respect, understanding, and unity among diverse groups. In this way, cultural dimensions contribute to the resilience and adaptive capacity of communities, ensuring that development is both inclusive and rooted in local realities.

However, cultural diversity is increasingly threatened by the forces of globalization and homogenization. Rapid economic development and urbanization can lead to the erosion of traditional practices and languages, as well as the marginalization of minority cultures. Sustainable development, therefore, must include strategies for protecting and revitalizing cultural heritage. Policies that support cultural education, heritage preservation, and the promotion of local arts and traditions are essential to ensuring that cultural diversity remains a vibrant component of social development (Naheed and Shooshtarian, 2022).

In addition, cultural considerations are vital when designing and implementing development projects. Policies that ignore local cultural contexts risk alienating communities and undermining the legitimacy of development efforts. By engaging local cultural leaders and stakeholders in the planning process, development initiatives can be tailored to respect and reflect community values. This culturally sensitive approach not only enhances the effectiveness of interventions but also builds trust between policymakers and the people they serve.

1.3 KEY INITIATIVES TO ACHIEVE SUSTAINABILITY

In the quest for a sustainable future, initiatives at various levels—from international agreements to local community actions—play a pivotal role in shaping policies and driving practical change. This section explores the diverse array of initiatives aimed at achieving sustainability by examining global frameworks, national policies, and innovative local projects. Together, these initiatives form a multi-layered strategy that not only addresses environmental degradation but also promotes social equity and economic resilience (Shrivastava, 2017).

1.3.1 International Frameworks and Agreements

Global challenges such as climate change, biodiversity loss, and resource depletion demand collective action. In response, international frameworks and agreements have been established to foster collaboration among nations and create binding targets for sustainable development. One of the earliest and most influential of these is the United Nations Framework Convention on Climate Change (UNFCCC), adopted in 1992. The UNFCCC provides the foundation for global climate action by setting out a framework for countries to monitor and reduce greenhouse gas emissions. Its enduring legacy lies in its role as a forum for negotiation and cooperation, paving the way for subsequent accords (Hermwille, Obergassel, Ott et al., 2017).

A landmark outcome of the UNFCCC is the Paris Agreement, adopted in 2015. This agreement represents a significant milestone in global efforts to mitigate climate change by establishing a common goal: to keep global temperature rise well below 2 degrees Celsius above pre-industrial levels, with a further aim to limit the increase to 1.5 degrees Celsius. The Paris Agreement not only commits signatory nations to emission reduction targets but also emphasizes the importance of adaptation, finance, and technology transfer. Its success depends on both robust national commitments and international cooperation, demonstrating that sustainability is a shared responsibility that transcends borders.

Another important international initiative is the Sendai Framework for Disaster Risk Reduction, adopted in 2015. Although not exclusively focused on environmental sustainability, the Sendai Framework is vital in enhancing resilience against natural hazards, which are increasingly exacerbated by climate change. By advocating for risk reduction measures and the integration of disaster risk management into national policies, this framework underscores the interconnected nature of sustainable development. It reminds us that the well-being of communities, hinges not only on reducing emissions but also on protecting lives and livelihoods from environmental hazards (Fletcher, Ripple, Newsome et al., 2024).

These international agreements serve as critical reference points for national and local sustainability initiatives. They provide both the impetus and the benchmarks for countries to develop their own policies and strategies. The legal and normative frameworks established by these agreements create a sense of urgency and accountability that drives further action at all levels of governance.

1.3.2 Global Goals and Targets

Building on the momentum of international agreements, the United Nations introduced the 2030 Agenda for Sustainable Development in 2015, a comprehensive framework centred around 17 SDGs. The SDGs offer a universal blueprint that covers a wide range of issues, including poverty eradication, quality education, clean energy, responsible consumption, and climate action. Each goal is accompanied by specific targets and indicators, ensuring that progress can be monitored and measured over time.

The strength of the SDGs lies in their integrative approach. They are designed to be indivisible, meaning that economic, social, and environmental objectives are seen as mutually reinforcing. For instance, achieving quality education (SDG 4) not only empowers individuals economically but also fosters awareness of environmental issues, which in turn supports sustainable consumption and production patterns (SDG 12). Similarly, promoting gender equality (SDG 5) has been shown to yield positive outcomes across health, education, and economic productivity.

The SDGs have galvanized action among governments, the private sector, civil society, and international organizations. They provide a common language and a shared set of targets that align diverse efforts toward a sustainable future. By setting clear priorities and measurable targets, the SDGs have become a cornerstone of global sustainability initiatives, driving policy reforms, mobilizing resources, and encouraging innovation across sectors.

In addition to their role as a goal-setting mechanism, the SDGs serve as a platform for international cooperation. The interconnected nature of the goals means that progress in one area often depends on or accelerates progress in another area. This interdependence reinforces the need for integrated strategies and collaborative action—a

FIGURE 1.3 Sustainable Development Goals.

recurring theme in sustainable development. The SDGs thus embody the vision of a world where progress is not achieved in isolation but through concerted, holistic efforts that span national boundaries (Van Tulder and van Mil, 2022).

1.3.3 National and Local Initiatives

While international frameworks and global goals provide an essential foundation, sustainable development ultimately depends on the actions taken at national and local levels. Governments and local authorities are tasked with translating global targets into practical policies and programmes that address the unique challenges of their regions.

One prominent example of national initiatives is the Green New Deal, a set of policy proposals aimed at addressing climate change and economic inequality, particularly in the United States. The Green New Deal advocates for a rapid transition to renewable energy, substantial investments in green infrastructure, and the creation of high-quality jobs. Its holistic approach seeks to remedy historical inequities while fostering a sustainable, resilient economy. By linking environmental goals with social justice, the Green New Deal illustrates how national policies can simultaneously address multiple dimensions of sustainability.

Other countries have pursued similar ambitious agendas. For instance, several European nations have developed comprehensive national sustainability strategies that integrate climate policies with economic reforms. These strategies often include significant investments in renewable energy, energy efficiency measures, and public transportation systems, coupled with initiatives to support social inclusion and job creation in emerging green sectors. Such policies underscore the importance of aligning economic development with environmental stewardship and social equity, demonstrating that sustainability can be a driver of innovation and growth (Widjaja, 2023).

At the local level, initiatives often emerge from grassroots movements and community-based organizations. Local governments, in collaboration with civil society, have implemented projects that range from urban greening and waste management to renewable energy cooperatives and community-supported agriculture. These initiatives are particularly effective because they are tailored to the specific needs and resources of the community, ensuring that solutions are both culturally relevant and contextually appropriate.

For example, eco-villages—intentional communities designed around sustainable living practices—exemplify how localized initiatives can create replicable models of sustainability. Eco-villages typically emphasize low-impact living, renewable energy, organic agriculture, and collective decision-making. They serve as living laboratories where innovative practices are tested and refined, offering valuable lessons that can be scaled up or adapted in different settings.

Local initiatives also play a crucial role in building social cohesion and resilience. Community-driven projects encourage citizen participation, empower marginalized groups, and foster a sense of ownership over local resources. Urban sustainability projects, such as the development of green spaces, sustainable public transport, and energy-efficient housing, not only contribute to environmental goals but also enhance the quality of life for residents. In doing so, they help to create more inclusive, vibrant, and resilient communities.

1.3.4 Cross-Cutting Strategies and Innovations

Beyond specific agreements and localized projects, a range of cross-cutting strategies and innovative approaches have emerged as key drivers of sustainability. These strategies cut across traditional sectoral boundaries, integrating technological innovation, market-based instruments, and community engagement to achieve sustainable outcomes.

One of the most influential cross-cutting strategies is the promotion of a circular economy. Unlike the traditional linear economy—which follows a "take-make-dispose" model—a circular economy emphasizes resource efficiency by reusing, recycling, and repurposing materials. This approach not only minimizes waste and reduces environmental impacts but also creates economic opportunities through the development of new industries and markets. Companies that embrace circular economy principles often find that they can achieve cost savings, improve supply chain resilience, and enhance their overall competitiveness in a rapidly changing global market.

Innovation in renewable energy is another critical cross-cutting strategy. Advances in solar, wind, and other forms of renewable energy technology have made clean energy more accessible and affordable than ever before. These innovations are transforming energy systems around the world, reducing reliance on fossil fuels and mitigating greenhouse gas emissions. Moreover, the deployment of renewable energy is frequently accompanied by improvements in energy efficiency and the development of smart grid technologies, which together contribute to a more resilient and sustainable energy infrastructure (Alotaibi, Abido, Khalid et al., 2020).

Market-based instruments, such as carbon pricing and emissions trading schemes, represent yet another innovative approach to achieving sustainability. By putting a price on carbon emissions, these instruments create economic incentives for reducing greenhouse gas emissions and investing in low-carbon technologies. The revenue generated from such mechanisms can be reinvested in sustainable projects, further reinforcing the virtuous cycle of sustainable development. These market-based approaches demonstrate how economic principles can be harnessed to drive environmental improvements, illustrating the potential for synergy between market forces and sustainability objectives.

Digital technologies and data analytics are also emerging as powerful tools for promoting sustainability. From smart cities and precision agriculture to real-time environmental monitoring, digital innovations enable more efficient resource management and more informed decision-making. For example, cities equipped with sensors and data platforms can optimize energy use, reduce traffic congestion, and improve public services—all of which contribute to a more sustainable urban environment. The integration of digital technology with sustainability initiatives is creating new opportunities for innovation, enhancing transparency, and empowering citizens to engage more effectively with their communities.

Another innovative approach involves public–private partnerships (PPPs), which bring together the strengths of government, business, and civil society. PPPs can mobilize significant resources, share risks, and accelerate the implementation of sustainable projects. Whether through the development of renewable energy infrastructure or the revitalization of urban spaces, these partnerships demonstrate that collaboration across sectors is essential for achieving complex sustainability goals. They also help bridge the gap between policy and practice, ensuring that innovative ideas are translated into tangible outcomes that benefit communities and the environment alike (Casady, Cepparulo, and Giuriato, 2024).

1.3.5 Bridging Initiatives Across Scales

One of the most promising aspects of sustainability initiatives is the ability to bridge actions across different scales. International frameworks set the agenda and provide a common set of goals, while national policies translate these goals into specific targets and regulatory measures. At the same time, local initiatives bring the human dimension into focus, ensuring that policies are responsive to community needs and conditions. This multi-scalar approach creates a feedback loop in which each level reinforces the others, leading to more coherent and integrated strategies for sustainable development.

For example, the international commitments made under the Paris Agreement and the SDGs serve as benchmarks that guide national legislation and local projects. When a country sets ambitious climate targets based on these global frameworks, it often follows with concrete policies—such as investments in renewable energy, energy efficiency standards, or urban greening initiatives—that are tailored to its local context. Local governments, in turn, act as the testing grounds for innovative approaches that can later be scaled up or adapted by other regions. This interconnected system underscores the importance of collaboration and coordination among all stakeholders, ensuring that sustainability initiatives are not isolated efforts but part of a comprehensive and dynamic process (Adanma and Ogunbiyi, 2024).

The interplay between different scales also emphasizes the need for effective governance and coordination mechanisms. Institutional frameworks that facilitate communication and collaboration among international organizations, national governments, and local authorities are critical for translating global commitments into real-world outcomes. Platforms for knowledge exchange, capacity-building initiatives, and shared monitoring systems help ensure that progress is tracked and that best practices are disseminated widely. By bridging the gap between different scales of action, sustainability initiatives can achieve a greater impact and foster a more resilient, adaptive development model.

1.4 MAJOR COMPONENTS OF SUSTAINABLE DEVELOPMENT

Sustainable development is a multifaceted endeavour that rests on the integration of several key components. Traditionally, the concept is described in terms of three primary pillars—economic, social, and environmental—which together create a framework for achieving long-term well-being. In addition, effective governance

TABLE 1.1
Components of Sustainable Development

Component	Description	Key Indicators
Economic	Growth that is inclusive and environmentally sound	GDP per capita, investment in green technologies
Social	Equity, community well-being, and quality of life	Education, healthcare access, income distribution
Environmental	Protection and sustainable management of natural resources	Air quality, biodiversity, water usage
Governance	Transparent, inclusive, and accountable institutions	Policy effectiveness, regulatory compliance

and institutional capacity are increasingly recognized as critical enablers of sustainable outcomes. This section examines each of these components in detail, discussing their individual contributions to sustainable development as well as the synergies that emerge when they are addressed holistically.

1.4.1 Economic Dimension

The economic dimension of sustainable development focuses on promoting growth that is not only robust but also environmentally sound and socially inclusive. Traditional economic models often emphasize rapid growth and increased production; however, such models can lead to resource depletion, environmental degradation, and widening social inequalities. In contrast, a sustainable economic model seeks to align economic activity with the finite nature of natural resources and the need for equitable benefit sharing.

One of the primary objectives of the sustainable economic approach is to transition from a linear, "take-make-dispose" economy toward a circular economy. In a circular economy, resources are used more efficiently by emphasizing reuse, recycling, and remanufacturing processes. This approach not only reduces waste and environmental impacts but also creates new opportunities for innovation and job creation. By closing the loop on resource use, economies can decouple growth from environmental degradation, ensuring that economic activities contribute to long-term prosperity without depleting the resources upon which future growth depends.

Investment in green technologies is another critical aspect of the economic dimension. Renewable energy, energy-efficient infrastructure, and low-carbon transportation systems are central to reducing greenhouse gas emissions and mitigating climate change. These investments are not only environmentally beneficial but also stimulate economic activity through the creation of new industries and employment opportunities. Governments and the private sector alike are increasingly recognizing the economic advantages of green investments, which can drive competitiveness and foster sustainable industrial growth.

Moreover, sustainable economic policies emphasize inclusive growth—ensuring that the benefits of development are distributed equitably across society. Policies designed to enhance access to education, healthcare, and financial services can empower marginalized populations and reduce income disparities. Inclusive growth is essential for maintaining social stability and preventing the negative social outcomes that often accompany rapid, uneven economic expansion. In this context, economic sustainability is not measured solely by GDP figures but also by indicators of social welfare, such as employment rates, income distribution, and access to essential services (Islam, 2024).

In summary, the economic component of sustainable development requires a paradigm shift from short-term profit maximization to long-term value creation. This includes adopting circular economy principles, investing in green technologies, and ensuring that growth is both inclusive and environmentally responsible. Such an integrated approach creates a resilient economic system capable of adapting to the challenges of a rapidly changing world.

1.4.2 Social Dimension

The social dimension of sustainable development is centred on human well-being, equity, and social cohesion. It recognizes that the success of any development strategy depends fundamentally on the quality of life of individuals and communities. Social sustainability addresses issues such as poverty, inequality, health, education, and cultural diversity—all of which are essential for building a resilient and harmonious society (Rogamelia and Setiawati, 2024).

At its core, social sustainability demands that development should be equitable. This means that both the benefits and burdens of economic growth and environmental protection must be distributed fairly. Social equity involves ensuring that all individuals, regardless of their background, have equal access to opportunities and resources. When disparities in income, education, or healthcare persist, they can lead to social unrest and undermine the overall stability of society. Policies that promote social inclusion—such as targeted social programmes, affirmative action, and community development initiatives—are therefore vital for achieving sustainable outcomes.

Another key aspect of the social dimension is the enhancement of quality of life and well-being. Beyond material wealth, well-being encompasses physical and mental health, education, housing, and access to social services. A sustainable society is one in which citizens can enjoy a high quality of life, free from the stress and uncertainty that often accompany social exclusion and economic deprivation. Investments in public health, education, and social infrastructure are critical for building such a society. For example, universal healthcare programmes and affordable housing initiatives not only improve individual well-being but also contribute to social stability and productivity.

Social cohesion, or the strength of relationships and trust among community members, is another important component. A cohesive society is better able to mobilize collective action, respond to crises, and adapt to changes. Community-based initiatives, participatory governance, and local cultural activities can strengthen social bonds and create an environment in which diverse groups work together toward common goals. This collective spirit is especially important in times of environmental or economic stress, where the ability to rely on strong social networks can mean the difference between resilience and collapse.

Cultural diversity also plays a crucial role in the social dimension of sustainability. A society that values and preserves its cultural heritage is more likely to foster creativity, innovation, and social inclusion. Traditional knowledge systems and cultural practices often contain sustainable principles that have been honed over centuries. Integrating such indigenous and local knowledge into modern sustainability practices not only enriches the development process but also ensures that solutions are culturally relevant and widely accepted.

Ultimately, the social component of sustainable development emphasizes that true progress must improve the lives of all people. This requires policies that address social inequities, promote inclusivity, and build the human and social capital needed for long-term resilience. When economic and environmental policies are designed with a focus on social well-being, the result is a more just and robust foundation for sustainable development (Dyakovaet, 2017).

1.4.3 Environmental Dimension

The environmental dimension of sustainable development is perhaps the most widely recognized and discussed. It is grounded in the understanding that natural ecosystems provide the essential services and resources that underpin all human activity. The primary goal of environmental sustainability is to protect and preserve these ecosystems for current and future generations.

Environmental sustainability involves a broad range of activities aimed at mitigating pollution, conserving biodiversity, and ensuring the responsible use of natural resources. At its core is the recognition that the environment has finite carrying capacity; overexploitation of resources can lead to irreversible damage, which in turn jeopardizes economic growth and social well-being. As such, environmental policies are designed to maintain the delicate balance between human needs and ecological integrity.

Key strategies within the environmental dimension include conservation, restoration, and the sustainable management of natural resources. Conservation efforts focus on preserving critical habitats, protecting endangered species, and maintaining biodiversity. Restoration initiatives, on the other hand, aim to rehabilitate degraded ecosystems—such as reforesting areas that have been deforested or cleaning up polluted waterways. Both types of initiatives are essential for ensuring that ecosystems remain robust and capable of providing services such as clean air, water, and fertile soil.

Climate change represents one of the most urgent challenges within the environmental realm. The rapid increase in greenhouse gas emissions due to industrial activity and deforestation has led to global warming and extreme weather events, threatening both natural and human systems. Mitigation strategies—such as reducing carbon emissions, transitioning to renewable energy, and enhancing energy efficiency—are therefore central to environmental sustainability. Adaptation measures, which help communities adjust to the unavoidable impacts of climate change, are also critical. These measures may include building flood defences, developing drought-resistant crops, and implementing early warning systems for natural disasters.

The environmental dimension also encompasses sustainable land use and water management practices. Urbanization and industrialization place enormous pressure on natural resources, leading to soil degradation, water scarcity, and habitat fragmentation. Integrated land use planning that considers environmental, social, and economic factors can help balance development needs with conservation goals. Likewise, sustainable water management practices—such as the efficient use of water resources, protection of watersheds, and treatment of wastewater—are essential for maintaining the quality and availability of this vital resource (Silve, 2023).

Overall, the environmental component of sustainable development emphasizes that economic growth and social progress must be achieved within the limits of our planet's ecosystems. This requires a proactive approach to conservation, innovative technologies for reducing environmental impacts, and policies that promote the sustainable use of natural resources. By protecting the environment, we ensure that future generations inherit a planet capable of supporting a high quality of life.

1.4.4 Governance and Institutional Dimension

Beyond the traditional three pillars, governance and institutional capacity are increasingly recognized as critical components of sustainable development. Effective governance refers to the frameworks, policies, and institutions that enable coordinated action among diverse stakeholders. It is the mechanism through which global agreements are translated into national policies and local practices, ensuring that SDGs are met.

Sound governance requires transparency, accountability, and inclusivity. Institutions that operate with these principles can build trust among citizens and foster cooperation across sectors. Transparent decision-making processes ensure that policies are open to public scrutiny, reducing the risk of corruption and enhancing policy legitimacy. Moreover, when governance structures are inclusive—allowing participation from various segments of society, including marginalized groups—they are more likely to design policies that reflect the needs and priorities of the broader community.

Institutional capacity is also essential for monitoring, evaluating, and adjusting sustainability initiatives. Robust institutions are capable of collecting data, analysing outcomes, and making evidence-based decisions that refine policies over time. For instance, national sustainability strategies often include detailed monitoring and evaluation frameworks that track progress against established indicators. These frameworks enable governments to identify challenges early, adjust strategies as needed, and ensure that the goals of sustainable development remain on track.

Intergovernmental organizations and PPPs also play a significant role in fostering governance for sustainability. International bodies such as the United Nations, the World Bank, and regional organizations help coordinate efforts across national borders and provide technical and financial support to countries in need. PPPs, by combining the resources and expertise of governments, businesses, and civil society, can accelerate the implementation of sustainable projects and ensure that policies are both innovative and practical (Mazutis and Sweet, 2022).

Furthermore, good governance extends to the legal and regulatory frameworks that underpin sustainable development. Environmental regulations, labour laws, and social protection policies are all part of the institutional fabric that supports sustainable practices. These regulations not only safeguard against practices that could harm the environment or social well-being but also create incentives for businesses and individuals to adopt sustainable behaviours. In this sense, governance serves as both a guide and an enforcer of sustainability standards, ensuring that all sectors work in harmony toward common goals.

In summary, the governance and institutional dimension is the backbone of sustainable development. It provides the structure within which economic, social, and environmental policies are developed and implemented. Strong, transparent, and inclusive institutions are essential for translating ambitious sustainability goals into practical outcomes that benefit all members of society.

1.4.5 Integration and Interdependencies

While each component of sustainable development is critical on its own, the true power of sustainability lies in the integration of these components. The economic, social, environmental, and governance dimensions are not separate silos; rather, they are interdependent systems that must be aligned to achieve a sustainable future.

The interdependencies among these dimensions become evident when considering the ripple effects of policy decisions. For example, a government policy that promotes renewable energy (an economic and environmental measure) can simultaneously create new jobs and reduce pollution, thereby enhancing social well-being. Similarly, policies aimed at reducing inequality (a social measure) can lead to more inclusive economic growth, which in turn generates resources for further environmental protection and improved governance.

Integrated strategies are often supported by cross-sectoral planning and multi-stakeholder collaboration. Mechanisms such as national sustainability councils, public–private partnerships, and community forums provide platforms where diverse stakeholders can discuss and align their objectives. Such collaborative approaches not only ensure coherence in policymaking but also build the resilience needed to adapt to emerging challenges—be they environmental shocks, economic downturns, or social upheavals.

Moreover, the adoption of comprehensive indicators that measure progress across all dimensions is essential for integrated planning. Traditional metrics, like GDP, are increasingly complemented by broader measures of well-being and environmental health. These include the Human Development Index (HDI), ecological footprint analysis, and composite sustainability indices that provide a more holistic view of progress. By integrating diverse indicators, policymakers can better understand the complex interplay between economic, social, environmental, and governance factors, leading to more informed decisions and effective interventions.

TABLE 1.2
Challenges to Sustainability

Category	Key Challenges
Economic	• Lack of Funding • High Costs • Economic Inequality
Social	• Limited Awareness • Resistance to Change • Social Inequality
Environmental	• Climate Change • pollution • Resource Depletion
Political	• Weak Policies • Lack of Cooperation • Corruption

In conclusion, the major components of sustainable development—economic, social, environmental, and governance—are deeply interconnected. A sustainable future requires that these dimensions should be addressed in tandem rather than in isolation. By recognizing and managing the interdependencies among these components, policymakers can develop strategies that are robust, adaptive, and capable of delivering long-term benefits for both current and future generations (Jarvis Lau, Cook et al., 2011).

1.5 CHALLENGES IN SUSTAINABILITY DEVELOPMENT

Sustainable development, while a critical goal for the future of humanity, faces numerous challenges that threaten its successful implementation. These challenges span environmental, social, economic, and political dimensions, creating a complex web of obstacles that must be addressed in order to build a more sustainable world. In this section, we delve into the key challenges in sustainability development by examining issues such as population growth, energy supply, water scarcity, global climate change, societal and political inaction, economic and technological complexities, and the dynamics of adaptive cycles. Understanding these challenges is essential for developing robust strategies to overcome them and achieve the goals of sustainable development.

1.5.1 Introduction to Challenges in Sustainability

The pursuit of sustainability confronts a myriad of barriers that are as diverse as they are interconnected. At a fundamental level, the challenge lies in balancing the immediate needs of a growing global population with the long-term imperative to preserve natural resources and maintain ecological balance. The interplay between rapid demographic expansion, escalating resource demands, and environmental degradation creates tensions that can derail progress toward sustainable development. Moreover, the persistence of outdated political and economic paradigms, combined with technological limitations and institutional inertia, further complicates efforts to implement sustainable practices. Drawing on insights provided by Bruce R. Conard (2013) and corroborated by subsequent reviews, this section examines these challenges in detail, highlighting their causes, implications, and potential avenues for mitigation (Şahin, Arslan, and Özdemir, 2024).

1.5.2 Population Growth

One of the most pressing challenges to sustainability is the rapid growth of the global population. Since the global population reached 7 billion in 2011, projections suggest that it will approach 9 billion by the mid-21st century and may stabilize between 10 and 11 billion by 2100. This exponential increase exerts enormous pressure on all forms of natural capital—food, water, energy, and land—as well as on social and economic infrastructures. The challenge is compounded by the uneven distribution of population growth and resource consumption; while developed regions such as North America and Europe have high per capita resource use and emissions, rapidly industrializing nations like China and India are experiencing surges in both population and pollution levels. Such disparities complicate global efforts to implement uniform sustainability measures, as regions with higher growth rates face more acute shortages and environmental degradation, whereas high-consumption regions are challenged by entrenched habits and infrastructural dependencies (Lemanski, 2023).

1.5.3 Energy Supply and Consumption

The global energy system is another critical area where challenges to sustainability are especially pronounced. The prevailing reliance on fossil fuels not only depletes

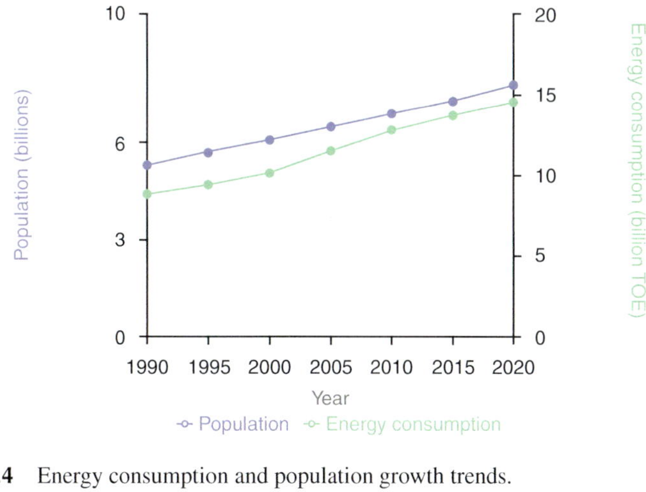

FIGURE 1.4 Energy consumption and population growth trends.

finite resources but also generates significant environmental damage through greenhouse gas emissions and other pollutants. Fossil fuels, by their nature, contribute to air and water pollution, acid rain, and a host of health issues among exposed populations. Transitioning to renewable energy sources is vital; however, such a shift requires massive investments in research, infrastructure, and technology. Furthermore, the energy efficiency of current systems must be improved, particularly in developed nations where per capita energy consumption remains disproportionately high.

The challenge is not merely one of replacing one energy source with another—it is also about reconfiguring entire energy networks to be more resilient, distributed, and adaptable to changing conditions. This calls for coordinated policy interventions and significant shifts in both consumer behaviour and industrial practices (Sima, Gheorghe, Subić et al., 2020).

1.5.4 Fresh Water Availability

Although water covers approximately 71% of the Earth's surface, only a tiny fraction of this water is fresh and readily accessible. Of the total freshwater available, about 90% is locked in glaciers and ice caps, leaving just 0.3% that can be easily harnessed for human use. The uneven distribution of freshwater resources, combined with rising demand driven by population growth, industrial activity, and agricultural expansion, poses a formidable challenge to sustainability. Over-extraction of groundwater, particularly for irrigation and industrial use, is leading to the rapid depletion of aquifers in many parts of the world. A notable example is the Ogallala aquifer in the central United States, where extraction rates far exceed natural recharge, with recovery estimated to take thousands of years. This unsustainable use of water resources not only endangers long-term agricultural productivity but also threatens the availability of potable water for millions of people. Addressing freshwater scarcity will require innovative water management practices, improved efficiency in usage, and the development of alternative water sources, including recycling and desalination technologies (Ghazi Rizvi, Shahid et al., 2022).

1.5.5 Global Climate Change

Climate change represents perhaps the most urgent and multifaceted challenge to sustainable development. The rising concentrations of greenhouse gases, particularly carbon dioxide (CO_2) and methane (CH_4), are driving global warming and causing widespread disruptions to weather patterns. Accelerated melting of polar ice caps, glaciers, and permafrost contributes not only to rising sea levels but also to the destabilization of critical climatic systems such as the Gulf Stream. The consequences of these changes are profound: coastal cities face increased flooding and storm surges, agricultural zones shift unpredictably, and natural ecosystems are forced to adapt or perish. The melting of permafrost, in particular, releases additional methane—a greenhouse gas far more potent than CO_2—thus creating a feedback loop that exacerbates global warming. Mitigation and adaptation strategies must therefore address both the reduction of emissions and the resilience of vulnerable communities to climate impacts. However, the scale and complexity of the climate challenge demand global cooperation and long-term commitment, factors that are often hindered by political and economic constraints (Bodin, 2017).

1.5.6 Societal and Political Inaction

Despite the mounting scientific evidence and increasingly dire consequences of unsustainable practices, societal and political inaction remains a pervasive barrier to progress. Human societies are often more responsive to immediate crises than to slow-moving, long-term threats. This short-termism is reflected in political and economic decision-making, where immediate gains and electoral cycles overshadow the need for sustainable policies. Conard (2013) highlights how entrenched interests and institutional inertia often stymie reform efforts, even when the benefits of sustainability are clear. Economic growth and short-term profit frequently take precedence over environmental stewardship and social equity, leading to policies that favour the status quo rather than transformative change. The lack of political will, coupled with a reluctance to invest in long-term infrastructure and research, hampers the development and implementation of policies that could significantly mitigate sustainability challenges. Overcoming this inertia will require not only robust advocacy and public awareness but also systemic changes in governance and economic incentives.

1.5.7 Economic and Technological Complexity

Modern economic and technological systems are characterized by an unprecedented level of complexity and interdependence. Global supply chains, intricate financial networks, and advanced technological infrastructures have made economies highly efficient, yet also increasingly vulnerable to disruptions. The focus on maximizing short-term economic performance often obscures the long-term environmental and social costs of these systems. Technological innovations, while essential for progress, can also introduce unforeseen risks. For example, reliance on digital networks and automation may lead to significant disruptions if systems fail or are compromised.

Furthermore, the drive for continuous innovation can lead to rapid obsolescence and increased waste, undermining efforts to create a circular economy. Balancing the benefits of technological advancement with its potential downsides is a central challenge for sustainability. This requires not only advancements in technology but also a reevaluation of economic models to incorporate long-term environmental and social costs into decision-making processes (Tseng and Hung, 2014).

1.5.8 Adaptive Cycles and Resilience

The concept of adaptive cycles, originally proposed by ecological theorists such as Holling, offers a useful framework for understanding the dynamic nature of sustainable systems. According to this concept, all complex systems—be they ecological, social, or economic—undergo cycles of growth, collapse, regeneration, and renewal. In the context of human societies, the current phase of rapid growth and increasing complexity heightens the risk of systemic collapse if adaptive measures are not taken. Adaptive cycles underscore the importance of building resilience into our systems, ensuring that they can absorb shocks and recover from disruptions. Resilience can be enhanced through diversification, redundancy, and flexibility in both technological and social systems. By understanding and planning for these adaptive cycles, policymakers can design strategies that not only prevent collapse but also create opportunities for renewal and innovation in the face of challenges.

1.5.9 Synthesis of Challenges

The challenges to sustainability are vast and interrelated, creating a landscape in which isolated solutions are insufficient. Instead, the complexity of these challenges requires an integrated approach that considers the interdependencies between population dynamics, energy use, water availability, climate change, and sociopolitical factors. Each challenge exacerbates the others: for example, rapid population growth intensifies pressure on water and energy systems, which in turn contribute to environmental degradation and climate change. Similarly, political and economic inaction can stall the development of necessary technologies and policies, leaving societies ill-prepared to deal with emerging crises.

Addressing these challenges will necessitate a multifaceted strategy that combines technological innovation, policy reform, and cultural change. Scientists and researchers play a critical role in identifying the mechanisms behind these challenges and in proposing evidence-based solutions. Likewise, political leaders and institutions must be willing to enact long-term policies that prioritize sustainability over short-term gains. Public awareness and engagement are also essential; when citizens are informed about the risks and consequences of unsustainable practices, they are more likely to support transformative changes (Odume, Akamagwuna, Ntlokot al., 2023)

1.5.10 Moving Forward: Strategies for Overcoming Challenges

The path toward sustainable development is not without obstacles, but acknowledging these challenges is the first step toward overcoming them. Several strategies emerge as particularly promising in addressing the multifaceted nature of sustainability challenges:

1. **Technological Innovation:** Accelerating the development and deployment of renewable energy sources, water-saving technologies, and waste reduction methods can help alleviate resource pressures. Innovation in digital technologies and data analytics also offers new ways to monitor and manage resource use efficiently.
2. **Policy Reform:** Governments must design and implement policies that integrate long-term sustainability goals into economic planning. This includes incentives for green technologies, regulatory frameworks to reduce environmental impacts, and measures to ensure social equity.

3. **Institutional Change:** Strengthening institutions and enhancing transparency in decision-making processes can overcome political inertia. Inclusive governance structures that empower communities and marginalized groups can foster broader support for sustainable initiatives.
4. **Public Engagement and Education:** Raising awareness about the challenges and consequences of unsustainable practices can mobilize public support for change. Education initiatives, community outreach, and participatory governance models are critical to building a resilient and informed society.
5. **International Cooperation:** Global challenges require global solutions. Enhanced international cooperation and the sharing of best practices can help bridge the gaps between developed and developing regions, ensuring that all nations can contribute to and benefit from sustainability efforts.

1.6 MEASURES OF SUSTAINABILITY

The measurement of sustainability is a multifaceted task that requires the assessment of economic, social, and environmental impacts. Unlike traditional metrics that focus solely on economic output, sustainability measures aim to capture a holistic picture of development. Various methods and tools have been developed to quantify and evaluate the performance of sustainability initiatives. This section explores these measures in detail, drawing on the work of Serzante and Khudozhnyk (2023) among other sources, and provides insights into the most appropriate methods for evaluating sustainability across different contexts.

1.6.1 Introduction to Measures of Sustainability

Measuring sustainability is inherently complex because it involves multiple dimensions that are often interrelated. Traditional economic indicators such as GDP

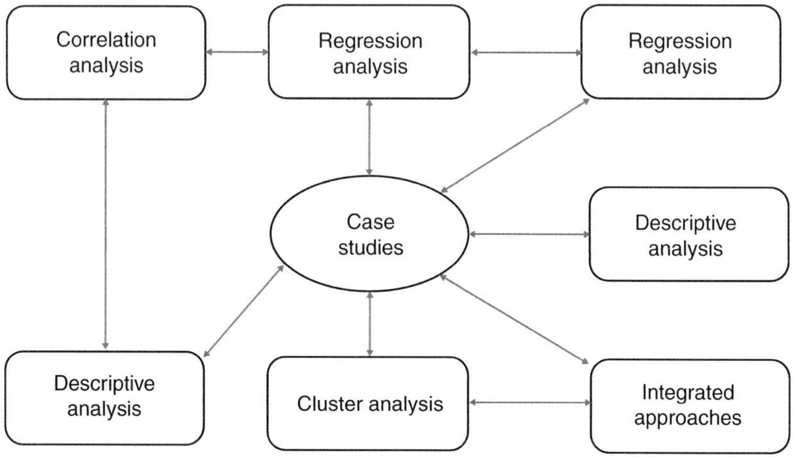

FIGURE 1.5 Methods of measure of sustainability.

do not account for the degradation of natural resources or social inequities. Therefore, new measurement approaches have emerged that integrate environmental, social, and economic data to provide a more comprehensive understanding of sustainability. These approaches not only enable policymakers and business leaders to track progress but also help identify areas where intervention is needed. As sustainability becomes a central goal for both public and private sectors, the development of robust, integrated measures is critical for ensuring that sustainability initiatives are effective, comparable, and continuously improved (Wang, Dabbaghjamanesh, Kavousi-Fard et al., 2024).

1.6.2 Correlation Analysis

One of the primary statistical methods employed in sustainability research is correlation analysis. This technique is used to assess the relationship between two or more variables. In the context of sustainability, correlation analysis can help determine how different sustainability factors relate to economic performance, environmental health, or social well-being. The correlation coefficient, which ranges from –1 to 1, indicates the strength and direction of the relationship. A coefficient of 1 denotes a perfect positive correlation, whereas –1 indicates a perfect negative correlation.

For instance, researchers might use correlation analysis to explore the relationship between investment in renewable energy and reductions in carbon emissions. A strong positive correlation would suggest that as investment increases, emissions decrease—a finding that can guide further policy development. This method is particularly useful for preliminary analysis, offering clear insights into potential causal relationships that can be explored more deeply with advanced techniques (Remeler and Van Ryzin, 2021).

1.6.3 Regression Analysis

While correlation analysis provides a snapshot of relationships between variables, regression analysis goes further by modelling these relationships and predicting outcomes. Regression analysis is a statistical technique that examines the impact of one or more independent variables on a dependent variable. In sustainability studies, regression analysis can be used to predict economic performance based on various sustainability initiatives or to determine how changes in environmental policies might affect social outcomes.

There are several forms of regression analysis. Simple regression involves one independent variable, while multiple regression considers several independent variables simultaneously. Panel data regression analysis, which incorporates data over time, is particularly valuable. For example, by analysing panel data, researchers can assess how a company's Environmental, Social, and Governance (ESG) disclosure performance evolves in response to changes in sustainability initiatives. The insights gained from regression analysis are crucial for developing strategies that enhance sustainability performance, as they help isolate the specific factors that drive success.

1.6.4 Case Study Method

The case study method is another important tool for measuring sustainability, particularly when a nuanced understanding of complex issues is required. This method involves an in-depth examination of a particular instance—such as a company's transition to sustainable practices—to draw broader insights. Case studies are highly valuable because they allow researchers to explore the context, challenges, and outcomes in detail, thereby generating qualitative insights that might be missed by purely quantitative methods.

For example, a case study of a manufacturing firm that successfully reduced waste through a circular economy model can provide valuable lessons for other businesses. Although the findings from a case study may not be generalizable to all contexts, they offer rich, contextualized evidence that can inspire similar initiatives elsewhere. The strength of the case study method lies in its ability to capture the complexities and subtleties of sustainability practices in real-world settings (Lim, 2024).

1.6.5 Descriptive Analysis

Descriptive analysis is fundamental to the measurement of sustainability because it involves summarizing and presenting data in a clear and concise manner. This method uses statistical measures such as frequency distributions, means, medians, and measures of variability to describe the characteristics of a dataset. In sustainability research, descriptive analysis can help identify patterns and trends—for example, the distribution of resource usage among different sectors or the prevalence of sustainable practices among small and medium enterprises (SMEs).

By applying descriptive statistics, researchers can identify outliers or anomalies that may require further investigation. For instance, descriptive analysis of sustainability practices in European SMEs has revealed that many firms rely heavily on internal funding for resource efficiency initiatives. Such insights can guide future research and policymaking by highlighting areas where targeted support may be needed (Janssen and Helbig, 2018).

1.6.6 Cluster Analysis

Cluster analysis is a technique used to group individuals or entities based on shared characteristics. In the context of sustainability, this method can segment markets, identify consumer preferences, or classify organizations based on their sustainability performance. By grouping similar data points together, cluster analysis helps reveal underlying patterns that might not be visible through other analytical methods.

For example, cluster analysis might be used to identify groups of companies that exhibit similar behaviours regarding energy consumption, waste management, or social responsibility practices. However, the validity of the results can be affected by the choice of variables, sample size, and the analytical techniques used. Therefore, it is essential to apply cluster analysis carefully and to interpret the findings in the context of other data sources and methods.

1.6.7 Integrated Approaches to Measuring Sustainability

Recognizing the inherent complexity of sustainability, there has been a growing trend toward integrated approaches that combine multiple methods and dimensions into a single framework. Integrated approaches are designed to capture the full spectrum of sustainability by considering economic, social, and environmental factors holistically. Rather than relying on a single indicator, integrated methods use composite indices and multi-criteria analysis to provide a more accurate and comprehensive picture of sustainability performance.

For instance, sustainability indices such as the Global Reporting Initiative (GRI) or the Sustainability Accounting Standards Board (SASB) frameworks combine quantitative data with qualitative assessments to evaluate an organization's overall sustainability. These integrated approaches not only enhance comparability between organizations and sectors but also help in benchmarking progress over time. The use of integrated methods is crucial for developing standardized and reliable sustainability criteria that can guide both policy and practice.

1.6.8 Global Distribution of Sustainability Research

The measurement of sustainability is not uniform across the globe, and research efforts have shown varying degrees of focus on different regions and sectors. Studies indicate that Asia and Europe have been the most intensively researched regions in terms of sustainability measurement. These regions have seen significant investments in sustainability research due to both environmental pressures and the availability of robust data.

In addition to geographic differences, there is notable variation across sectors. The secondary sector—including manufacturing, construction, and industrial activities—has received considerable attention because of its direct impact on resource use and environmental degradation. SMEs have also emerged as a prominent focus area, given their significant role in the global economy and their potential to adopt innovative sustainable practices. This focus on SMEs highlights the need for more nuanced research that considers the diverse capacities and challenges faced by companies of varying sizes.

1.6.9 Quantitative and Qualitative Methods

Sustainability measurement employs both quantitative and qualitative methods, each offering distinct advantages. Quantitative methods provide objective, numerical assessments that are essential for benchmarking and statistical analysis. Techniques such as correlation and regression analysis, as well as descriptive and cluster analyses, offer valuable insights into the relationships between sustainability initiatives and performance outcomes.

On the other hand, qualitative methods provide depth and context that numbers alone cannot capture. Through interviews, focus groups, and case studies, qualitative research explores the subjective experiences, cultural factors, and organizational dynamics that influence sustainability. The combination of quantitative and qualitative approaches—often referred to as mixed-methods research—yields a more comprehensive understanding of sustainability dynamics. This integrated approach allows researchers to validate quantitative findings with qualitative insights, ensuring that the measures of sustainability are both rigorous and contextually relevant (Lim et al., 2024).

1.6.10 Implications for Policy and Business Practice

The various methods and tools discussed above have significant implications for both policy and business practice. For policymakers, robust measures of sustainability are essential for setting targets, monitoring progress, and designing effective interventions. For instance, integrated sustainability indices can help governments identify priority areas for investment and regulatory reform, while regression and correlation analyses can provide evidence for the economic benefits of sustainable practices.

Businesses, too, can leverage these measurement tools to enhance their sustainability performance. By using descriptive and cluster analyses, companies can benchmark their performance against industry standards and identify areas for improvement. Case studies and qualitative research provide actionable insights that can guide strategic decision-making. Moreover, by adopting integrated approaches, organizations can present a comprehensive picture of their sustainability efforts, thereby improving transparency and stakeholder engagement.

1.6.11 Future Directions in Sustainability Measurement

As sustainability challenges evolve, so too must the methods used to measure them. The future of sustainability measurement lies in the further integration of advanced technologies, such as big data analytics, artificial intelligence, and remote sensing. These tools can greatly enhance the precision and timeliness of sustainability assessments. For example, real-time data from sensors and satellite imagery can provide immediate feedback on environmental conditions, enabling rapid responses to emerging issues.

Furthermore, the continued development of standardized frameworks and indicators will be essential for ensuring consistency and comparability in sustainability reporting. International initiatives that promote data sharing and collaborative research can help harmonize measurement approaches across regions and sectors. This, in turn, will empower policymakers and business leaders to make more informed decisions, ultimately driving progress toward a more sustainable future (Dushkova and Ivlieva, 2024).

1.7 CONCLUSION

This chapter has provided a comprehensive exploration of sustainable development by examining its conceptual framework, social dimensions, key initiatives, major components, challenges, and methods of measurement. We began by defining sustainability and tracing its evolution from the seminal Brundtland Commission Report to the ambitious goals of the 2030 Agenda. The discussion then delved into the social factors that underpin sustainable development—highlighting issues such as social equity, community cohesion, participatory governance, and quality of life. We further examined the range of initiatives from global agreements and national policies to innovative local projects that aim to create a sustainable future. In addressing the challenges—from population growth and energy constraints to climate change and political inaction—we underscored the complexity of the sustainability agenda. Finally, the chapter detailed various quantitative and qualitative methods for measuring sustainability, emphasizing the need for integrated, holistic approaches that capture economic, social, and environmental performance.

This integrated review of sustainable development highlights that achieving sustainability is not the result of isolated actions but the outcome of coordinated, long-term strategies that address interdependent issues. By aligning policies, fostering innovation, and ensuring inclusive participation at all levels, societies can overcome the multifaceted challenges of our time and pave the way for a resilient, equitable, and sustainable future.

REFERENCES

Adanma, U. M., and Ogunbiyi, E. O. (2024). Evaluating the effectiveness of global governance mechanisms in promoting environmental sustainability and international relations. *Finance & Accounting Research Journal,* 6(5), 763–791.

Ajirotutu, R. O., Adeyemi, A. B., Ifechukwu, G. O., Iwuanyanwu, O., Ohakawa, T. C., and Garba, B. M. P. (2024). Future cities and sustainable development: Integrating renewable energy, advanced materials, and civil engineering for urban resilience. *International Journal of Sustainable Urban Development*, *12*(02), 235–250.

Al-Mansoori, F., and Hamdan, A. (2023). Integrating indigenous knowledge systems into environmental education for biodiversity conservation: A study of sociocultural perspectives and ecological outcomes. *AI, IoT and the Fourth Industrial Revolution Review*, 13(7), 61–74.

Alotaibi, I., Abido, M. A., Khalid, M., and Savkin, A. V. (2020). A comprehensive review of recent advances in smart grids: A sustainable future with renewable energy resources. *Energies*, 13(23), 6269.

Binagwaho, A., Bonciani Nader, H., Brown Burkins, M., Davies, A., Hessen, D. O., Mbow, C. et al. (2022). *Knowledge-driven Actions: Transforming higher education for global sustainability: Independent expert group on the universities and the 2030 agenda.* UNESCO Publishing.

Bodin, Ö. (2017). Collaborative environmental governance: Achieving collective action in social-ecological systems. *Science*, 357(6352). DOI: 10.1126/science.aan1114

Casady, C. B., Cepparulo, A., and Giuriato, L. (2024). Public-private partnerships for low-carbon, climate-resilient infrastructure: Insights from the literature. *Journal of Cleaner Production*, 143338.

Chiroli, D. M. D. G., Menezes, M. G., Zola, F. C., Aragão, F. V., de Almeida, R. D., and Tebcherani, S. M. (2023). Integrating resilience and sustainability: A systematic analysis of resilient cities using ISO 37123. *International Journal of Disaster Risk Reduction,* 96, 103960.

Conrad, C. P. (2013). The solid Earth's influence on sea level. *Bulletin*, *125*(7–8), 1027–1052

David, L. O., Aigbavboa, C., Nula, N., and Adepoju, O. O. (2024). *Sustainable Synergy*. Springer Nature.
Dushkova, D., and Ivlieva, O. (2024). Empowering communities to act for a change: A review of the community empowerment programs towards sustainability and resilience. *Sustainability*, 16(19), 8700.
Dyakova, M. (2017). *Investment for health and well-being: A review of the social return on investment from public health policies to support implementing the sustainable development goals by building on health 2020*, NCBI.
Eizenberg, E., and Jabareen, Y. (2017). Social sustainability: A new conceptual framework. *Sustainability*, 9(1), 68.
Fitzgibbons, J., and Mitchell, C. L. (2021). Inclusive resilience: Examining a case study of equity-centred strategic planning in Toronto, Canada. *Cities*, 108, 102997.
Fletcher, C., Ripple, W. J., Newsome, T., Barnard, P., Beamer, K., Behl, A. et al. (2024). Earth at risk: An urgent call to end the age of destruction and forge a just and sustainable future. *PNAS Nexus*, 3(4), 106.
Florini, A., and Pauli, M. (2018). Collaborative governance for the sustainable development goals. *Asia & the Pacific Policy Studies*, 5(3), 583–598.
Ghazi, Z. M., Rizvi, S. W. F., Shahid, W. M., Abdulhameed, A. M., Saleem, H., and Zaidi, S. J. (2022). An overview of water desalination systems integrated with renewable energy sources. *Desalination,* 542, 116063.
Hariram, N. P., Mekha, K. B., Suganthan, V., and Sudhakar, K. (2023). Sustainalism: An integrated socio-economic-environmental model to address sustainable development and sustainability. *Sustainability*, 15(13), 10682.
Hassan, S. J., and Umar, O. M. (2024). Actualizing the sustainable development goals 2030: The role of environmental education in poverty eradication and tackling insecurity in Nigeria. *Scholar J-Science and Education*, 2(1), 150–167.
Hermwille, L., Obergassel, W., Ott, H. E., and Beuermann, C. (2017). UNFCCC before and after Paris–What's necessary for an effective climate regime?. *Climate Policy*, 17(2), 150–170.
Hosseini, S. H. (2023). The well-living paradigm: Reimagining quality of life in our turbulent world. *Discover Global Society*, 1(1), 19.
Islam, M. R. (2024). Empowering marginalized communities: A Bangladesh perspective. In *Family planning and sustainable development in Bangladesh: Empowering marginalized communities in Asian contexts* (pp. 67–84). Emeralds, Emerald Publishing Limited.
Janssen, M., and Helbig, N. (2018). Innovating and changing the policy-cycle: Policy-makers be prepared!. *Government Information Quarterly*, 35(4), S99–S105.
Jarvis, A., Lau, C., Cook, S., Wollenberg, E., Hansen, J., Bonilla, O., and Challinor, A. (2011). An integrated adaptation and mitigation framework for developing agricultural research: Synergies and trade-offs. *Experimental Agriculture*, 47(2), 185–203.
Kuhlman, T., and Farrington, J. (2010). What is sustainability?, *Sustainability*, 2(11), 3436–3448.
Lang, D. J., Wiek, A., Bergmann, M., Stauffacher, M., Martens, P., Moll, P. et al. (2012). Transdisciplinary research in sustainability science: practice, principles, and challenges. *Sustainability Science,* 7, 25–43.
Lemanski, C. (2023). Broadening the landscape of post-network cities: A call to research the off-grid infrastructure transitions of the non-poor. *Landscape Research*, 48(2), 174–186.
Lim, W. M. (2024). What is qualitative research? An overview and guidelines. *Australasian Marketing Journal*, *37*(2), 264619.
Mazutis, D., and Sweet, L. (2022). The business of accelerating sustainable urban development: A systematic review and synthesis. *Journal of Cleaner Production*, 357, 131871, ISSN 0959-6526. https://doi.org/10.1016/j.jclepro.2022.131871.

Naheed, S., and Shooshtarian, S. (2022). The role of cultural heritage in promoting urban sustainability: A brief review. *Land*, 11(9), 1508.

Odume, O. N., Akamagwuna, F. C., Ntloko, P., Dallas, H. F., Nnadozie, C. F., and Barber-James, H. M. (2023). A trait database for southern African freshwater invertebrates. *African Journal of Aquatic Science*, *48*(1), 64–70

Pahl-Wostl, C., Odume, O. N., Scholz, G., De Villiers, A., and Amankwaa, E. F. (2023). The role of crises in transformative change towards sustainability. *Ecosystems and People*, 19(1), 2188087.

Remler, D. K., and Van Ryzin, G. G. (2021). *Research methods in practice: Strategies for description and causation*. Sage Publications.

Rogamelia, R., and Setiawati, R. (2024). Effective strategies for strengthening harmonious families in Bandar Lampung during the era of society transformation 5.0. *RADEN INTAN: Proceedings on Family and Humanity*, 1(1), 466–472.

Şahin, E., Arslan, N. N., and Özdemir, D. (2024). Unlocking the black box: An in-depth review on interpretability, explainability, and reliability in deep learning. *Neural Computing and Applications, 37*(2):859–965.

Schaltegger, S., Beckmann, M., and Hockerts, K. (2018). Collaborative entrepreneurship for sustainability. Creating solutions in light of the UN sustainable development goals. *International Journal of Entrepreneurial Venturing,* 10(2), 131–152.

Selvakumar, P. (2025). The role of technology in enhancing social sustainability. In *Enhancing social sustainability in manufacturing supply chains* (pp. 275–302). IGI Global Scientific Publishing.

Serzante, M., and Khudozhnyk, A. (2023). Reviewing sustainability measurement methods for enterprises. *Sustainability*, *15*(21), 15514.

Shrivastava, P., and Guimarães-Costa, N. (2017). Achieving environmental sustainability: The case for multi-layered collaboration across disciplines and players. *Technological Forecasting and Social Change*, 116, 340–346.

Silva, J. A. (2023). Wastewater treatment and reuse for sustainable water resources management: A systematic literature review. *Sustainability*, 15(14), 10940.

Sima, V., Gheorghe, I. G., Subić, J., and Nancu, D. (2020). Influences of the industry 4.0 revolution on the human capital development and consumer behavior: A systematic review. *Sustainability*, 12(10), 4035.

Tseng, S. C., and Hung, S. W. (2014). A strategic decision-making model considering the social costs of carbon dioxide emissions for sustainable supply chain management. *Journal of Environmental Management,* 133, 315–322.

Van Tulder, R., and van Mil, E. (2022). *Principles of sustainable business: Frameworks for corporate action on the SDGs*. Routledge

Vargas, C. M., and Cooper, P. J. (2024). *Implementing sustainable development: From global policy to local action*. Rowman & Littlefield.

Wang, B., Dabbaghjamanesh, M., Kavousi-Fard, A., and Yue, Y. (2024). AI-enhanced multi-stage learning-to-learning approach for secure smart cities load management in IoT networks. *Ad Hoc Networks,* 164, 103628.

Widjaja, G. (2023). Economic development transformation with environmental vision: Efforts to create sustainable and inclusive growth. *Kurdish Studies*, 11(2), 3154–3177.

Williams, D. R., Mohammed, S. A., Leavell, J., and Collins, C. (2010). Race, socioeconomic status, and health: Complexities, ongoing challenges, and research opportunities. *Annals of the New York Academy of Sciences*, 1186(1), 69–101.

2 Sustainable Development Goals (SDGs 6, 11, 12, 13, 14, and 15) and Waste Management

Adhith Krishna Erath Shine

2.1 INTRODUCTION

Waste management has become one of the most paramount environmental challenges of the 21st century. With the exponential increase in population and rapid expansion of industrialization and urbanization, there has been an unprecedented increase in waste generation. The 2024 Global Waste Management outlook published by the United Nations Environment Programme and International Solid Waste Association projects an increase of 80.9% in global municipal waste generation to 3.8 billion metric tons by 2050. This is coupled with an increase in the cost for global waste management to $640.3 billion after factoring in the hidden costs due to pollution, climate change, and poor health caused by poor waste disposal practices. This increase poses significant environmental, social, and economic challenges, as poor waste management contributes to pollution, public health hazards, and ecological degradation (Kumar et al., 2017).

This issue is not limited to developed nations. Many low- and middle-income countries across the seven continents, each with their own unique economic, cultural, and environmental contexts, struggle with inadequate waste management. This leads to severe environmental consequences, such as water contamination, air pollution, and biodiversity loss. Mismanaged waste, particularly plastic waste and e-waste, has found its way into landfills, rivers, and oceans, contaminating soil and water and severely affecting the ecosystems that thrive off them. Inefficient management of waste in landfills and more primitive methods of waste management such as incineration leads to an increased emission of greenhouse gases, which in turn exacerbates climate change and global warming.

The United Nations Sustainable Development Agenda (SDA), formulated in 2015, provides the framework to tackle this global crisis through sustainable consumption, resource efficiency, and environmental protection. Each sustainable development goal emphasizes the importance of sustainable waste management practices in ensuring a cleaner, healthier, and more sustainable future. This chapter will explore the intricate

DOI: 10.1201/9781003637677-2

relationship between waste management and the Sustainable Development (SDGs), highlighting key challenges, successful global initiatives, and actionable solutions (Ferronato et al., 2019).

2.2 SUSTAINABLE DEVELOPMENT AGENDA

The SDA is a globally recognized framework for achieving social, economic, and environmental sustainability. It was adopted by all United Nations member nations in 2015 as a part of the 2030 Agenda for Sustainable Development. It outlines 17 SDGs which work in tandem with each other to address the world's most pressing challenges, including poverty, inequality, climate change, and environmental degradation. It aims to ensure a better future for all by integrating sustainability into governmental policies, business practices, and everyday life.

2.2.1 Origins of the Sustainable Development Agenda

2.2.1.1 Brundtland Report, London (1987)

A resolution adopted in the 38th session of the United Nations General Assembly in 1983 led to the formation of an independent body called the World Commission on Environment and Development. As per the study of the Brundtland Report (McChensey, 1991), the establishment of this commission led to a significant departure from the conventional approach used previously to tackle environmental issues. As the orthodox agenda, for which most part only focused on highlighting the "effects" of the aforementioned environmental issues, was highly criticized, an alternative agenda was formulated, which in turn set its primary focus on the causes. The commission also radically changed the methodology in which data was obtained to garner more public input regarding these environmental issues. They pushed the discussion from the jurisdiction of scientific and government agencies and introduced it to the public. Between 1985 and 1987, the commission conducted meetings and public hearings in nine countries across North and South America, Europe, Asia, and Africa, where hundreds of organizations and individuals presented testimonies and reports on the difficulties they encountered. They went as far as to visit sites of specific environmental concern to validate these testimonies and to add their own findings to the same. In 1987, the commission finally released their report entitled "Our Common Future," which laid the foundations for the concept of sustainable development for the modern era. The report was then presented to a plethora of presidents, prime ministers, nongovernmental organizations, and press of more than 100 nations for further refinement before being presented in a special plenary session of the United Nations in October 1987 by Gro Harlem Bruntland, the then prime minister of Norway, which led to it being eponymously named the Bruntland Report.

> Bruntland Report coined the widely accepted definition for 'sustainable development' as being "Development which meets the needs of the present, without compromising the ability of future generations to meet their own needs."

It was the foremost report to highlight the interconnected nature of economic growth, environmental protection, and social equity and further emphasized that sustainability cannot be achieved without addressing these issues at their roots. It cautioned about the irreversible damage that could be caused due to unchecked industrialization and resource exploitation and stressed on the need for policies that balance economic progress with ecological conversation. It also urged long-term planning and called for global cooperation, recognizing that environmental challenges transcended national borders and requires coordinated international action. It laid the foundation for major global initiatives that followed, such as the 1992 Earth Summit, Agenda 21, Kyoto Protocol, and finally the Sustainable Development Agenda.

2.2.1.2 Agenda 21, Rio De Janeiro (1992)

Agenda 21 was the comprehensive action plan adopted at the 1992 Earth Summit (United Nations Conference on Environment and Development—UNCED) in Rio de Janeiro. It was developed as a blueprint to help achieve sustainable development globally by addressing social, economic, and environmental issues. Although non-binding the document served as the guiding framework for governments, local authorities, businesses, and general public to integrate sustainability into their policies and practices (United Nations, 1992).

Sustainable development encompasses multiple dimensions, including social and economic aspects that aim to alleviate poverty, promote sustainable consumption, and ensure the inclusion of marginalized groups in decision-making processes. It also involves the conservation and management of natural resources through biodiversity protection, sustainable land use, efficient water and energy management, pollution control, and climate change mitigation. Furthermore, it stresses the importance of partnerships among governments, private sectors, nongovernmental organizations, and communities to advance sustainability efforts. Effective implementation of these goals relies on mechanisms such as technology transfer, capacity building, adequate financial resources, and strong institutional frameworks (United Nations, 1992).

Agenda 21 helped lay the foundation for the present-day SDA, influencing major sustainability policies and agreements. Its emphasis on integrating sustainability into governance helped shape future international frameworks, such as the Millennium Development Goals (MDGs) in 2000 and the 2030 Agenda for Sustainable Development in 2015.

Its key contributions include promoting local level sustainability initiatives, referred to as Local Agenda 21 (LA21) initiatives worldwide. Cities and municipalities worldwide adopted their own sustainable development strategies, demonstrating the need for localized solutions to combat global challenges. This decentralized approach remains central in modern sustainability practices and policies. It also brought forward multilateral international cooperation and thereby helped establish various international climate and environmental agreements. It influenced modern-day treaties such as the Kyoto Protocol (1997) and the Paris Agreement (2015), both of which focus on reducing carbon footprint and addressing climate change—one of the core issues at the heart of the present-day SDA. Furthermore, its principles of sustainable consumption, resource efficiency, and social equity directly shaped

SDGs such as SDG 11 (Sustainable Cities), SDG 12 (Responsible Consumption and Production), and SDG 13 (Climate Action). Agenda 21 played a critical role in the evolution of the SDA by fostering a holistic and participatory approach to the domain of sustainability.

2.2.1.3 Millennium Development Goals—MDGs (2000)

The MDGs Sachs, J. D. (2012) were a set of eight objectives established at the turn of the millennium in 2000 as a part of the United Nations Millennium Declaration. It was aimed at addressing pressing global issues that plagued the international community through global cooperation in order to ensure equitable sustainable development worldwide. The eight MDGs, established in 2000, represented a global commitment to addressing some of the world's most pressing development challenges. These goals aimed to eliminate extreme poverty and hunger, achieve universal primary education, empower women and promote gender equality, reduce child mortality, improve maternal health, combat HIV/AIDS, malaria, and other diseases, ensure environmental sustainability, and foster a global partnership for development. Collectively, they provided a framework for international cooperation and measurable progress in human development up to 2015 (United Nations, 2008). The impact of MDGs had a substantial impact on sustainable development globally and brought remarkable changes with its 15-year time period, from 2000 to 2015. By the 2015 deadline, significant progress was observed across the world; however, this was observed to be uneven across regions with many developing nations showcasing improvement while conflict-affected and lesser developed countries lagged behind.

Global efforts under the MDGs also helped curb the spread of various diseases such as HIV/AIDS, malaria, and tuberculosis. Goal 7 pertaining to environmental sustainability saw mixed results. Biodiversity, climate change, and deforestation increased at an alarming rate while access to potable water increased. A crucial

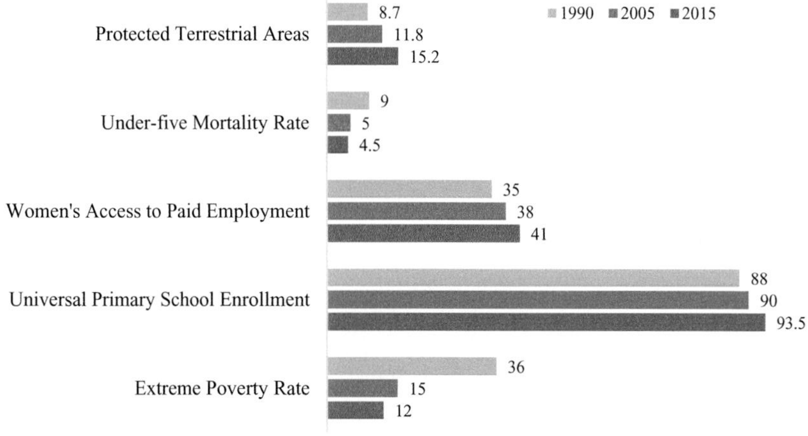

FIGURE 2.1 Illustration of percentage change in various global development parameters (United Nations, 2008).

failure of the MDGs was their narrow focus and lack of emphasis on long-term sustainability. These shortcomings led to the development and adoption of the SDGs in 2015, which built upon the foundations of its predecessor while being broader, integrated, and widely applicable.

2.2.1.4 Rio +20 Conference (2012)

The 20th Anniversary of the original 1992 Earth Summit was commemorated by the United Nations Conference on Sustainable Development (UNCSD), also known as Rio +20 Conference. The aim of the conference was to review the global progress in sustainable development and to establish renewed commitment to sustainability. The conference called for the development and adoption of a wide range of measures and actions in order to shape a better future for the people and the planet. As per the official document associated with the Rio+20 summit, In *The Future We Want*, Paul (2012) outlines a comprehensive set of actions aimed at advancing sustainable global development. These include initiating the process for establishing the SDGs, emphasizing the role of the green economy to achieve sustainability, and strengthening the United Nations Environment Programme (UNEP) along with the creation of a high-level political forum on sustainable development. The document also advocates for corporate sustainability reporting, the development of metrics that go beyond the gross domestic product (GDP) to measure national well-being, and the formulation of a strategy for financing sustainable development. Additionally, it introduces a framework for sustainable consumption and production, highlights the importance of promoting gender equality, and calls for stronger engagement of civil society and integration of scientific input into policymaking (Paul, 2012). The 2012 Rio +20 conference brought about three pivotal changes in the approach toward sustainability. It promoted the involvement of businesses, nongovernmental organizations, indigenous communities, and civil society in decision-making processes. This shift ensured that sustainable development was no longer solely the responsibility of governments but required multi-stakeholder engagement.It also established the High-Level Political Forum on Sustainable Development (HLPF), which became the primary platform for monitoring and reviewing progress toward sustainability goals globally. The HLPF continues to play a crucial role in tracking the implementation of the SDGs and encouraging accountability among nations. Finally, the Rio +20 summit also called for ocean governance, which led to stronger policies on marine ecosystem conservation and sustainable fisheries. This directly influenced SDG 14 (Life below Water), which is intended to help protect marine diversity and combat ocean pollution. The conference revitalized global sustainability efforts and bridged the gap between previous agendas and ambitious development goals and played a critical role in the formation of the Sustainable Development Agenda for 2030.

2.2.1.5 UN General Assembly (2015)

A more ambitious and comprehensive framework, the 2030 Agenda for Sustainable Development, replaced the Millennium Development Goals at the UN Headquarters during the UN General Assembly held in New York, September 25–27, 2015. It introduced 17 goals and 169 targets, aimed to guide global development efforts for a period of 15 years up to 2030. These 17 goals, known as the SDGs, were established to address a broad range of issues and were integrated into all aspects of governance, encouraging governments, businesses, and individuals to align their policies and practices with sustainable development practices. The 2030 Agenda also emphasized equity and inclusivity to ensure that marginalized and vulnerable populations, including women, people with disabilities, and indigenous communities, were at the center of developmental policies. The agenda also brought forth the requirement of detailed data collection and reporting to monitor progress across different groups worldwide instead of relying on national averages. The approach toward environmental issues were also subjected to great change, with greater importance deemed toward it via clear targets and strengthened support toward the Paris Agreement on Climate Change, which was finalized later in 2015.

2.2.2 Key Principles of the Sustainable Development Agenda

The SDA is grounded in several key principles that guide its implementation across all nations, recognizing the global nature of sustainability challenges. Universality is a core tenet, marking a shift from earlier development frameworks by applying the SDGs to both developed and developing countries alike (Kanie and Biermann, 2017). The agenda emphasizes integration and indivisibility, underscoring the need to balance and harmonize economic, social, and environmental dimensions, as progress in one area can significantly affect outcomes in others (Nilsson et al., 2016). Central to the SDA is the commitment to "leaving no one behind," which focuses on eradicating poverty and inequality by ensuring that marginalized populations—such as women, indigenous communities, people with disabilities, and the poor—have equitable access to resources and opportunities (United Nations, 2015). Multi-stakeholder participation is also essential, highlighting the importance of collaboration among governments, the private sector, civil society, academia, and international organizations in achieving the goals (Kroll et al., 2019). The SDA further promotes evidence-based decision-making and accountability, relying on mechanisms such as Voluntary National Reviews (VNRs) and the HLPF to ensure transparency and track progress through reliable data (United Nations, 2017). Finally, the agenda stresses sustainability and resilience, advocating for long-term strategies to address climate change, disaster risk reduction, and resource conservation to safeguard the well-being of future generations (Sachs et al., 2019).

2.2.3 Pillars of the Sustainable Development Agenda

Sustainability is a multidimensional concept encompassing economic, social, environmental, and governance dimensions, all of which are essential for achieving long-term development goals. Economic sustainability involves fostering inclusive and sustained economic growth, creating decent jobs, ensuring fair wages, and promoting innovation through green economy principles, sustainable trade, renewable energy, and digital infrastructure (Sachs et al., 2019). Social sustainability emphasizes equity, inclusivity, and human well-being, focusing on poverty eradication, access to education and healthcare, gender equality, human rights, and the preservation of cultural and indigenous knowledge (Barraclough and Stewart, 2019). Environmental sustainability centers on the responsible management of natural resources through climate action, biodiversity conservation, circular economy practices, and renewable energy adoption to minimize ecological degradation (Rockström et al., 2009). Governance and institutional sustainability, while often less highlighted, is critical for enabling these efforts through transparent institutions, anti-corruption measures, evidence-based policymaking, strong legal frameworks, and global cooperation, ensuring that sustainability initiatives are implemented and monitored effectively (Meadowcroft, 2007).

2.3 COMPARING THE WASTEWATER TREATMENT METHODOLOGIES

One of the most fundamental resources present in our planet is water, with it being essential for both life and economic development. However, in the 21st century, population growth, industrialization, and urbanization have increased the strain on freshwater resources, leading to extensive water pollution. Sustainable wastewater management is paramount for ensuring public health safety, protecting ecosystems and meeting SDGs. The aim of wastewater management is to remove contaminants and reclaim the water for reuse, helping mitigate water scarcity and environmental degradation.

Efficient wastewater management not only helps decrease environmental pollution but also plays a key role in addressing water shortages and helps promote a circular economy at local, regional, and international levels. A comprehensive understanding of wastewater treatment systems is crucial to develop strategies to optimize resource use while maintaining ecological sustainability. The diverse wastewater treatment systems that have been developed to cater to various regions and industries as an outcome of effective policies, infrastructural development, and technological advancements will be dealt with in this section.

2.3.1 Principles of Wastewater Management

The core principle of wastewater treatment is to improve the physical, biological, and chemical attributes of polluted water before discharging it into the environment. The core principles of wastewater management are the removal of physical contaminants, biological degradation of organic matter, chemical neutralization and removal of hazardous substances, sustainability, and resource recovery. Effective wastewater management also follows the principle of pollution prevention and minimization at the source. Industries and municipalities are encouraged to implement best management practices to reduce wastewater generation, such as water-efficient appliances, green infrastructure, and industrial process modifications. By controlling pollution at the source, the burden on water treatment facilities is minimized, improving efficiency. Different treatment processes can be characterized into various different subsets on the basis of treatment stage, mechanisms, and type and origin of wastewater.

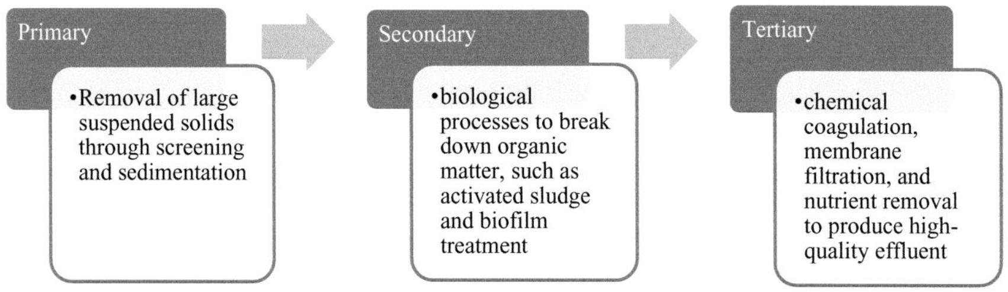

FIGURE 2.2 Illustration of the various stages of wastewater treatment (Tchobanoglous et al., 2003).

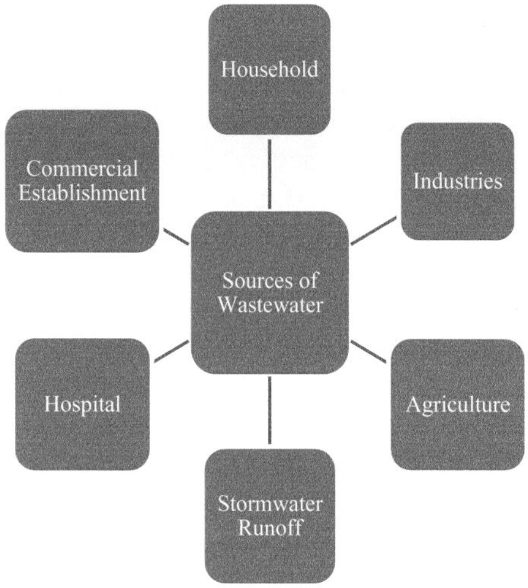

FIGURE 2.3 Illustration of various sources of wastewater (Verstraete, 2009).

2.3.2 Types of Wastewater Management Systems

2.3.2.1 Centralized Wastewater Management System

In major cities, urban areas, and industrial zones, there exist large-scale facilities that operate by collecting wastewater through extensive sewer networks and channeling it to a central treatment facility. The water is then made to undergo multiple stages of purification before being discharged into the environment or reused.

One of the biggest advantages of centralized wastewater treatment is its ability to treat large volumes of wastewater effectively. The system provides complete removal of pollutants, including organic content, heavy metals, and toxic pathogens, through a combination of mechanical, biological, and chemical treatment. The initial treatment stage targets the removal of large solids and grit through screening and sedimentation processes. Secondary treatment follows with biological processes, such as the activated sludge process, decomposing organic pollutants through microbial activity.

The final stage of treatment, referred to as tertiary treatment, involves advanced filtration, chemical disinfection, and nutrient removal to ensure that the effluent discharged into the environment is within environmental guidelines (Massoud et al., 2009).

In spite of their effectiveness, centralized wastewater treatment systems are plagued by many limitations. One of the biggest shortcomings involves the cost of developing and maintaining infrastructure. Development and operation of large-scale treatment facilities demand high capital investment, thus limiting such systems for economically disadvantaged communities. Centralized treatment facilities also demand high energy input, thus contributing to greenhouse gas emissions. Another limitation is the vulnerability of centralized systems to disruption, such as natural disasters or failure of infrastructure, which may lead to mass contamination if not handled appropriately.

In a bid to promote the sustainability of centralized wastewater treatment, various strategies have been implemented. Incorporation of energy recovery technologies, such as anaerobic digestion, helps to reduce the carbon footprint of treatment facilities by generating biogas as a renewable source of energy. Similarly, wastewater reuse and recycling schemes help to recycle treated effluent for reuse in irrigation, industrial processes, and even drinking water supplies in some cases. Through incorporation of sustainable practices, centralized wastewater treatment systems can become a key factor in promoting water security and conservation of the environment.

2.3.2.2 Decentralized Wastewater Management System

Decentralized wastewater treatment systems offer localized solutions to wastewater management for areas where large-scale centralized infrastructure is not feasible. Decentralized systems are especially useful for rural areas, small municipalities, and areas with difficult geographical conditions that make the establishment of large-scale sewage systems difficult.

Decentralized wastewater treatment encompasses a variety of methodologies, such as septic tanks, constructed wetlands, and biofiltration systems. Of these, septic tanks are among the most common decentralized alternatives, especially in residential use. They function through sedimentation and anaerobic digestion processes to degrade organic matter, thereby lowering the pollutant load before the effluent is discharged into a drainage field. Constructed wetlands are another popular approach, utilizing the natural filtration properties of aquatic plants and microbial communities to eliminate contaminants from wastewater. These systems are energy-efficient and utilize minimal energy, making them a popular choice for sustainable water management practices. One of the primary advantages of decentralized wastewater treatment systems is their built-in flexibility. Unlike centralized treatment plants, decentralized systems do not need extensive sewer infrastructures, making them more viable for remote locations. Moreover, these systems can be tailored to meet the specific needs of a community or industry, thereby ensuring maximum efficiency and environmental impact. The decentralized treatment approach also minimizes the likelihood of large-scale system failure, given that each unit operates independently (Massoud et al., 2009).

Nonetheless, decentralized wastewater treatment systems also face some limitations. Many small-scale systems need regular maintenance to function efficiently, and poor maintenance can result in environmental pollution and public health concerns. Furthermore, the treatment capacity of decentralized systems is often limited, making them unsuitable for areas with high population densities or large industrial wastewater discharges. To ensure maximum efficiency in decentralized wastewater treatment, various innovations, such as membrane bioreactors and electrocoagulation technologies, are being researched. Advanced treatment technologies maximize pollutant removal efficiency while at the same time ensuring low operating costs. By investing in decentralized technologies and integrating them with sustainable water management, communities are able to develop resilience against water scarcity and pollution threats. Decentralized systems are best suited to rural or remote communities with little access to large-scale infrastructure. These include septic tanks, constructed wetlands, and composting toilets, which all employ natural processes for water purification. Decentralized treatment systems offer significant advantages, including lower capital outlays, the capacity to adapt to local conditions, and low reliance on extensive networks of pipelines.

Constructed wetlands are an effective decentralized treatment approach that involves the use of aquatic plants and microbial communities to remove contaminants. A diagrammatic representation of constructed wetlands would show wastewater flowing through different wetland zones, thus demonstrating the efficiency of pollutant removal. These wetlands act as natural biofilters, removing nitrogen, phosphorus, and heavy metals effectively from wastewater prior to discharge into surrounding water bodies. Another prominent decentralized approach involves the use of anaerobic digestion in small-scale wastewater treatment systems. Anaerobic digestion produces biogas, which can be used as a renewable energy source while at the same time minimizing sludge volumes. This approach is of particular advantage to rural communities and industries that seek to reduce their use of fossil fuels.

2.3.2.3 Industrial Wastewater Management

Industrial wastewater management entails the treatment and disposal of wastewater produced as a result of manufacturing, processing, and production processes. Textile, chemical, pharmaceutical, food processing, and metal working industries all produce huge amounts of wastewater with a range of contaminants such as heavy metals, organic compounds, and toxic chemicals. Industrial wastewater management entails the application of advanced treatment technologies to remove the unique contaminants in the effluent. Pre-treatment is normally the initial step, with processes such as oil and grease removal, pH adjustment, and sedimentation to settle the solids. Primary treatment is then followed, normally with coagulation and flocculation to clump suspended solids for easy removal. Secondary treatment employs biological processes to break down organic contaminants, with processes such as activated sludge systems, trickling filters, and sequencing batch reactors. Highly contaminated wastewater industries need tertiary treatment, with advanced filtration, adsorption, and chemical oxidation to produce high-quality effluent (Fu and Wang, 2011).

One of the main issues with industrial wastewater is the potential to be environmentally detrimental if not managed appropriately. Untreated or semi-treated wastewater discharges can cause water pollution, damaging aquatic life and contaminating drinking water sources. To reduce these risks, industries are increasingly adopting zero liquid discharge (ZLD) systems, where wastewater is fully treated and recycled, with discharge of liquid waste completely removed. Water recycling and resource recovery technologies such as reverse osmosis and membrane distillation are also being utilized to enhance water efficiency in industrial processes.

Regulatory compliance is an essential part of industrial wastewater management. Governments and environmental agencies set stringent discharge standards and monitoring to prevent any pollution. Industries that fail to meet the standards face the risk of penalty and damage to their reputation. In response to these challenges, industries are adopting cleaner production strategies, environmentally friendly wastewater treatment technologies, and low-impact closed-loop water systems. Advancements in industrial wastewater treatment are increasingly making it possible to recover valuable resources from wastewater, including metals, nutrients, and energy. By integrating circular economy strategies into industrial water management, organizations can reduce waste, lower costs, and enhance global sustainability initiatives.

Industrial wastewater is chemically concentrated and therefore requires special treatment. Treatments range from pre-treatment on the industrial site, chemical neutralization, and reuse of water to achieve minimum environmental discharge. Industrial wastewater treatment complexity depends on the nature of the industry, for instance, textile, pharmaceutical, food processing, and heavy metal industries.

2.3.2.4 Agriculture Wastewater Management

Management of agricultural wastewater is important in the prevention of water pollution from farm, livestock, and irrigation runoff. Agricultural farming releases wastewater containing fertilizers, pesticides, sediments, and organic matter, which, if not managed, can pollute water bodies.

The most outstanding source of agricultural wastewater is irrigation runoff. Irrigation water beyond the requirements is carried by runoff with dissolved agrochemicals and nutrients into adjacent rivers and lakes, posing problems such as eutrophication and algal blooms. Irrigation techniques such as drip irrigation and controlled drainage reduce runoff and increase the efficiency of water use. Livestock wastewater such as manure, pathogens, and ammonia is another source of significant magnitude. Concentrated Animal Feeding Operation (CAFO) wastewater must be treated using anaerobic lagoons, constructed wetlands, or biogas digesters to avert pollution and tap valuable resources in the form of methane gas.

Recovery of nutrients is an inherent principle of sustainable management of agricultural wastewater. Technologies such as struvite precipitation and biological nutrient removal enable the recovery of phosphorus and nitrogen from wastewater, which can be used as a fertilizer. Apart from minimizing the environmental footprint of agricultural discharges, this promotes circular economy patterns in agriculture as well. Control of agricultural wastewater is challenging due to the extensive scope of pollution sources from farms and the inability to implement a common treatment approach. In contrast to industrial wastewater, which is treated at a common plant, agricultural wastewater is distributed across extensive spatial areas, thus making decentralized and landscape-based treatment more important.

To meet these challenges, farmers and policymakers are embracing integrated water resource management (IWRM) practices that integrate best irrigation, land use, and conservation practices. Constructed wetlands, buffer strips, and vegetative filter strips are being used more frequently to capture and treat runoff before it reaches natural water bodies. Precision agriculture practices like soil moisture monitoring and variable-rate fertilization also assist in maximizing the utilization of resources and minimizing the generation of excess wastewater.

By embracing sustainable agricultural wastewater management practices, the agricultural sector can minimize its environmental impact while sustaining productivity. Interdisciplinary collaboration among farmers, researchers, and regulators will be essential in formulating innovative solutions that balance agricultural production with water conservation and ecosystem protection.

2.4 SUSTAINABLE DEVELOPMENT GOALS

The SDGs are an all-round world initiative targeting fundamental environmental, social, and economic issues. Agreed to by all UN member states in September 2015, the 2030 Agenda for Sustainable Development launched 17 SDGs and 169 targets aimed at developing a more sustainable, fairer, and more prosperous world by 2030. They draw on the successes and learnings of the MDGs, broadening their focus to incorporate sustainability, climate action, economic growth, and social inclusion.

Fundamentally, the SDGs seek to reconcile three aspects of sustainability: economic growth, social justice, and environmental conservation. They understand that poverty alleviation and economic growth need to go hand in hand with efforts to combat climate change, inequality reduction, and the protection of natural resources. In contrast to the MDGs, where developing countries were mostly targeted, the SDGs target every country universally, stressing that sustainability is everybody's global responsibility (Sachs et al., 2019).

2.4.1 SDG 6—Clean Water and Sanitation

Sustainable Development Goal 6 (SDG 6) seeks to ensure the availability and sustainable management of water and sanitation for all. Water is one of the most basic natural resources, crucial for human existence, economic growth, and environmental sustainability. In spite of its essentiality, billions of individuals across the globe do not have access to safe drinking water, proper sanitation, and good hygiene facilities. By UN Water, it is estimated that around 2.2 billion individuals have no safely managed drinking water services, and 4.2 billion individuals do not have access to safely managed sanitation facilities. The effects of poor water and sanitation services include the spread of waterborne diseases, lack of hygiene, food insecurity, and environmental degradation.

SDG 6 acknowledges that clean water and sanitation are human rights, and attaining this goal is critical to enhancing health, poverty reduction, and sustainable economic development. It is also crucial for the attainment of other SDGs, such as SDG 3 (Good Health and Well-being), SDG 11 (Sustainable Cities and Communities), and SDG 13 (Climate Action). Water scarcity, lack of sanitation, and climate-related water threats compromise human health and economic productivity and render SDG 6 an international priority.

2.4.1.1 Targets and Indicators of SDG 6

SDG 6 is divided into eight targets that address various aspects of water management, sanitation, and ecosystem conservation. These targets provide a comprehensive framework for improving water security and sustainability worldwide.

6.1 —Achieve Universal and Equitable Access to Safe and Affordable Drinking Water
- This target aims to provide safe, affordable, and reliable drinking water to all people, particularly marginalized communities in developing regions.
- Progress is measured by the proportion of the population using safely managed drinking water services.

6.2 —Achieve Access to Adequate and Equitable Sanitation and Hygiene for All
- Focuses on eliminating open defecation and ensuring everyone has access to basic sanitation facilities such as toilets and handwashing stations.
- Emphasizes the needs of women, girls, and vulnerable populations, as inadequate sanitation disproportionately affects their health and safety.

6.3 —Improve Water Quality by Reducing Pollution
- Targets the reduction of untreated wastewater discharge, hazardous chemical contamination, and nutrient pollution in freshwater sources.
- Encourages countries to promote wastewater treatment, recycling, and safe water reuse to improve water quality.

6.4 —Increase Water-Use Efficiency and Ensure Sustainable Withdrawals
- Aims to address water scarcity by improving efficiency in agriculture, industry, and domestic water use.
- Reducing overextraction of groundwater and implementing water conservation strategies is a key part of this target.

6.5 —Implement Integrated Water Resources Management
- Promotes holistic approaches to water management that consider the interconnectedness of rivers, lakes, groundwater, and ecosystems.
- Encourages international cooperation on transboundary water resources, ensuring shared water bodies are sustainably managed.

6.6 —Protect and Restore Water-Related Ecosystems
- Focuses on safeguarding wetlands, rivers, lakes, and aquifers, which are crucial for maintaining biodiversity and water availability.
- Encourages countries to implement policies that prevent habitat destruction and promote ecosystem restoration.

6.A —Expand International Cooperation and Capacity-Building
- Calls for increased financial and technical support to help developing countries improve water infrastructure and management.

6.B —Support Local Participation in Water and Sanitation Management
- Emphasizes the importance of community-led initiatives in ensuring sustainable water management and sanitation services.

2.4.1.2 Challenges to Achieving SDG 6

In spite of international efforts, serious obstacles impede progress toward SDG 6. These obstacles are especially acute in developing countries, where infrastructure, finance, and governance problems make it hard to deliver clean water and sanitation services (Sachs et al., 2019).

- More than 2.3 billion individuals reside in water-scarce nations, where water demand outstrips supply.
- Climate change exacerbates water scarcity by modifying rainfall patterns, enhancing the frequency of droughts, and triggering glacier melting.
- Increased global temperatures result in increased evaporation rates, further depleting available freshwater supplies.
- Rural areas and urban slums often have inadequate water and sanitation infrastructure, resulting in the use of unsafe water sources.
- Poor governance and lack of funds restrict developing nations from constructing and maintaining water treatment plants.
- Privatization of water services in certain areas renders clean water unaffordable to low-income groups.
- Industrial effluent, agricultural runoff, and inadequate wastewater treatment cause serious water pollution.
- Un-treated sewage and chemical waste pollute rivers, lakes, and groundwater, making water undrinkable.
- Waterborne diseases like cholera, dysentery, and typhoid are spread due to consumption of polluted water.
- Women and girls in most developing nations experience serious impediments to accessing secure sanitation facilities.
- Insufficient hygienic toilets in schools compel numerous girls to withdraw, impacting their education and lives ahead.

2.4.2 SDG 11—Sustainable Cities and Communities

Sustainable Development Goal 11 (SDG 11) seeks to make cities and human settlements inclusive, safe, resilient, and sustainable. The world's population keeps increasing, and urbanization is happening at an unprecedented level. Over half of the world's population is now living in urban areas, and by 2050, the figure is likely to increase to almost 70%. Though urbanization offers economic benefits, innovation, and cross-cultural exchange, it also poses major problems, such as congestion, pollution, poor infrastructure, and exposure to climate change (Satterthwaite, 2010).

SDG 11 recognizes that sustainable urbanization is fundamental to the realization of more comprehensive development objectives. Cities are centers of economic growth and innovation but also devours of huge amounts of resources and producers of most of the world's carbon emissions and waste. Making cities inhabitable, environmentally sustainable, and inclusive is essential for the welfare of both present and future generations.

This objective is interrelated with several other SDGs, such as SDG 6 (Clean Water and Sanitation), SDG 9 (Industry, Innovation, and Infrastructure), SDG 13

(Climate Action), and SDG 15 (Life on Land). Without good urban planning and governance, cities will perpetually face social inequalities, environmental degradation, and economic instability. SDG 11 outlines a pathway for the development of smart, sustainable, and resilient cities that promote economic development without causing extensive environmental damage.

2.4.2.1 Targets and Indicators of SDG 11

SDG 11 consists of 10 key targets, each addressing a different aspect of sustainable urbanization. These targets provide a framework for improving urban resilience, housing accessibility, and environmental sustainability worldwide (Geissdoerfer, 2017).

11.1—Ensure Access to Adequate, Safe, and Affordable Housing and Basic Services
- Aims to eliminate slums and informal settlements by providing affordable housing and basic infrastructure, such as clean water, sanitation, and electricity.
- Measured by the proportion of urban populations living in slums or inadequate housing.

11.2—Provide Access to Sustainable Transport Systems
- Focuses on expanding public transportation, improving road safety, and reducing traffic congestion.
- Encourages non-motorized transport options, such as cycling and walking, to reduce carbon emissions.

11.3—Enhance Inclusive and Sustainable Urbanization
- Promotes participatory urban planning where communities have a voice in city development.
- Aims to integrate social, economic, and environmental considerations in urban planning policies.

11.4—Strengthen Efforts to Protect Cultural and Natural Heritage
- Seeks to preserve historical sites, traditional architecture, and biodiversity in urban areas.
- Recognizes that cultural and natural heritage are vital for tourism, education, and community identity.

11.5—Reduce the Impact of Natural Disasters and Climate Change on Cities
- Focuses on improving urban resilience to floods, earthquakes, and extreme weather events.
- Encourages the development of disaster response strategies and early warning systems.

11.6—Reduce Environmental Impact of Cities
- Targets air quality improvement and waste management by reducing pollution from industries and transportation.
- Encourages sustainable waste disposal, recycling programs, and green infrastructure.

11.7—Provide Universal Access to Safe, Inclusive, and Accessible Public Spaces
- Promotes green spaces, parks, and recreational areas to improve urban livability.
- Ensures that women, children, elderly people, and those with disabilities can safely access public spaces.

11.A—Support Positive Economic, Social, and Environmental Links Between Urban and Rural Areas
- Encourages integrated regional development planning to connect cities with surrounding rural communities.
- Recognizes that rural development is essential for reducing migration pressure on urban centers.

11.B—Increase Urban Resilience and Disaster Risk Reduction Strategies
- Calls for local governments to adopt climate resilience policies and build disaster-resistant infrastructure.
- Encourages collaborative governance between national and local governments.

11.C—Support Least Developed Countries in Building Sustainable and Resilient Urban Infrastructure
- Seeks to increase international aid and investment in sustainable infrastructure projects in developing nations.

2.4.2.2 Challenges to Achieving SDG 11

In spite of advancement in sustainable urbanization, some challenges prevent the realization of SDG 11, especially in developing nations whose urban infrastructure is underfinanced and underdevelopment (Geissdoerfer, 2017).

More than one billion people currently live in slums or informal settlements without access to essential services, highlighting the growing inequality in urban environments. As urbanization continues to outpace the development of adequate housing, affordable shelter remains unattainable for many low-income households. This rapid expansion also contributes to severe traffic congestion in most cities, driving up air pollution and carbon emissions. Inadequate public transportation systems further compel reliance on private vehicles, worsening environmental degradation. Climate change-related threats such as sea level rise, heatwaves, and intense storms pose significant risks to urban infrastructure, particularly in coastal and flood-prone regions, where displacement and economic loss are increasingly common. Additionally, inefficient waste collection and recycling practices contribute to widespread environmental contamination, while industrial activities release pollutants into the air, water, and soil, posing serious health risks to urban populations.

2.4.3 SDG 12—Responsible Consumption and Production

Sustainable Development Goal 12 (SDG 12) seeks to ensure sustainable consumption and production patterns. With growing global populations and economies, the need for natural resources, energy, and products has grown exponentially. The existing rate of resource extraction, production, and generation of waste is unsustainable, causing extreme environmental degradation, such as deforestation, loss of biodiversity, pollution, and climate change. The shift toward a circular economy, where resources are utilized optimally, waste is reduced, and materials are recycled, is critical to realizing sustainable development.

SDG 12 encourages sustainable consumption and production through the reduction of food loss and waste, increased energy efficiency, sustainable business practices, and sustainable management of natural resources. This objective is key to achieving other SDGs, such as SDG 13 (Climate Action), SDG 14 (Life Below Water), and SDG 15 (Life on Land), since unsustainable patterns of consumption are directly linked to environmental degradation. Ensuring that economic growth does not harm the environment is central to this objective and is a key element in long-term global sustainability.

2.4.3.1 Targets and Indicators of SDG 12

SDG 12 consists of 11 targets that focus on resource efficiency, sustainable business practices, and waste reduction. These targets address critical areas of production and consumption to create a more sustainable economic system (Traverso and Mankaa, 2023).

12.1—Implement the 10-Year Framework of Programs (10YFP) on Sustainable Consumption and Production
- Calls for all countries, especially developing nations, to integrate sustainable production policies into their national strategies.

12.2—Achieve Sustainable Management and Efficient Use of Natural Resources
- Focuses on reducing overexploitation of forests, minerals, fossil fuels, and freshwater to ensure long-term availability.

12.3—Halve Global Food Waste at the Retail and Consumer Levels
- Aims to reduce food loss along production chains and prevent wastage at the household and commercial levels.

12.4—Achieve Environmentally Sound Management of Chemicals and Waste
- Seeks to minimize the release of hazardous chemicals and pollutants into air, water, and soil to protect human and environmental health.

12.5—Substantially Reduce Waste Generation through Prevention, Reduction, Recycling, and Reuse
- Encourages countries to promote recycling programs, circular economy policies, and waste management systems.

12.6—Encourage Companies to Adopt Sustainable Practices and Sustainability Reporting
- Calls for businesses to implement corporate social responsibility (CSR) policies and transparent sustainability reporting.

12.7—Promote Sustainable Public Procurement Practices
- Encourages governments to prioritize environmentally friendly products and services in their procurement policies.

12.8—Ensure That People Have the Information and Awareness for Sustainable Development and Lifestyles
- Promotes education, behavioral change, and awareness campaigns to encourage sustainable consumption habits.

12.A—Support Developing Countries in Strengthening Scientific and Technological Capacity for Sustainable Consumption and Production

- Provides financial and technical support for sustainable industrialization in low-income nations.

12.B—Develop and Implement Tools to Monitor Sustainable Tourism

- Encourages eco-friendly tourism practices that benefit local communities and protect the ecosystem.

12.C—Rationalize Inefficient Fossil Fuel Subsidies

- Calls for the eliminations of fossil fuel subsidies that promote overconsumption and environmental damage.

2.4.3.2 Challenges to Achieving SDG 12

In spite of attempts to ensure sustainable consumption and production, various challenges are hampering global progress toward SDG 12. These challenges have their origins in economic systems, consumerism, and weak regulatory frameworks.

The global economy is currently consuming natural resources at a rate 1.75 times faster than the Earth can regenerate, placing immense pressure on ecological systems (Traverso and Mankaa, 2023). Irresponsible exploitation of forests, minerals, and fossil fuels is accelerating biodiversity loss and ecosystem degradation. Additionally, about one-third of all food produced—equivalent to over 1.3 billion tons—is wasted annually due to inefficiencies in storage, transportation, and consumer behavior, exacerbating both food insecurity and environmental damage. Corporate focus on short-term profits often overshadows commitments to environmental and social sustainability, while greenwashing misleads consumers and undermines genuine progress. Poor management of electronic waste, plastics, and industrial chemicals continues to pollute air, land, and water, particularly in developing nations lacking effective waste management infrastructure. Moreover, annual government subsidies exceeding $500 billion for fossil fuels hinder the competitiveness of renewable energy sources. This inefficiency, coupled with high energy use in buildings, transport, and industry, contributes significantly to greenhouse gas emissions (Lathabai and Nedungadi, 2024; Marcos et al., 2023).

2.4.4 SDG 13—Climate Action

Sustainable Development Goal 13 (SDG 13) seeks to take action to end climate change and its effects. One of the most serious problems facing the world today is climate change, with increasing global temperatures, weather-related disasters, and natural resource degradation posing dangers to human health, the environment, and economic security. The climate of the Earth is altering at a record pace through human activity, mainly the combustion of fossil fuels, deforestation, and unsustainable land use. The Intergovernmental Panel on Climate Change (IPCC) predicts that unless urgent action is taken, warming will pass 1.5°C above pre-industrial levels, with catastrophic and irreversible effects.

Climate change touches all areas of life, including food security and water supply on the one hand, and health and economic productivity on the other hand. Sea-level rise endangers coastal communities, while heatwaves, droughts, and hurricanes are increasing in frequency and severity. Developing countries, although being the least contributor to greenhouse gases, are generally the most at risk from climate-related disasters owing to limited means, poor infrastructure, and reliance on climate-sensitive sectors like agriculture.

SDG 13 identifies the imperative for climate action for sustainable development and interlinkages to other goals including SDG 6 (Clean Water and Sanitation), SDG 7 (Affordable and Clean Energy), SDG 11 (Sustainable Cities and Communities), and SDG 15 (Life on Land). Global effort must be initiated on a urgent basis to slow down the rates of greenhouse gases emission, strengthen the resilience toward climate change, and shift toward low-carbon economies.

2.4.4.1 Targets and Indicators of SDG 13

SDG 13 consists of five key targets that focus on mitigation, adaptation, policy implementation, and education to address climate change. These targets provide a roadmap for reducing emissions, strengthening climate resilience, and promoting environmental sustainability.

13.1—Strengthen Resilience and Adaptive Capacity to Climate-Related Hazards and Natural Disasters

- Calls for climate-resilient infrastructure, early warning systems, and disaster risk reduction strategies to minimize climate-related damages.

13.2—Integrate Climate Change Measures into National Policies, Strategies, and Planning

- Encourages governments to align their economic and development policies with climate action and reduce reliance on fossil fuels.

13.3—Improve Education, Awareness, and Human and Institutional Capacity on Climate Change Mitigation and Adaptation

- Promotes climate literacy, public engagement, and research on sustainable solutions for climate action.

13.A—Implement the Commitment by Developed Countries to the UN Climate Finance Goal of Mobilizing $100 Billion Annually

- Developed nations must provide financial assistance to developing countries to help them mitigate and adapt to climate change.

13.B—Promote Climate Resilience in Least Developed Countries and Small Island Developing States

- Focuses on supporting vulnerable communities in building resilience to climate change through technology, funding, and policy development.

2.4.4.2 Challenges to Achieving SDG 13

In spite of growing knowledge and policy pledges, a number of challenges impede world progress toward the attainment of SDG 13. These are economic, political, and technological challenges that hamper climate action (Zhou and Wang, 2023).

Global carbon dioxide (CO_2) emissions continue to rise due to industrial activities, transportation, and large-scale deforestation, while the global energy system remains heavily dependent on fossil fuels—coal, oil, and gas—which account for over 80% of energy consumption. Although renewable energy sources like wind and solar are expanding, they still represent a small fraction of the energy mix, and most countries lack the infrastructure and financial resources to make a rapid transition away from fossil fuels. Climate change impacts are disproportionately severe in poorer nations, where droughts, floods, and hurricanes lead to displacement, food insecurity, and economic hardship. Small Island Developing States (SIDS) are especially vulnerable to rising sea levels and climate-related disasters. Despite commitments by wealthier countries to provide $100 billion annually to support climate action in developing nations, this pledge remains underfunded and largely unrealized. Furthermore, many countries lack comprehensive climate policies and enforcement mechanisms to meaningfully reduce emissions. Ongoing deforestation for agriculture, mining, and urban growth continues to reduce the planet's carbon absorption capacity, while desertification and soil degradation diminish agricultural productivity, exacerbating food scarcity and economic instability.

2.4.5 SDG 14—Life Below Water

Sustainable Development Goal 14 (SDG 14) seeks to conserve and manage sustainably the oceans, seas, and marine resources for sustainable development. Oceans cover over 70% of the Earth's surface, have a significant function in climate regulation, and accommodate marine biodiversity, livelihoods, and global food security. Nonetheless, pollution, overfishing, habitat degradation, and climate change pose significant threats to the health of the world's oceans, jeopardizing marine life and the millions of people who make their living off them.

The oceans take in close to 30% of CO_2 emissions resulting from human activities and serve to abate climate change. Yet ocean acidification as a result of excess CO_2 destroys marine life, impacting coral reefs, shellfish, and fishery. Meanwhile, plastic trash, oil spillage, and chemical pollution persist in polluting marine life, while illegal and overfishing empty fish stock, undermining food security and economic stability.

SDG 14 calls for ocean conservation, sustainable fishing, reduced pollution, and global cooperation to restore and maintain ocean ecosystems. The goal has a direct correlation with SDG 2 (Zero Hunger), SDG 12 (Responsible Consumption and Production), and SDG 13 (Climate Action) because healthy oceans play a critical role in food security, economic life, and climate resilience.

2.4.5.1 Targets and Indicators of SDG 14

SDG 14 includes ten key targets focused on marine ecosystem preservation, sustainable fisheries, pollution control, and international partnerships. These targets aim to reduce human impact on the oceans and promote long-term sustainability (Cormier and Elliott, 2017).

14.1—Reduce Marine Pollution

- Aims to prevent and significantly reduce marine pollution, particularly from land-based activities, including plastic waste, sewage, and agricultural runoff.

14.2—Protect and Restore Marine Ecosystems

- Focuses on sustainably managing and protecting marine and coastal ecosystems to prevent environmental degradation and promote ecosystem resilience.

14.3—Minimize Ocean Acidification

- Calls for urgent action to combat ocean acidification, which threatens marine biodiversity, coral reefs, and shellfish populations.

14.4—Ensure Sustainable Fishing and Restore Fish Stocks

- Aims to regulate overfishing, illegal, unreported, and unregulated (IUU) fishing, and destructive fishing practices to restore depleted fish stocks.

14.5—Conserve at Least 10% of Coastal and Marine Areas

- Encourages governments to designate marine protected areas (MPAs) to safeguard vulnerable marine habitats and biodiversity.

14.6—Prohibit Harmful Fisheries Subsidies

- Calls for the elimination of government subsidies that contribute to overfishing and illegal fishing activities.

14.7—Increase Economic Benefits from Sustainable Marine Resources

- Aims to enhance the economic benefits of sustainable fisheries, aquaculture, and marine tourism, particularly in small island developing states (SIDS) and least developed countries (LDCs).

14.A—Support Marine Research and Technology Transfer

- Promotes scientific research and marine technology to enhance knowledge of ocean ecosystems and support sustainable management practices.

14.B—Provide Access to Marine Resources for Small-Scale Fishers

- Encourages governments to support artisanal and small-scale fisheries by ensuring fair access to marine resources and markets.

14.C—Strengthen International Laws for Ocean Governance

- Calls for global cooperation and the implementation of international agreements to improve marine conservation efforts and sustainable ocean management.

2.4.5.2 Challenges to Achieving SDG 14

Although increased awareness of the importance of conserving oceans exists, there are various challenges preventing the achievement of SDG 14. Such challenges stem from economic demands, governance, and environmental alterations within marine ecosystems.

Each year, over 8 million tons of plastic waste enter the ocean, causing severe harm to marine ecosystems and species, while microplastics infiltrate the food chain, posing risks to both marine organisms and human health. Approximately one-third of global fish stocks are overfished, a situation worsened by illegal, unreported, and unregulated fishing. Bycatch—the unintentional capture of non-target marine species—further accelerates biodiversity loss. Climate change has intensified ocean warming, leading to coral bleaching, sea-level rise, and significant habitat degradation. Ocean acidification disrupts shellfish reproduction and undermines the structural integrity of coral reefs, threatening entire marine ecosystems. Coastal degradation due to rapid urbanization, tourism, and industrial activity has led to the loss of critical habitats such as mangroves, seagrasses, and coral reefs. Additionally, unsustainable aquaculture practices contribute to coastal water contamination through antibiotic use and nutrient overload. Many nations lack effective policy frameworks and enforcement mechanisms to combat marine pollution and illegal fishing, while conflicting priorities between economic development and environmental conservation hinder timely action (Cormier and Elliott, 2017).

2.4.6 SDG 15—Life on Land

Sustainable Development Goal 15 (SDG 15) aims to conserve, restore, and promote sustainable use of terrestrial ecosystems, manage forests sustainably, combat desertification, and stop biodiversity loss. Land ecosystems offer critical services like clean air, water regulation, food production, and climate regulation, but human activities have led to extreme environmental degradation. Deforestation, desertification, land degradation, and loss of biodiversity are proceeding at an alarming rate, which poses a threat to global sustainability.

Forests occupy about 31% of the planet's surface and are home to more than 80% of the world's terrestrial species, but deforestation for agriculture, forestry, and infrastructure development still destroys habitats. Land degradation also impacts about 3.2 billion people, lowering crop productivity and increasing food insecurity, particularly in developing nations. Loss of biodiversity, mainly caused by habitat loss, pollution, and climate change, upsets ecosystems and lowers the Earth's ability to support life.

SDG 15 is interconnected with SDG 13 (Climate Action), SDG 6 (Clean Water and Sanitation), and SDG 12 (Responsible Consumption and Production) because land preservation is vital to ensure climate resilience, water security, and sustainable management of resources. Unless drastic action is taken to recover lost land, safeguard forests, and maintain biodiversity, the coming generations will have to contend with devastating ecological emergencies.

2.4.6.1 Targets and Indicators of SDG 15

SDG 15 includes 12 key targets aimed at reducing deforestation, restoring degraded ecosystems, protecting biodiversity, and strengthening environmental policies. These targets outline the necessary steps to ensure sustainable land management and ecosystem conservation (Gao and Zhang, 2019).

15.1—Conserve and Restore Terrestrial and Freshwater Ecosystems
- Calls for the protection and restoration of forests, wetlands, and drylands to maintain ecosystem services.

15.2—Promote Sustainable Forest Management and Halt Deforestation
- Encourages reforestation, afforestation, and responsible forestry practices to prevent biodiversity loss.

15.3—Combat Desertification and Restore Degraded Land
- Aims to reverse land degradation by promoting sustainable agricultural and land-use practices.

15.4—Ensure Conservation of Mountain Ecosystems
- Seeks to protect mountain biodiversity, water sources, and fragile ecosystems from human-induced damage.

15.5—Reduce Degradation of Natural Habitats and Halt Biodiversity Loss
- Targets the reduction of species extinction caused by deforestation, pollution, and land conversion.

15.6— Ensure Fair and Equitable Sharing of Genetic Resources
- Promotes fair access to genetic resources and benefits derived from biodiversity, particularly for indigenous communities.

15.7—Combat Poaching and Illegal Wildlife Trafficking
- Strengthens laws and enforcement to prevent illegal hunting, trade, and exploitation of endangered species.

15.8—Prevent the Introduction and Spread of Invasive Species
- Aims to control and eradicate invasive species that threaten ecosystems and agricultural lands.

15.9—Integrate Ecosystem and Biodiversity Values into Planning and Development
- Encourages countries to incorporate environmental protection into national economic policies.

15.A—Mobilize Financial Resources for Biodiversity Conservation
- Calls for increased funding and investment in conservation projects.

15.B—Promote Sustainable Land Use Practices
- Encourages sustainable agriculture, agroforestry, and land management techniques.

15.C—Support Local Communities in Preventing Poaching and Trafficking
- Strengthens community-based initiatives and conservation efforts to protect wildlife and prevent illegal activities.

2.4.6.2 Challenges to Achieving SDG 15

In spite of increasing awareness, various obstacles are in the way of achieving SDG 15. These obstacles are the result of economic stress, poor governance, and climate change effects on land ecosystems.

Over 10 million hectares of forest are lost annually due to logging, agriculture, and infrastructure expansion, with illegal logging and ineffective land management further accelerating habitat destruction and carbon emissions. Nearly 75% of the planet's terrestrial surface has been degraded through overgrazing, soil erosion, and unsustainable farming practices. This widespread land degradation, particularly in Africa and Asia, is contributing to desertification, reduced agricultural productivity, and forced migration. The Earth is currently experiencing a Sixth Mass Extinction, with approximately one million species at risk due to human activities such as habitat destruction, overhunting, and climate change. The illegal wildlife trade, estimated at over $20 billion annually, drives species like elephants, rhinos, and pangolins toward extinction. Conservation efforts are often undermined by weak enforcement mechanisms and corruption. Moreover, rising temperatures and extreme weather events are damaging key ecosystems—including forests, wetlands, and grasslands—compromising their roles in carbon storage and water regulation (Gao and Zhang, 2019).

2.5 CONTRIBUTION OF WASTEWATER MANAGEMENT TO SDGS

Wastewater management is key to the attainment of several SDGs as it helps to address water scarcity, pollution, health hazards, and environmental degradation. More than 380 billion cubic meters of wastewater is produced annually worldwide, with most of it going untreated or mismanaged. If wastewater is treated, reused, and managed properly, it is a valuable resource instead of a pollutant. Sustainable wastewater management contributes to SDG 6 (Clean Water and Sanitation), SDG 3 (Good Health and Well-being), SDG 11 (Sustainable Cities and Communities), SDG 12 (Responsible Consumption and Production), SDG 13 (Climate Action), SDG 14 (Life Below Water), and SDG 15 (Life on Land).

Treatment, recycling, and environmentally friendly disposal of wastewater avoid water pollution, protect freshwater resources, ensure hygiene, maintain sustainable agriculture, and curb greenhouse gas emissions. Moreover, recent advances in technologies for treating wastewater, energy reuse, and recovering nutrients additionally facilitate sustainability. Here, how various SDGs can be attained by wastewater management, and how such management must be part of national policy if a better, more sustainable world is desired, are elucidated (Obaideen and Al-Gheethi, 2022).

One of the most straightforward contributions of wastewater management to SDG 6 is to provide universal access to clean water and sanitation. Effective wastewater treatment and reuse enhance water quality through decreased pollution, the elimination of dumping, and reduced releases of hazardous chemicals, while more of the treated wastewater is reused. Today, 80% of global wastewater is released untreated into rivers, lakes, and groundwater. Sustainable management of wastewater can boost the amount of safely treated water available for recycling in irrigation, industry, and even drinking water supplies. Singapore and Israel have already made successful large-scale programs of wastewater recycling, thereby dramatically decreasing their dependency on freshwater sources (Kopecká et al., 2024).

Availability of clean water and sanitation underlies public health. Lack of proper wastewater management causes extreme health hazards such as polluted drinking water, epidemics of waterborne diseases, and exposure to harmful chemicals. SDG 3 calls for the prevention of waterborne disease and the achievement of sustainable water resources. Wastewater carries bacteria, viruses, and parasites that pollute drinking water supplies, particularly in developing nations with inadequate sanitation infrastructure. Annually, 1.8 million people succumb to waterborne disease, and most are located in poor countries. Investing in wastewater treatment plants, decentralized sanitation facilities, and sewer systems helps lower the risks of illnesses and promotes improved health within the communities.

Urbanization is fast escalating, with more than 4.4 billion citizens dwelling in cities today. This growth causes high levels of strain on the water and sanitation facilities. SDG 11 aims to make cities inclusive, resilient, and sustainable, and wastewater management is central to mitigating pollution and enhancing urban water cycles. Most megacities experience severe water scarcity, exacerbated by unregulated wastewater discharge. Effective wastewater treatment keeps urban water bodies from being waste dumps, maintaining ecosystems and enhancing the quality of life for city dwellers. Metropolises such as Tokyo, Los Angeles, and Copenhagen have embraced wastewater recycling systems to develop closed water cycles that minimize groundwater withdrawal and protect aquifers.

Circular economy promotes effective wastewater management that minimizes waste and encourages efficiency in the use of resources. SDG 12 calls for minimizing waste, enhancing recycling, and ensuring sustainable industrial processes. Industries produce a high amount of toxic wastewater that has chemicals, heavy metals, and organic contaminants. Installation of on-site wastewater treatment, water recycling plants, and zero-liquid discharge (ZLD) technologies can significantly save water and reduce pollution. Circular wastewater management entails the recovery of nutrients (phosphorus and nitrogen) from sewage, which can be employed as fertilizers in agriculture. This minimizes the use of synthetic fertilizers, curtailing land and water pollution.

Wastewater treatment has far-reaching consequences for climate change mitigation and adaptation. SDG 13 demands that greenhouse gas emissions are reduced and climate resilience is improved. Conventional wastewater treatment is a source of greenhouse gas emissions, primarily through methane (CH_4) and nitrous oxide (N_2O) emissions from unreclaimed sewage. Anaerobic digestion and biogas recovery combined with the application of energy-efficient treatment technologies can transform wastewater plants into net energy producers instead of being net energy consumers. For instance, the Strass wastewater treatment plant in Austria produces more energy than it uses through the utilization of biogas from sewage sludge. Scaling up such projects can contribute to emissions reduction and the development of renewable energy sources.

Wastewater pollution poses a significant risk to marine and terrestrial ecosystems. SDG 14 and SDG 15 aim to safeguard aquatic and land ecosystems from human activities. Untreated wastewater releases nutrients into rivers and oceans, leading to eutrophication and dead zones. This causes oxygen depletion, massive fish kills, and habitat loss. Excess nutrients are removed by advanced wastewater treatment before entering aquatic ecosystems. Wastewater can also be reused for irrigation in dry climates, lessening the demand on freshwater resources and enhancing soil health. In Israel, nearly 90% of treated wastewater is used for irrigation, making it a global leader in sustainable water reuse.

2.6 CONCLUSION

Wastewater management is a key pillar of sustainability that is strongly interlinked with the achievement of several SDGs. Effective treatment and recycling of wastewater not only ensures access to clean water and sanitation (SDG 6) but also improves public health (SDG 3), sustainable urban development (SDG 11), responsible production patterns (SDG 12), climate resilience (SDG 13), and marine and terrestrial ecosystem protection (SDGs 14 and 15). Shift toward sustainable wastewater management systems, driven by the strength of technology, regulatory innovation, and global collaboration, is essential to reducing environmental degradation, preserving natural resources, and developing resilient communities. In the context of growing severity of water scarcity and rising environmental pressures, embracing a circular strategy to wastewater—marked by waste minimization and resource recovery—will be essential to securing a sustainable future. Incorporation of wastewater management into national and global sustainability agendas is not only an environmental imperative but also a socioeconomic imperative, ensuring equitable access to clean water, preventing pollution, and improving ecosystems for future generations.

REFERENCES

Barraclough, L., and Stewart, F. (2019). Social sustainability: A review and critique. *International Journal of Sustainable Development, 22*(2), 101–117. https://doi.org/10.1504/IJSD.2019.100226

Ferronato, N., and Torretta, V. (2019). Waste mismanagement in developing countries: A review of global issues. *International Journal of Environmental Research and Public Health, 16*(6), 1060. https://doi.org/10.3390/ijerph16061060

Fu, F., and Wang, Q. (2011). Removal of heavy metal ions from wastewaters: A review. *Journal of Environmental Management, 92*(3), 407–418.

Gao, Y., and Zhang, Y. (2019). Measuring SDG 15 at the county scale: Localization and practice in China. *ISPRS International Journal of Geo-Information, 8*(11), 515. https://doi.org/10.3390/ijgi8110515

Geissdoerfer, M., Savaget, P., Bocken, N. M. P., and Hultink, E. J. (2017). The circular economy: A new sustainability paradigm? *Journal of Cleaner Production, 143*, 757–768. https://doi.org/10.1016/j.jclepro.2016.12.048

Ian, M. (1991). *The Brundtland report and sustainable development in New Zealand*. Lincoln University and University of Canterbury. Centre for Resource Management.

Kanie, N., and Biermann, F. (Eds.). (2017). *Governing through Goals: Sustainable Development Goals as Governance Innovation*. MIT Press.

Kopecká, R., Hrad, M., and Huber-Humer, M. (2024). The role of the waste sector in the sustainable development goals and the IPCC assessment reports. Österreichische Wasser- und Abfallwirtschaft, 76(5), 300–307.

Kroll, C., Warchold, A., and Pradhan, P. (2019). Sustainable development goals (SDGs): Are we successful in turning trade-offs into synergies? *Palgrave Communications, 5*, 140. https://doi.org/10.1057/s41599-019-0335-5

Kumar, S., Smith, S. R., Fowler, G., Velis, C., Kumar, S. J., Arya, S., Rena Kumar, R., Cheeseman, C., and Kumar, S. (2017). Challenges and opportunities associated with waste management in India. *Royal Society Open Science, 4*(3), 160764. https://doi.org/10.1098/rsos.160764

Lathabai, H., and Nedungadi, P. (2024). Sustainable development goal 12 and its synergies with other SDGs: A scientometric analysis. *Discover Sustainability, 5*(1), 1–20. https://doi.org/10.1007/s43621-024-00289-0

Leal Filho, W. (Ed.). (2018). *Handbook of sustainability science and research*. Springer. https://doi.org/10.1007/978-3-319-63007-6

Marcos, A., Hartmann, P., and Barrutia, J. M. (2023). Toward the implementation of SDG12 to ensure sustainable consumption and production patterns: opportunities and neglected issues. In *Handbook of Sustainability Science in the Future: Policies, Technologies and Education by 2050* (pp. 1353-1376). Cham: Springer International Publishing.

Massoud, M. A., Tarhini, A., and Nasr, J. A. (2009). Decentralized approaches to wastewater treatment and management: Applicability in developing countries. *Journal of Environmental Management, 90*(1), 652–659. https://doi.org/10.1016/j.jenvman.2008.07.001

Meadowcroft, J. (2007). National sustainable development strategies: Features, challenges and reflexivity. *European Environment, 17*(3), 152–163. https://doi.org/10.1002/eet.449

Nations. (1992). *Agenda 21: Programme of Action for Sustainable Development.* United Nations Conference on Environment and Development (UNCED), Rio de Janeiro, Brazil. https://sustainabledevelopment.un.org/content/documents/Agenda21.pdf

Nilsson, M., Griggs, D., and Visbeck, M. (2016). Map the interactions between sustainable development goals. *Nature, 534*(7607), 320–322. https://doi.org/10.1038/534320a

Nwachukwu, C. P., Umobi, C. O., Nwaorgu, V. C., Okwuosa, F. C., and Offoma, S. O. (2026). Impacts of dumpsite leachate on groundwater quality in Awka, Anambra State, Nigeria and environs: A physicochemical and heavy metal analysis. *UNIZIK Journal of Agricultural and Biosystems Engineering*, *1*(1), 93–103.

Obaideen, K., Shehata, N., Sayed, E. T., Abdelkareem, M. A., Mahmoud, M. S., and Olabi, A. G. (2022). The role of wastewater treatment in achieving sustainable development goals (SDGs) and sustainability guideline. *Energy Nexus*, 7, 100112.

Paul, J. (2012). The future we want: Key outcomes from the UN conference on sustainable development (Rio+20). *Natural Resources Forum, 36*(4), 233–237. https://doi.org/10.1111/1477-8947.12005

Rockström, J., Steffen, W., Noone, K., Persson, Å., Chapin, F. S., Lambin, E. F. et al. (2009). A safe operating space for humanity. *Nature, 461*(7263), 472–475. https://doi.org/10.1038/461472a

Rezouga, F., Hamdi, M., and Sperandio, M. (2009). Variability of kinetic parameters due to biomass acclimation: case of para-nitrophenol biodegradation. *Bioresource technology*, *100*(21), 5021–5029.

Sachs, J. D. (2012). From millennium development goals to sustainable development goals. *The Lancet*, *379*(9832), 2206–2211.

Sachs, J. D., Schmidt-Traub, G., Kroll, C., Lafortune, G., and Fuller, G. (2019). *Sustainable Development Report 2019.* Bertelsmann Stiftung and Sustainable Development Solutions Network (SDSN). www.sdgindex.org/reports/sustainable-development-report-2019/

Satterthwaite, D. (2010). Urbanization and its implications for food and farming. *Philosophical Transactions of the Royal Society B: Biological Sciences, 365*(1554), 2809–2820. https://doi.org/10.1098/rstb.2010.0136

Tchobanoglous, G., Burton, F. L., and Stensel, H. D. (2003). Wastewater engineering: Treatment and reuse. *Journal of Environmental Engineering*, *129*(5), 456–465. https://doi.org/10.1061/(ASCE)0733-9372(2003)129:5(456)

Traverso, M., and Mankaa, R. N. (2023). The sustainable development goals. In *The Palgrave Handbook of Global Sustainability* (pp. 1255–1277). Springer. https://doi.org/10.1007/978-3-031-01949-4_83

United Nations. (2008). *The Millennium Development Goals Report 2008.* United Nations. www.un.org/millenniumgoals/2008highlevel/pdf/newsroom/mdg%20reports/MDG_Report_2008_ENGLISH.pdf

United Nations. (2015). *Transforming Our World: The 2030 Agenda for Sustainable Development.* https://sdgs.un.org/2030agenda

United Nations. (2017). *Voluntary National Reviews: Synthesis Report 2017.* https://hlpf.un.org

Verstraete, W., Van de Caveye, P., and Diamantis, V. (2009). Maximum use of resources present in domestic "used water". *Bioresource Technology*, *100*(23), 5537–5545. https://doi.org/10.1016/j.biortech.2009.05.039

Zhou, D., and Wang, Y. (2023). Sustainable development goal 13: Recent progress and challenges. *Environmental Science and Pollution Research, 30*(5), 12345–12360. https://doi.org/10.1007/s11356-023-00287-4

Zulfiqar, K., and Butt, M. J. (2021). Preserving community's environmental interests in a meta-ocean governance framework towards sustainable development goal 14: A mechanism of promoting coordination between institutions responsible for curbing marine pollution. *Sustainability*, *13*(17), 9983.

3 Adopting Lean Principles for Waste Management

Abhyudaya Sinha and Aravind Raj Sakthivel

3.1 INTRODUCTION TO WASTE MANAGEMENT

3.1.1 Understanding Waste

Manufacturing companies today operate under a multitude of pressures that shape their strategies and day-to-day operations. These pressures include stringent customer requests for timely yet low-cost deliveries, the necessity for continuous new product development and innovation, and the challenges posed by small lot production and product variety. Additionally, there is an increasing expectation for zero defects, high product quality, and even customization options tailored to individual customer needs. In this competitive landscape, any failure to meet these standards can have dire consequences. Companies may face monetary sanctions or charges due to noncompliance with regulations, delays in delivery schedules, or deficiencies in product quality and reliability (Agrogiannis, 2015). Such repercussions not only affect a company's bottom line but can also damage its reputation in the marketplace, highlighting the critical need for effective operational strategies that align with these demands.

At the centre of improving operational efficiency is the concept of waste. The basic definition of waste encompasses any human activity that consumes resources without creating value. However, perceptions of waste can vary significantly among individuals within an organization. For instance, a forward-thinking employer might define waste as "unused employee creativity," emphasizing the potential contributions of their workforce that go untapped. In contrast, an employee may view waste as "underutilized people," focusing on the inefficiencies stemming from not fully engaging staff capabilities. This divergence in definitions underscores the complexity of waste as a concept and its implications for manufacturing processes.

Activities that do not contribute value to the process or the final product are known as non-value adding (NVA) activities. In manufacturing contexts, these activities are frequently regarded as waste. NVA activities utilize time, financial resources, and materials without enhancing the quality or functionality of the end product.

In manufacturing specifically, waste refers to any activity or process that fails to add value to the final product from the customer's perspective. This includes various

DOI: 10.1201/9781003637677-3

inefficiencies—such as redundant quality checks, excessive handling of materials, or unnecessary delays—that ultimately lead to increased costs and lower productivity. The role of industrial engineering becomes crucial here; it emphasizes the identification and elimination of such waste to streamline operations and maximize profitability. By adopting methodologies like Lean Manufacturing or Six Sigma, companies can systematically address these inefficiencies, transforming their production processes into more agile and responsive systems (Singh, 2020).

Recognizing and addressing waste is not merely an operational concern; it is fundamental to sustainable business practices and achieving operational excellence. By minimizing waste, companies can enhance their resource allocation, reduce costs, and improve overall product quality factors that are increasingly important in today's environmentally conscious market. Moreover, fostering a culture that encourages continuous improvement allows organizations to adapt more readily to changing customer demands and market conditions. As such, addressing waste becomes a strategic imperative that not only drives economic performance but also positions companies favourably for future growth.

Effective waste management is integral to achieving operational efficiency in manufacturing. By optimizing the use of resources, companies can significantly reduce costs associated with waste disposal and resource procurement. For instance, implementing recycling programmes for materials such as metals and plastics not only minimizes waste but also allows manufacturers to reclaim valuable resources, thereby lowering the need for new materials. This practice contributes to a circular economy, where resources are reused and repurposed, enhancing overall sustainability and reducing environmental impact.

An effective waste management strategy is also predicated on employee engagement. Educating personnel on optimal waste reduction methodologies cultivates a culture of sustainability throughout the organization. When employees comprehend the importance of waste minimization and possess the requisite knowledge to enact efficacious procedures, they become proactive contributors to the attainment of operational efficiency. Soliciting employee feedback may generate novel concepts for the continued reduction of waste and the enhancement of processes.

3.1.2 Identification and Classification of Waste

Waste plays a pivotal role in overall resource consumption, negatively impacting sustainability and efficiency across various sectors. The presence of waste can lead to numerous detrimental consequences, including reduced productivity, profit losses, customer dissatisfaction, and safety concerns. Originating from the Toyota Production System (TPS) in the 1980s, the classification of waste has evolved significantly. TPS introduced the concept of the "Seven Wastes" (or *Muda*), which serves as a foundational principle in lean manufacturing (Sutrisno, 2018). This framework assists businesses in identifying inefficiencies across diverse sectors such as manufacturing, services, product development, and software development. Over time, these principles have expanded to encompass considerations at both the company-wide and supply chain levels.

Historically defined by Taiichi Ohno in the 1980s, the seven types of waste identified in lean manufacturing are presented next.

3.1.2.1 Overproduction

Overproduction occurs when products are manufactured in excess or ahead of demand. This type of waste ties up valuable resources, including materials, labour, and storage space, without adding value to the final product. The consequences of overproduction include increased lead times, excessive storage costs, and a heightened risk of obsolescence. The TPS advocates for "Just in Time" (JIT) manufacturing, which emphasizes producing only what is required at the moment. By minimizing excess inventory through JIT practices, companies can enhance responsiveness to customer demands and reduce costs associated with maintaining surplus stock.

3.1.2.2 Waiting

The waste of waiting arises whenever products are not moving or being processed. In traditional batch-and-queue manufacturing systems, products can spend over 99% of their lifecycle waiting to be processed due to poor material flow or long production runs. This idle time can result from equipment breakdowns, material shortages, or inefficient scheduling. According to the Theory of Constraints by Eliyahu Goldratt, an hour lost in a bottleneck process results in a loss to the entire factory's output that cannot be recovered. To mitigate waiting waste, it is essential to streamline processes so that one operation could directly feed into the next, thereby enhancing overall productivity.

3.1.2.3 Transportation

Waste in transportation encompasses the superfluous movement of materials from one process to another, where such movement does not contribute to the product's value. This excess transportation may result in elevated costs and a higher risk of product damage due to handling. Moreover, inefficient material flows can create delays and disrupt production schedules. To reduce transportation waste, organizations should analyse product flows and optimize plant layouts to minimize unnecessary movement. By strategically positioning workstations and equipment closer together, companies can enhance operational efficiency and reduce logistical costs.

3.1.2.4 Over/Inappropriate Processing

Inappropriate processing occurs when expensive or high-precision equipment is utilized for tasks that could be performed with simpler tools or methods. This type of waste often leads to inefficient plant layouts and encourages high asset utilization while failing to deliver value. To address inappropriate processing, companies should invest in smaller, more flexible equipment and create manufacturing cells that facilitate better workflow. By simplifying processes and utilizing appropriate technology for specific tasks, organizations can enhance productivity while reducing unnecessary costs.

3.1.2.5 Inventory

Excess inventory, also referred to as Work in Progress (WIP), is a direct consequence of overproduction and waiting. High levels of inventory can obscure underlying issues on the shop floor, making it difficult to identify and resolve problems effectively. Excess inventory increases lead times and consumes valuable floor space while inhibiting communication among team members. To combat this waste, companies should strive for seamless flow between work centres by implementing pull systems that align production with actual customer demand.

3.1.2.6 Motion

Unnecessary motion encompasses any excessive movement by workers or machines that does not contribute to the production process. This includes actions such as bending, stretching, walking excessively, or searching for tools and materials. Such movements not only waste time but also pose health and safety risks for employees. To minimize unnecessary motion, organizations should analyse workflows and redesign job tasks with ergonomics in mind. By optimizing workstations and eliminating redundant movements, companies can enhance worker efficiency and well-being.

3.1.2.7 Defects

Defects refer to quality issues that necessitate rework or result in scrap materials, both of which incur significant costs for organizations. The costs associated with defects can include quarantining inventory, re-inspecting products, rescheduling production efforts, and capacity loss due to downtime caused by quality issues. Defects not only impact profitability but also harm customer relationships and brand reputation. To minimize defects, companies should emphasize continuous process improvement (CPI) initiatives and engage employees in quality management efforts. By fostering a culture focused on quality at every stage of production, organizations can significantly reduce defects while improving overall performance.

3.1.3 Origins of Waste

The origins of waste in industrial operations can be traced back to systemic inefficiencies that permeate various aspects of manufacturing and production processes. An interplay of factors such as poor process design, lack of standardized operating procedures, and inadequate workforce training cumulatively contributes to waste, invariably impacting organizational performance.

At the heart of numerous waste-related challenges lies suboptimal process design. Inefficient workflows often culminate in bottlenecks, causing certain production stages to lag behind, leading to idle resources and time. For instance, a flawed assembly line design necessitating back-and-forth movement for workers exemplifies how design inefficiencies can prolong cycle times and increase unnecessary motion. The lack of a holistic approach to process design can create a ripple effect throughout the production system, exacerbating other forms of waste such as waiting and overproduction (Leksic, 2020)

Furthermore, the absence of standardized operating procedures (SOPs) leads to the introduction of various inconsistencies in task execution, leading to process variability and potential errors. A contributing factor to this issue is the insufficient training and lack of clear guidelines for the workforce. This lack of guidelines complicates training efforts for new employees, resulting in increased defect rates and misalignment within teams. Effective training programmes aim to provide the employees with the necessary skills and knowledge required to perform tasks efficiently. Moreover, without an understanding of lean principles and waste reduction strategies, employees may be able to identify improvement opportunities within their workflows, perpetuating inefficiencies.

External factors, such as misaligned supply chains, further exacerbate waste. Discrepancies in supply chain synchronization create material flow disruptions, leading to overproduction or excess inventory. When suppliers fail to deliver materials timely or there is a supply–demand mismatch, the entire production process witnesses' inefficiencies, causing increased lead times and operational costs. Moreover, ineffective communication within organizations significantly contributes to waste. Misunderstandings and misaligned efforts arise when critical information regarding production schedules, inventory levels, or quality standards is not effectively shared among teams (Harish, 2015). For example, an unawareness of production target changes due to poor communication can result in overproduction and surplus inventory within departments.

Identifying these waste origins is paramount to implementing effective waste reduction strategies. Through meticulous process analyses using methodologies like root cause analysis or value stream mapping, organizations can pinpoint inefficiencies. Understanding waste origins allows for targeted interventions addressing the underlying issues rather than just symptoms, fostering comprehensive waste reduction strategies enhancing operational efficiency and sustainability.

3.1.4 The Snowball Effect of Waste

The impact of waste in manufacturing extends far beyond immediate inefficiencies, manifesting as a compounding phenomenon often referred to as the *snowball effect*. This term encapsulates the notion that minor inefficiencies can accumulate over time, leading to significant operational and financial challenges that can hinder a company's competitive edge. The snowball effect illustrates how seemingly isolated instances of waste can interact and amplify each other, creating a cascade of negative consequences that are difficult to reverse once they gain momentum.

For instance, excessive inventory is a prime example of how waste can snowball. When a company produces more goods than necessary, it not only ties up capital but also necessitates additional resources for storage, handling, and management. This accumulation of inventory can lead to increased costs associated with warehousing and labour, as well as the risk of obsolescence, where products become outdated or unsellable. As inventory levels rise, the complexity of managing these resources also increases, often resulting in inefficient practices that further exacerbate the problem. Consequently, this excess inventory creates a burden on cash flow and diverts attention from more productive activities (Taher, 2024).

Similarly, defects in products represent another critical area where the snowball effect manifests. When defects occur, they can lead to customer dissatisfaction and warranty claims, which not only incur direct costs but also damage a company's reputation in the marketplace. Over time, this erosion of trust can result in lost sales and diminished customer loyalty. The repercussions of defects extend beyond immediate financial implications; they can instigate a cycle of rework and additional inspections that further strain resources and divert focus from core production activities. As companies grapple with the fallout from defects, they may find themselves entangled in a web of inefficiencies that stifles innovation and growth.

Recognizing the interconnected nature of various types of waste is critical for achieving lean efficiency. The snowball effect emphasizes that addressing one type of waste can have cascading benefits across the organization. Moreover, the snowball effect underscores the importance of fostering a culture of continuous improvement within organizations. By promoting awareness among employees regarding the implications of waste and encouraging proactive measures to identify, target, and eliminate inefficiencies, companies can create a self-reinforcing cycle of improvement. When employees are empowered to contribute to waste reduction initiatives, they become more engaged and invested in their work, leading to higher morale and productivity levels.

3.2 LEAN PRINCIPLES

3.2.1 Introduction to Lean Manufacturing

The foundational principles of lean manufacturing have a rich historical context that traces back centuries, gaining significant prominence through the works of influential figures such as Benjamin Franklin. In his writings in *Poor Richard's Almanack*, Franklin emphasized the importance of minimizing waste to enhance profitability, a notion that he elaborated further in his essay *The Way to Wealth*. This early recognition of waste reduction as a means to improve economic outcomes laid the groundwork for future developments in manufacturing efficiency. Later, Frederick Winslow Taylor formalized these ideas in his seminal 1911 work, *Principles of Scientific Management*, where he introduced systematic methodologies aimed at optimizing productivity. Taylor posited that management should rigorously evaluate and adopt superior work methods based on empirical evidence, thereby establishing a precedent for modern process improvement strategies.

As the early 20th century unfolded, American industrialists such as Henry Ford began to leverage these foundational principles to counter competition from lower-cost international labour. In the foreword to Taylor's *Shop Management*, Henry Towne underscored the critical role of productivity enhancements in maintaining economic competitiveness and supporting high wages. Ford's innovations in assembly line production exemplified the application of lean concepts long before the term "lean" was coined, demonstrating how efficiency could be achieved through standardized processes and reduced cycle times.

While Taylor's contributions laid the groundwork for efficiency in manufacturing, it was Shigeo Shingo and Taiichi Ohno of Toyota Motor Corporation who significantly shaped lean manufacturing into its contemporary form. Drawing inspiration from Taylor's work, Shingo and Ohno developed the TPS, which focused on addressing inefficiencies and fostering continuous improvement, known as Kaizen. In the 1930s, Kiichiro Toyoda identified flaws within Toyota's production processes, particularly concerning resource wastage due to defective castings. By meticulously analysing each stage of production, Toyota introduced Kaizen teams that played a pivotal role in refining TPS. In the aftermath of the Second World War, Japan's limited economic circumstances led Ohno to develop a "pull" system, in which production was guided by genuine customer demand instead of arbitrary targets. This groundbreaking strategy reduced overproduction and established the groundwork for what would eventually be known as JIT manufacturing.

The term "lean manufacturing" emerged in modern discourse through John Krafcik's 1988 article, *Triumph of the Lean Production System*, which was informed by his experiences with Toyota's practices. Krafcik noted that lean systems achieved superior productivity and quality with fewer resources, emphasizing the significance of a skilled workforce, streamlined product design, and robust supplier networks. The concept gained further traction through the influential works of James Womack, Daniel T. Jones, and Daniel Roos in their seminal book *The Machine That Changed the World* (1990) and its follow-up *Lean Thinking* (1996). They articulated five core principles essential to lean manufacturing: specifying value from the customer's perspective; mapping the value stream for each product; creating continuous value flow; enabling customer-driven production through a pull system; and striving for perfection through continuous improvement.

Lean manufacturing fundamentally revolves around reducing waste and enhancing value: achieving more with less by reducing labour, space, time, and resources while closely aligning with customer needs. Womack and Jones encapsulated this ethos by defining lean as "a way to do more and more with less and less." Lean practices emphasize streamlined resource utilization, waste minimization through the minimization of NVA activities, and enhancements in productivity, quality, cost efficiency, and lead times. TPS serves as an archetype for these principles; initially tailored for automotive manufacturing, they have since been adopted across diverse industries including electronics, ceramics, machinery, and wood processing (Singh, 2020).

Moreover, lean manufacturing is frequently complemented by other methodologies such as Six Sigma, which enhances quality while simultaneously reducing costs through systematic process improvement. Together, these approaches foster the development of efficient and high-performing production systems that underscore lean's role as a cornerstone of world-class manufacturing.

3.2.2 Why Lean?

Implementing lean principles provides a myriad of benefits, encompassing cost reduction, enhanced quality, and accelerated response times to market demands. A quintessential example is Toyota, which transformed automobile production through TPS. The application of these lean principles enabled Toyota to minimize waste throughout its operations, resulting in reduced production cycles, elevated vehicle quality, and diminished costs (Agrogiannis, 2015). This methodology not only bolstered operational efficiency but also afforded a substantial competitive edge, facilitating the company's dominance in the global automotive sector.

Another pivotal advantage of lean is the reduction of lead times and inventory levels. By adopting JIT practices, companies like Dell Computers could tailor products to individual customer orders while maintaining minimal inventory. This adaptability allowed them to swiftly address market demands, curtail storage expenses, and mitigate the risk of obsolete inventory.

Moreover, lean principles cultivate an ethos of collaboration and accountability, empowering employees at all tiers to take ownership of continuous improvement initiatives. In organizations such as Amazon, this strategy has been pivotal in refining operations and enhancing customer satisfaction. Employees are encouraged to identify inefficiencies, propose remedies, and implement modifications, fostering a dynamic and responsive work environment (Soltani, 2023).

3.2.3 The Concept of Muda

Muda, a Japanese term meaning "waste", is central to lean manufacturing. It refers to any activity or process that does not add value to the final product from the customer's viewpoint. *Muda* is classified into two categories: necessary but NVA activities and pure waste. Necessary NVA activities include actions such as quality inspections or equipment maintenance, which do not directly enhance the product but are essential for maintaining operational standards. Pure waste, on the other hand, consists of activities that consume resources without any benefit, such as overproduction or unnecessary transportation (Han, 2012).

The concept of *Muda* is often examined alongside two other critical terms: *Mura* (unevenness) and *Muri* (overburden). These three concepts form the foundation of waste elimination in lean manufacturing. For instance, *Mura* can arise from inconsistent production schedules, leading to idle time in one area and overwork in another. This imbalance often results in *Muri*, where machines or employees are overburdened, increasing the likelihood of errors and breakdowns. Together, they create a cycle that perpetuates *Muda*.

Addressing *Muda* requires a systematic approach. For example, in the automotive industry, Toyota implemented the JIT system to eliminate overproduction waste by producing only what is needed, when it is needed. Similarly, by redesigning workflows and introducing cellular manufacturing, companies can reduce motion waste and create more efficient production processes. Analysing *Muda* in conjunction with *Mura* and *Muri* ensures that efforts to reduce waste do not inadvertently create new inefficiencies, fostering a balanced and efficient production system.

3.2.4 Principles of Lean Manufacturing

The principles of lean manufacturing establish a structured framework designed to eliminate waste and optimize processes within organizations. By adhering to these principles, businesses can achieve streamlined, efficient, and customer-focused operations that enhance overall performance. The five fundamental principles, as articulated by Womack et al. (1990), serve as the cornerstone of lean thinking and operational excellence, which are discussed next.

3.2.4.1 Specify Value

A clear and precise definition of value from the customer's perspective is essential for any organization seeking to implement lean practices. Value is meaningful only when it reflects a specific product—whether a good, service, or a combination thereof—that meets customer needs at an appropriate price and within a timely manner. While the creation of value is the primary objective of production, many producers encounter difficulties in accurately defining it from the standpoint of their customers. This challenge necessitates a profound understanding of customer preferences and requirements to ensure that products align with market demands (Wu, 2013).

3.2.4.2 Identify the Value Stream

The value stream encompasses the complete set of actions required to deliver a product to the customer, including problem-solving tasks that span concept development, detailed design, and production launch; information management tasks that cover order-taking, scheduling, and delivery; and physical transformation tasks that encompass the journey from raw materials to the finished product in the customer's hands. Mapping the value stream is instrumental in exposing inefficiencies and waste—referred to as Muda—thereby revealing substantial opportunities for improvement. Organizations must strive to identify the entire value stream for each product or product family to uncover hidden inefficiencies and redundancies that may compromise operational effectiveness.

3.2.4.3 Create Flow

Achieving flow requires the elimination of interruptions, inefficiencies, and bottlenecks within processes. This step often necessitates significant shifts in traditional organizational thinking and processes; however, it is fundamental to reducing lead times, enhancing productivity, and improving overall operational efficiency. A continuous flow of work not only accelerates production cycles but also fosters an environment where resources are utilized more effectively.

3.2.4.4 Establish a Pull System

The introduction of flow subsequently paves the way for a pull system, wherein production is driven by actual customer demand rather than arbitrary targets. This approach ensures that only the required products are produced in desired quantities and at precisely the right time. The implementation of pull systems minimizes overproduction, reduces inventory waste, and increases responsiveness to fluctuating customer demands. Lean systems enable rapid transitions from concept to delivery, significantly shortening design, production, and fulfilment timelines.

3.2.4.5 Perfection

The pursuit of perfection represents the culmination of lean thinking. As organizations refine their understanding of value, optimize their value streams, establish flow, and implement pull systems, they enter a continuous cycle of improvement. This virtuous circle reduces effort, time, cost, and errors while delivering products that increasingly align with customer needs. Far from being an unattainable goal, perfection in lean thinking embodies a dynamic process that fosters ongoing innovation and improvement across all facets of an organization.

The consequences of adopting lean principles are significant. By focusing on these five fundamental principles—defining value, mapping the value stream, ensuring smooth flow, creating pull systems, and pursuing perfection—organizations can eradicate waste, improve process efficiency, and provide enhanced value to customers. Moreover, the interconnectedness of these principles promotes a culture of continuous improvement within organizations, driving them towards a lasting competitive edge in an ever-evolving marketplace.

Kaizen, a Japanese term meaning "continuous improvement," is a fundamental philosophy that aligns seamlessly with the principles of lean manufacturing. This approach emphasizes the importance of incremental, ongoing enhancements across all areas of an organization, fostering a culture where every employee is encouraged to identify inefficiencies and propose solutions. By integrating Kaizen into the framework of lean manufacturing, organizations can systematically eliminate waste and optimize processes, thereby enhancing overall value delivery to customers. The iterative nature of Kaizen complements the lean principles of specifying value, identifying the value stream, ensuring flow, establishing pull systems, and striving for perfection. As organizations engage in regular Kaizen activities, they cultivate an environment of continuous learning and improvement that not only addresses current operational challenges but also positions them for sustained competitive advantage in an ever-evolving marketplace.

3.2.5 Advantages and Disadvantages

Lean manufacturing is a methodical approach which aims at eliminating waste and optimizing efficiency, and presents both advantages and disadvantages that vary based on its implementation context. This analysis explores its benefits and challenges in a formal and research-oriented manner (Kumar, 2024).

3.2.5.1 Advantages

Cost and Time Efficiency: Lean manufacturing significantly reduces operational costs by enhancing efficiency. Streamlined workflows, optimized resource allocation, and minimized production and storage inefficiencies contribute to substantial cost savings for businesses across various industries. Furthermore, lean methodologies reduce lead times, facilitating quicker delivery of products and services. In addition to improving customer satisfaction, this also lessens the demand for excessive inventory and staffing, leading to a more efficient and cost-reducing workforce (Buer, 2021).

Environmentally Friendly: Lean manufacturing emphasizes waste elimination in processes, materials, and energy use, positioning it as a sustainable approach. By reducing energy consumption and fuel requirements, lean practices contribute to lower greenhouse gas emissions, supporting environmental sustainability objectives. Additionally, lean systems often incorporate energy-efficient machinery and environmentally conscious practices, such as recycling and waste reduction, providing both ecological and financial benefits.

Enhanced Customer Satisfaction: By improving operational efficiency and reducing waste, lean manufacturing ensures that customers receive high-quality products or services at competitive prices within shorter timeframes. Consistent and reliable delivery fosters customer trust and loyalty, which are essential for long-term business success. Satisfied customers are more likely to become repeat buyers and advocates for the brand, driving organic growth through positive word-of-mouth and referrals.

3.2.5.2 Disadvantages

Employee Safety and Well-being: Despite its focus on efficiency, lean manufacturing can inadvertently compromise employee welfare. The relentless emphasis on streamlining processes and eliminating waste often creates a high-pressure environment with minimal room for error. Employees may face increased workloads and stress levels, potentially leading to burnout or decreased morale. Comparisons have been drawn between lean practices and 19th-century scientific management techniques, which prioritized efficiency over worker well-being and were widely criticized during the labour reform movements of the early 20th century.

Impacts on Future Development: A notable criticism of lean manufacturing is its focus on immediate efficiency, which can sometimes come at the expense of long-term innovation and growth. By cutting resources or departments deemed "non-essential" to current operations, companies risk eliminating areas that could contribute to future development or diversification. This myopic approach may impede a business's capacity to adjust to evolving market conditions or to invest in research and development initiatives that guarantee long-term success.

Difficulties in Standardization: Although lean manufacturing being more of a cultural philosophy than a fixed methodology, it presents challenges in standardization. The lack of a universally accepted framework for lean implementation can lead to inconsistencies across organizations. This variability may cause confusion or scepticism, with some perceiving lean as a vague or loosely defined strategy rather than a robust and structured system.

3.3 ELIMINATING WASTE WITH LEAN PRINCIPLES

3.3.1 Small Changes, Major Impact

Lean principles underscore the transformative potential of small, incremental improvements that can yield substantial long-term outcomes for organizations. One of the most effective methodologies within this framework is the 5S approach—Sort, Set in Order, Shine, Standardize, and Sustain—which significantly enhances workplace organization and operational efficiency. For instance, a manufacturing facility that adopts the 5S methodology often experiences a marked reduction in time spent searching for tools and materials. A case study involving **Harley-Davidson** illustrates this point: by implementing the 5S system alongside Total Productive Maintenance (TPM), the company reported a 25% increase in productivity due to improved organization and reduced downtime.

Cellular manufacturing is another lean principle that can dramatically enhance productivity while minimizing waste. In this model, workstations are arranged in a sequence that supports smooth workflow, thereby reducing motion waste. **John Deere**, for example, adopted cellular manufacturing techniques in its production processes, resulting in a 30% reduction in cycle times and a significant decrease in transportation waste within the facility. By organizing workstations to facilitate a continuous flow of materials and information, John Deere has been able to optimize its operations while simultaneously enhancing product quality.

The concept of Kaizen, which we talked about earlier, is fundamental to lean and emphasizes the importance of fostering a culture where employees at all levels are encouraged to identify inefficiencies and propose enhancements. This philosophy is not merely about addressing immediate issues; rather, it aims to establish long-term goals that drive ongoing progress. For instance, **Coca-Cola** has integrated Kaizen principles into its operations since the early 1990s, focusing on waste elimination across various processes. As a result of these efforts, Coca-Cola achieved a 20% reduction in water usage and a 15% decrease in energy consumption, demonstrating how continuous improvement initiatives can lead to significant resource savings and enhanced sustainability.

Moreover, organizations that embrace lean principles often set ambitious long-term goals aimed at achieving operational excellence. The pursuit of perfection is an ongoing journey rather than a final destination; it requires commitment to continual evaluation and refinement of processes. For example, **Intel** has implemented lean methodologies within its semiconductor manufacturing operations, focusing on reducing cycle times and improving product quality through continuous innovation. By systematically applying Kaizen principles across its production lines, Intel has achieved an impressive 40% reduction in manufacturing cycle times, illustrating how sustained efforts towards improvement can yield substantial competitive advantages.

3.3.2 Correlating Waste Types to Lean Principles

Eliminating waste in manufacturing through lean principles can be effectively illustrated with quantitative case studies from various companies that have successfully addressed each type of waste. These examples not only demonstrate the practical application of lean methodologies but also highlight measurable outcomes that underscore their effectiveness.

Overproduction is a significant issue in many industries, leading to excess inventory and associated costs. **Zara**, the global fashion retailer, has mastered the art of minimizing overproduction through its agile supply chain (Vergara, 2016). By producing limited quantities of designs based on real-time customer preferences, Zara achieves an impressive inventory turnover rate of 12 times per year, compared to the industry average of 3 to 4 times (Pedrinazzi, 2020). This strategy results in only 10% of unsold inventory annually, significantly lower than the typical 17% to 20% seen in the fashion sector. Zara's ability to adapt quickly to market trends allows it to maintain a competitive edge while minimizing waste (Costa, 2017).

Inventory waste, characterized by excess stock that ties up capital, can be effectively managed through innovative business models. Dell Computers exemplifies this with its build-to-order approach. By allowing customers to customize their PCs online and commencing assembly only after receiving an order, **Dell** has kept its inventory levels remarkably low. This model has enabled Dell to maintain an average inventory turnover ratio of approximately 50 times per year, compared to the industry average of around 10 times. This method not only reduces storage costs but also enhances customer satisfaction by delivering personalized products promptly.

The issue of **defects** in manufacturing can lead to substantial rework and material waste. Honda has implemented Poka-yoke systems—error-proofing mechanisms designed to minimize defects—resulting in significant quality improvements. For instance, **Honda**'s use of automated sensors on assembly lines has reduced defect rates by approximately 30%, translating into savings of millions of dollars annually due to decreased rework and scrap costs. This proactive attitude not only enhances product quality but also fosters customer trust and loyalty.

Motion waste, which involves unnecessary movements by workers or machines, can be minimized through efficient workplace design. **Unilever** has successfully tackled this issue by optimizing its manufacturing facilities. By employing ergonomic principles and strategically placing tools and materials within easy reach, Unilever has reported a 20% reduction in worker movement, leading to increased productivity and decreased worker fatigue. This optimization not only enhances operational efficiency but also contributes to a safer working environment.

In addressing **over-processing**, **Starbucks** has standardized beverage preparation across its locations, streamlining operations and reducing redundancy. This standardization has led to a 15% decrease in service times, enabling baristas to serve customers more efficiently while maintaining product quality. As a result, Starbucks has enhanced customer satisfaction and increased throughput during peak hours, demonstrating how lean practices can lead to tangible improvements in service delivery.

Reducing the **waiting** waste, which is characterized by idle time for workers or machines, was effectively addressed by **Boeing** through the implementation of a moving assembly line for aircraft production. This innovation has resulted in a 25% reduction in lead times, allowing GE to deliver aircraft more quickly while optimizing resource utilization. The steady flow of work minimizes delays and enhances overall productivity, showcasing the benefits of lean methodologies in complex manufacturing environments.

Finally, reducing **transportation** waste is crucial for enhancing supply chain efficiency. **Procter & Gamble** (P&G) improved its transportation logistics by strategically placing distribution centres closer to key markets, resulting in a 30% reduction in transit times and a corresponding decrease in logistics costs. This optimization not only enhances service reliability but also contributes to lower carbon emissions associated with transportation activities.

3.3.3 Implementing Lean Principles

Implementing lean manufacturing represents a continuous and dynamic process that necessitates ongoing evaluation, adjustment, and refinement. The inherent nature of lean processes demands regular modifications to work practices, and continuous monitoring is critical to ensure that improvements are sustained over time. Establishing a comprehensive "lean toolbox," which encompasses various tools and techniques, can assist in streamlining management systems; however, it is essential to recognize that lean serves more as a guiding philosophy than as a rigid set of standardized procedures (Bhasin, 2006).

3.3.3.1 Establishing Clear and Specific Goals

The initial phase of implementing lean manufacturing involves the establishment of clear objectives, which must be effectively communicated to all members of the organization. It is imperative to ascertain the overarching purpose of adopting lean processes: Is the goal to optimize workflow for expedited product delivery? Is the intention to enhance profitability through efficiency gains? Or is the objective to cultivate greater organizational agility? Regardless of the specific aim, it is crucial that these goals are clearly defined. Clear objectives provide a roadmap that guides the organization, motivates employees, and aids in overcoming potential obstacles encountered during the transformation process (Huang, 2022). By articulating these goals, leadership ensures that all organizational members comprehend the broader vision and align their efforts accordingly.

3.3.3.2 Designing a Simplified Manufacturing System

A fundamental principle of lean manufacturing is the simplification of processes. By deconstructing manufacturing systems into their most basic components, organizations can more readily identify inefficiencies and areas ripe for improvement. Simplification facilitates more effective monitoring and enhancement of individual processes while enabling the identification and elimination of waste, an essential tenet of lean philosophy (Wahab, 2013). A streamlined system is critical for an organization's long-term success, as it enhances agility and responsiveness to fluctuating market demands.

3.3.3.3 Implementing Design Improvements Continuously

Identifying opportunities for improvement alone is insufficient; organizations must ensure that enhancements are systematically integrated into designs, procedures, and processes. Implementation should be approached strategically, emphasizing incremental, continuous changes rather than large-scale disruptions. Gradual improvements allow organizations to monitor each change's impact and make necessary adjustments, thus fostering a more adaptable system. Establishing key performance metrics to track improvement success is essential, as it enables organizations to quantify progress and assess various strategies' effectiveness.

A pragmatic approach involves initiating lean practices within a single team or department before expanding these practices organization-wide. This pilot phase allows companies to test methodologies, refine processes, and establish a robust foundation for broader implementation. In larger organizations, forming a cross-functional pilot group with representatives from various teams can effectively introduce lean principles (Saad, 2023). These representatives should ideally be influential employees, such as senior members or informal leaders, who can act as change agents promoting lean practices within their respective teams while ensuring consistency across the organization.

3.3.3.4 Seek Employee Buy-In

Successful implementation of lean manufacturing hinges on securing employee support and engagement. Without staff buy-in, lean initiatives are unlikely to succeed. Given that lean processes impact all business areas, it is essential for management to collaborate with employees in identifying and addressing root causes of inefficiencies. A pivotal aspect of this collaboration is clear communication; employees must understand what lean manufacturing entails, its rationale for introduction, and how it will benefit both the organization and themselves personally. This approach fosters a sense of ownership in the process, crucial for overcoming resistance to change.

Furthermore, leaders should adopt a participatory approach to change management rather than imposing lean practices from above. Guiding the organization through transition fosters a culture of shared leadership, allowing team members to assume greater responsibility and actively contribute to continuous improvement efforts. Lean processes focus on delivering superior customer value by eliminating waste; thus, employee engagement is vital in driving this transformation. Involving staff in developing a "lean plan," where they can provide feedback and suggestions, cultivates a collaborative environment responsive to employee needs.

An important consideration is addressing human factors influencing resistance to change—particularly if employees perceive lean implementation as an additional burden or threat to job security. To mitigate such resistance, providing coaching and training across all levels ensures employees possess the requisite knowledge and skills for effectively understanding and implementing lean concepts (Deshmukh, 2022). By making lean manufacturing a collective responsibility while promoting open communication, organizations can foster a sense of shared ownership that contributes significantly to the long-term success of lean initiatives.

3.4 STRATEGIES FOR LEAN WASTE REDUCTION

Strategies aimed at reducing lean waste are crucial for developing more efficient and cost-effective production processes. By methodically identifying and eradicating waste, organizations can improve their overall productivity and provide increased value to their customers. These strategies not only contribute to minimizing resource consumption but also promote a culture of sustainable operational excellence.

3.4.1 Control Charts

Control charts are a fundamental statistical tool employed in the realm of quality control and process management. Their primary function is to monitor and measure the stability of processes within a production system, thereby facilitating the identification of variations that may indicate underlying issues or opportunities for enhancement. By systematically tracking data over time, control charts provide valuable insights into process performance, enabling organizations to maintain high standards of quality and efficiency (Sundar, 2014).

3.4.1.1 Functionality

Control charts operate by plotting data points collected from a specific process on a graph. This graphical representation includes upper and lower control limits, which delineate the acceptable range of variation for the process being monitored. These control limits are statistically determined based on historical data and reflect the expected variability inherent in the process. When data points fall within these limits, the process is considered stable and predictable. Conversely, if data points exceed these limits or exhibit patterns indicative of instability—such as trends or cycles—this serves as a critical signal for managers to initiate further investigation.

The ability of control charts to highlight deviations from expected performance is vital for effective process management. By identifying out-of-control conditions early, organizations can implement corrective actions before minor issues escalate into significant problems. This tactic not only helps in maintaining product quality but also enhances overall operational efficiency.

3.4.1.2 Benefits

Control charts offer a clear visual representation of process stability, enabling managers and team members to quickly assess process health. This graphical format facilitates immediate recognition of trends, shifts, or unusual patterns. They endorse a culture of continuous improvement by reassuring data-driven discussions about performance and identifying areas for enhancement.

3.4.2 Kanban Boards

Kanban boards are a pivotal tool within the framework of lean management, specifically designed to visualize and optimize workflows in various production systems. The primary objective of Kanban boards is to facilitate efficient work management by providing a clear and immediate visual representation of the status of tasks. This enables team members to quickly grasp what is currently in progress, what requires attention, and what has been completed. By enhancing visibility into the workflow, Kanban boards play an important role in improving productivity and efficiency (Kanawade, 2020).

3.4.2.1 Structure

A typical Kanban board is structured into columns that represent different stages of a process. Commonly, these stages include "To Do," "In Progress," and "Done," although organizations can customize these categories to align with their specific workflows. Work items are depicted as cards or sticky notes, which move across the board as tasks progress through the various stages of completion. This dynamic movement not only illustrates the flow of work but also helps teams manage tasks effectively by providing a tangible representation of workload distribution.

The flexibility of Kanban boards allows them to be applied across a wide range of workflows, from manufacturing processes to project management scenarios (Rahman, 2013). This adaptability makes Kanban an invaluable tool for organizations seeking to streamline operations and enhance collaboration among team members.

3.4.2.2 Benefits

Kanban boards offer significant advantages by promoting transparency and accountability within teams. By providing a clear visual representation of ongoing work, team members can easily track progress and identify potential bottlenecks, facilitating proactive problem-solving. This visibility allows teams to address issues before they escalate into significant challenges. Additionally, Kanban boards enable better workload management by allowing teams to adjust task assignments based on current capacity and priorities. This adaptability is particularly beneficial in fast-paced environments where priorities may frequently shift (Agarwal, 2020). This iterative approach empowers team members, leading to higher stages of engagement and job satisfaction.

3.4.3 5S Methodology

The 5S methodology is a systematic approach designed to organize and standardize the workplace, thereby promoting cleanliness, efficiency, and safety. Originating from Japan, this methodology is integral to lean manufacturing and continuous improvement practices. The five principles of 5S—*Seiri* (Sort), *Seiton* (Set in order), *Seiso* (Shine), *Seiketsu* (Standardize), and *Shitsuke* (Sustain)—collectively focus on creating a well-ordered environment where tools, materials, and information are readily accessible, enabling employees to work more effectively and efficiently (Thakur, 2016).

3.4.3.1 Five Principles of 5S

The implementation of the 5S system involves guiding teams through a series of structured steps aimed at optimizing the workplace. The first principle, **Sort**, involves identifying and eliminating unnecessary items from the workspace. This step is crucial as it helps to declutter the environment, allowing employees to focus on essential tools and materials that contribute directly to their work processes.

Following Sort is **Set in order**, which emphasizes organizing the remaining items in a manner that enhances accessibility and efficiency. This principle advocates for arranging tools and materials logically so that employees can easily locate what they need without wasting time searching. Visual aids such as labels or colour-coding can be employed to facilitate this organization.

The third principle, **Shine**, focuses on maintaining cleanliness in the workplace. This step goes beyond mere tidiness; it encourages teams to adopt regular cleaning practices that promote a safe working environment. A tidy workspace not only boosts employee morale but also minimizes the likelihood of accidents and mistakes.

Standardize is the fourth principle, which involves establishing standardized processes for maintaining the first three S's. This includes creating guidelines and checklists that ensure everyone adheres to the organizational practices established during Sort, Set in order, and Shine. Standardization is vital for ensuring consistency across teams and shifts.

Finally, **Sustain** emphasizes the importance of maintaining the gains achieved through the previous steps over time. This principle encourages a culture of continuous improvement where all team members actively participate in upholding the standards set by 5S. Regular audits and training sessions can reinforce these practices and ensure long-term adherence.

3.4.3.2 Benefits

The 5S methodology significantly enhances workplace efficiency and safety by creating a more organized environment, reducing waste and inefficiencies. This improved accessibility saves time and minimizes frustration, leading to higher employee satisfaction. Additionally, a well-organized workspace fosters continuous improvement, as employees engaged in regular 5S activities identify opportunities for enhancement (Rodríguez, 2020). This proactive mindset contributes to productivity gains and workplace safety by eliminating unnecessary items and maintaining cleanliness. Standardization ensures consistent best practices, reducing process variability, errors, and enhancing quality outputs. Consequently, employees are better equipped to perform their tasks effectively within a standardized system.

3.4.4 Multi-Process Handling

Multi-process handling is a critical concept in production and operations management that involves the simultaneous management and control of multiple processes. This capability is essential for ensuring efficient resource allocation and maintaining a smooth production flow across various stages of operation. In environments where processes are interdependent or where multiple tasks must be managed concurrently, effective multi-process handling becomes vital for optimizing throughput and enhancing overall operational performance.

3.4.4.1 Coordination of Resources and Tasks

In practice, multi-process handling necessitates the careful coordination of resources, time, and tasks. Managers and operators are tasked with monitoring several processes simultaneously, ensuring that each one progresses according to established plans and timelines. This requires a comprehensive understanding of how different processes interact and depend on one another, as well as the ability to anticipate potential challenges that may arise during production.

To facilitate this coordination, organizations often employ specialized software or tools designed to monitor processes in real time. These utilities offer crucial visibility into the state of individual processes, allowing managers to make well-informed determinations related to resource allocation and the ranking of tasks. For instance, if one process is lagging behind, managers can quickly identify this issue and reallocate resources or adjust schedules to mitigate delays. This real-time monitoring capability is crucial for maintaining operational efficiency and responsiveness in dynamic production environments.

3.4.4.2 Benefits

Effective multi-process handling offers numerous benefits, such as reducing bottlenecks within the production system, which enhances overall production flow and minimizes idle time for resources. This leads to improved efficiency. Additionally, it maximizes resource utilization by ensuring that all resources—human, technological, and material—are effectively employed across various tasks, which is crucial in competitive markets where efficiency impacts profitability. Multi-process handling also contributes to faster production cycles by streamlining workflows and reducing delays between stages, allowing companies to expedite their time-to-market. Moreover, it fosters a culture of collaboration among team members, as coordinating various processes helps employees understand how their roles contribute to organizational goals. This collaborative environment encourages communication and teamwork, essential for driving continuous improvement initiatives.

3.4.5 Poka-Yoke

Error proofing, commonly referred to as Poka-Yoke, is a vital technique employed in production and quality management aimed at preventing mistakes or defects during the manufacturing process. The core objective of Poka-Yoke is to design systems that automatically correct or prevent errors, thereby ensuring that work is performed accurately and consistently without variation. This proactive approach to quality control is essential for enhancing operational efficiency and maintaining high standards in product quality.

3.4.5.1 Mechanisms of Error Proofing

The implementation of error-proofing techniques involves the use of various mechanisms designed to detect and mitigate potential errors before they result in defects. These mechanisms can take several forms, including physical devices, visual cues, and software-based systems. For instance, on an assembly line, a sensor may be installed to verify that parts are positioned correctly before the next step in the process can occur. This not only ensures that components are assembled accurately but also prevents the progression of errors down the line.

Visual cues also play a significant role in error proofing. These can include colour-coded labels or markings that guide workers in performing tasks correctly. By providing clear visual instructions, organizations can reduce the risk of human error stemming from misinterpretation or oversight. Additionally, software-based systems can be employed to alert workers if a critical step has been skipped or if an incorrect action is taken during the production process. Such alerts serve as immediate feedback mechanisms, allowing for timely corrections before defects occur (Rahardjo, 2023).

3.4.5.2 Benefits

Implementing error-proofing techniques, also known as Poka-Yoke, brings significant benefits to organizations by minimizing defects in the production process. This proactive approach improves output quality, reduces the need for extensive quality control checks, and lowers rework and scrap rates. As a result, companies save time, resources, and enhance overall productivity by focusing on value-added activities. Additionally, higher quality outputs lead to increased customer satisfaction, fewer complaints, and enhanced brand loyalty, which are crucial for sustaining a positive reputation and long-term success in a competitive market (Sujatha, 2013). Furthermore, error-proofing fosters a culture of continuous improvement within organizations by empowering employees to identify and address potential errors, thereby prioritizing quality and boosting employee morale and ownership of work processes.

3.4.6 Rank Order Clustering

Rank Order Clustering is a valuable lean tool utilized in the analysis and optimization of production flow by systematically grouping similar tasks or processes. This methodology is instrumental in identifying relationships and dependencies among various tasks, thereby enabling organizations to enhance overall flow and minimize inefficiencies. By focusing on the interconnections between tasks, Rank Order Clustering facilitates a more coherent and efficient production environment.

3.4.6.1 The Process of Rank Order Clustering

The implementation of Rank Order Clustering begins with a thorough analysis of relevant data pertaining to tasks and processes within the production system. This data-driven approach involves evaluating each task's significance or impact on production outcomes. Once this analysis is complete, tasks are ranked according to their importance, allowing organizations to prioritize those that contribute most significantly to operational efficiency.

Following the ranking phase, tasks that exhibit similar characteristics or functions are clustered together. This clustering process is critical as it enables the optimization of resources and streamlining of operations. By grouping related tasks, organizations can reduce unnecessary movement and delays that often occur when similar activities are dispersed throughout the production environment. This strategic organization not only enhances workflow but also fosters a more cohesive approach to task management.

3.4.6.2 Benefits

Rank Order Clustering significantly enhances production flow by effectively grouping related tasks, minimizing unnecessary movements, and reducing delays in the production process. This leads to a smoother, faster production cycle and improved responsiveness to market demands. Additionally, it helps identify areas for resource optimization, resulting in substantial cost savings through streamlined operations and reduced waste. By maximizing resource utilization, companies enhance productivity while minimizing operational costs. Moreover, Rank Order Clustering fosters a culture of continuous improvement as teams analyse and cluster tasks, deepening their understanding of workflows and interdependencies. This promotes ongoing evaluation and refinement of processes, contributing to sustained operational excellence.

3.4.7 Single-Point Scheduling

Single-Point Scheduling is a strategic tool designed to optimize the scheduling of tasks and resources by concentrating on one task or process at a time. The primary objective of this approach is to simplify the scheduling process and ensure that resources are fully dedicated to completing a single task before transitioning to the next (Hicks, 2007). This focused methodology is particularly beneficial in environments where managing multiple competing priorities can lead to confusion and inefficiencies.

3.4.7.1 The Process of Single-Point Scheduling

In practice, Single-Point Scheduling involves prioritizing and scheduling one task or operation at any given moment. By directing attention exclusively to a single task, organizations can prevent the system from becoming overwhelmed with simultaneous demands. This singular focus allows for more effective resource allocation, as all available resources, whether human, technological, or material, can be concentrated on completing the designated task.

This method also minimizes the potential for mistakes that often arise when managing multiple tasks concurrently. When employees are not distracted by competing priorities, they can maintain higher levels of concentration and accuracy in their work. This enhanced clarity improves individual performance and simultaneously promotes a more efficient organizational workflow.

3.4.7.2 Benefits

Single-Point Scheduling enhances focus and resource allocation by dedicating resources to one task at a time, avoiding over-scheduling and production bottlenecks. This focused approach reduces idle time, improves production timelines, and allows swift responses to market changes. It fosters a culture of accountability among team members, leading to higher quality outputs and increased employee satisfaction (Panigrahi, 2023). Additionally, it supports continuous improvement initiatives, encouraging teams to evaluate and refine processes, driving operational efficiency and innovation. Overall, Single-Point Scheduling maximizes productivity while minimizing operational costs.

3.4.8 Total Productive Maintenance

TPM is a proactive maintenance strategy aimed at enhancing the reliability, availability, and performance of equipment within an organization. The primary goal of TPM is to maximize equipment effectiveness while minimizing unplanned downtime, thereby ensuring that machines operate in optimal condition. This comprehensive approach to maintenance not only addresses the technical aspects of machinery but also fosters a culture of continuous improvement and employee engagement.

3.4.8.1 Core Principles of TPM

TPM encompasses several key components that work synergistically to improve equipment performance. One of the fundamental aspects of TPM is the implementation of regular inspections and preventive maintenance schedules. These proactive measures are designed to identify potential issues before they escalate into significant problems or breakdowns. By conducting routine checks and maintenance activities, organizations can ensure that their equipment remains in peak operating condition.

Another critical element of TPM is the participation of employees at all levels, particularly operators and maintenance teams. This collaborative approach encourages operators to take ownership of their equipment by actively participating in its upkeep. Operators are often the first to notice irregularities or signs of wear and tear. Therefore, their insights are invaluable in identifying and resolving issues promptly. By fostering a sense of responsibility among employees, organizations can create a culture where everyone is invested in maintaining high equipment performance.

3.4.8.2 Benefits

Implementing TPM enhances operational efficiency by reducing unplanned downtime through proactive maintenance, improving workflow continuity, and increasing productivity. It extends equipment lifespan, reducing costs associated with premature replacements. TPM also boosts product quality by ensuring reliable and consistent equipment performance, leading to fewer defects (Bhamu, 2014). Additionally, it fosters continuous improvement, encouraging regular evaluation and refinement of maintenance processes for sustained operational excellence.

3.4.9 Single-Minute Exchange of Die (SMED)

SMED is a critical lean manufacturing tool designed to significantly reduce the time required for changeovers between different manufacturing processes or setups. The primary objective of SMED is to streamline these changeovers, thereby minimizing downtime and enhancing overall production efficiency. By focusing on reducing the time lost during transitions, SMED enables manufacturers to operate more flexibly and responsively in today's dynamic market environment (Ondra, 2022).

3.4.9.1 The Process of SMED

The implementation of SMED involves a systematic analysis of the steps involved in a changeover. This analysis typically begins with a detailed observation of the current changeover process to identify all tasks and their respective durations. Once the steps are mapped out, the next phase is to categorize them into two groups: internal and external tasks. Internal tasks are those that can only be performed when the machine is stopped, whereas external tasks can be conducted while the machine is still running (Ribeiro, 2022).

The goal of SMED is to minimize internal tasks and convert as many external tasks as possible to parallel activities. For instance, preparation work such as gathering tools or materials can be performed while the machine is still operational, thereby reducing the total downtime associated with changeovers (Moreira, 2011). Furthermore, implementing quick-change tooling and standardized procedures can facilitate faster transitions between setups (Che Ani, 2014). Training operators in these optimized procedures enhances their efficiency and confidence during changeovers, further reducing time lost.

3.4.9.2 Benefits

Adopting the SMED methodology reduces changeover time, enabling rapid transitions between processes or products and improving production flexibility. This efficiency supports JIT production, enhances overall productivity, and leads to cost savings (Toki, 2023). Moreover, SMED fosters a culture of continuous improvement, empowering teams to optimize processes and collaborate effectively.

3.4.10 Value Stream Mapping (VSM)

VSM serves as an effective lean methodology employed to illustrate the progression of materials and data across the entirety of the production cycle. The principal aim of VSM is to discern areas of operational deficiency, process bottlenecks, and prospects for value augmentation or waste reduction. Through the provision of an exhaustive depiction of the manufacturing workflow, VSM empowers organizations to acquire a deeper understanding of their operational activities and to facilitate judicious decision-making in the pursuit of enhancement (Batwara, 2023).

3.4.10.1 The Process of Value Stream Mapping

The VSM process begins with mapping out each step of the production journey, from the intake of raw materials to the delivery of finished products. This mapping exercise involves documenting every activity involved in the production process, including both value-adding and non-value-adding activities (Patil, 2021). Activities that directly contribute to the production of goods or the provision of services that fulfil customer requirements are classified as value-adding. Conversely, NVA activities are those that do not augment the value of the product or service and can be regarded as redundant (Joshi, 2012).

By capturing this detailed information, VSM provides a clear picture of the entire value stream. It highlights how materials and information flow through various stages of production, allowing teams to visualize where delays, redundancies, or unnecessary steps may occur (Setiawan, 2021). This holistic view is crucial for identifying areas that require attention and improvement.

3.4.10.2 Benefits

VSM helps organizations identify waste in processes, such as excessive inventory and long lead times. It improves understanding of material and information flow, enabling workflow optimization and bottleneck elimination. VSM fosters collaboration and continuous improvement among cross-functional teams. Its visual nature aids communication and alignment on process improvement goals (Salwin¸ 2021).

3.5 CASE STUDY: ADVANCED LEAN OPTIMIZATION IN A GERMAN AUTOMOBILE MANUFACTURER'S INDIAN FACILITY

3.5.1 Background

A major German automobile manufacturer operating in India initiated a strategic intervention to optimize its Body Shop's rear architecture assembly line, accommodating production ramp-up for a new chassis model. This project aimed to identify and mitigate process inefficiencies by reducing NVA activities and enhancing station-level synchronization. Applying rigorous lean manufacturing methodologies, the initiative aligned with the company's global commitment to operational excellence, efficiency, and premium quality.

3.5.2 Objective

The main objective was to systematically identify, quantify, and eliminate NVA activities on the rear architecture line. The project focused on aligning actual and ideal CGIS-defined cycle times, improving ergonomic conditions for operators, and implementing long-term workflow optimizations. Following core lean manufacturing principles, the initiative targeted waste reduction, emphasizing motion, transportation, and waiting.

3.5.3 Methodology

The project was executed in two principal phases:

1. Identification and Quantitative Analysis of NVAs: A dual-method time study evaluated station-level performance. CGIS system data provided baseline theoretical cycle times, which were cross-referenced with empirical observations of operator activities across multiple shifts to capture performance variability under diverse operational conditions.
2. Optimization Ideation and Implementation: Data-driven insights from the time study informed targeted interventions to rectify identified inefficiencies. These efforts included workstation reconfigurations, ergonomic adjustments, tool repositioning, and process enhancements aimed at mitigating NVA impacts.

Data aggregation and consolidation were performed using advanced Excel functions and R-based statistical modelling to generate visual tools—time comparison graphs, deviation analyses, and performance distribution charts—providing granular insights into station-level inefficiencies and guiding subsequent improvement efforts.

3.5.4 Findings

The time study illuminated significant inefficiencies across multiple stations, mapped to the lean manufacturing framework's seven wastes:

- Motion Waste: At station one of the final stations, operators experienced frequent ergonomic strain due to the low placement of spot-welding gun switches. Repeated bending motions increased cycle time and operator fatigue. Relocating switches to a more accessible platform eliminated this waste, enhancing ergonomic conditions and output.

- Transportation Waste: Another similar station experienced workflow disruptions due to disorganized tool storage, prolonging tool retrieval and impeding continuity. Standardizing toolbox organization reduced delays, minimized cycle time variations, and improved station efficiency.
- Waiting Waste: Inefficient guiderail systems for MER carrier movement caused material handling delays and periodic operator idling. Optimized guiderail configurations improved material flow, reduced waiting times, and enhanced production consistency.

The findings were consolidated into a comprehensive report and presented to the Body Shop's leadership. The presentation detailed time discrepancies, operational bottlenecks, and station-specific improvements tailored to each inefficiency.

3.5.5 Implementation and Results

Several Critical Optimizations Were Implemented Based on the Findings

- Ergonomic Reconfigurations: Repositioning spot-welding gun switches at the final station eliminated excessive bending motions, reducing cycle time deviations and improving operator comfort and safety.
- Toolbox Standardization: Introducing standardized tool storage systems streamlined retrieval processes, improved station efficiency, and reduced unnecessary motion.
- Guiderail and Carrier Flow Enhancements: An analysis of MER carrier movement patterns led to optimized guiderail configurations, reducing material transport delays and alleviating operator waiting times.

These interventions resulted in reduced cycle time deviations and measurable improvements in overall line efficiency. Ergonomic adjustments also fostered a safer, more operator-centric work environment.

3.6 Conclusion and Future Scope

This case study demonstrates the transformative impact of structured time studies and targeted optimizations within lean manufacturing frameworks. Addressing motion, transportation, and waiting waste in the rear architecture assembly process led to substantial operational enhancements and positioned the manufacturer's Indian facility for sustained excellence.

Continuous improvement initiatives, including periodic time studies and iterative process refinements, will remain critical as production volumes increase and the rear architecture line adapts to additional models. Automating repetitive tasks offers potential for further reducing manual interventions, enhancing precision, and supporting long-term operational fluidity. The operational improvements realized on the rear architecture line have strengthened the manufacturer's ability to adapt to evolving production demands. This case study emphasizes the vital role of data-driven process optimization in lean manufacturing, ensuring that waste-free practices remain central to operational philosophy.

REFERENCES

Agarwal, S., and Agarwal, A. (2020). Uses, advantages and opportunities of Kanban methods in mechanical engineering and product manufacturing. *International Journal of Scientific and Research Publications (IJSRP)*, *10*(1), p9704.

Agrogiannis, Serafim, and Christos Agrogiannis. "A critical review and evaluation of the lean concept and corporate social responsibility/sustainability: Investigating their interrelation and contribution in terms of business competitive positioning." 1, (2015): 2–167.

Batwara, Amber, Vikram Sharma, Mohit Makkar, and Antonio Giallanza. "Towards smart sustainable development through value stream mapping–a systematic literature review." *Heliyon* 9, no. 5 (2023).

Bhamu, Jaiprakash, and Kuldip Singh Sangwan. "Lean manufacturing: Literature review and research issues." *International Journal of Operations & Production Management* 34, no. 7 (2014): 876–940.

Bhasin, Sanjay, and Peter Burcher. "Lean viewed as a philosophy." *Journal of Manufacturing Technology Management* 17, no. 1 (2006): 56–72.

bin Che Ani, Mohd Norzaimi, and Mohd Sollahuddin Solihin Bin Shafei. "The effectiveness of the single minute exchange of die (SMED) technique for the productivity improvement." *Applied Mechanics and Materials* 465 (2014): 1144–1148.

Buer, Sven-Vegard, et al. "The complementary effect of lean manufacturing and digitalisation on operational performance." *International Journal of Production Research* 59, no. 7 (2021): 1976–1992.

Carrizo Moreira, António, and Gil Campos Silva Pais. "Single minute exchange of die: A case study implementation." *Journal of Technology Management & Innovation* 6, no. 1 (2011): 129–146.

Costa, Ana Mafalda Ricardo Morgado. *Zara's case study: the strategy of the fast fashion pioneer*. MS thesis, 2017.

De la Vega-Rodríguez, Marina, et al. "Lean manufacturing: A strategy for waste reduction." *New Perspectives on Applied Industrial Tools and Techniques* (2017): 153–174.

Deshmukh, Manish, et al. "Study and implementation of lean manufacturing strategies: A literature review." *Materials Today: Proceedings* 62 (2022): 1489–1495.

Han, Sangwon, SangHyun Lee, and Feniosky Peña-Mora. "Identification and quantification of NVA effort from errors and changes in design and construction projects." *Journal of Construction Engineering and Management* 138, no. 1 (2012): 98–109.

Harish, K. A., and M. Selvam. "Lean wastes: A study of classification from different categories and industry perspectives." *The Asian Review of Civil Engineering* 4, no. 2 (2015): 7–12.

Hicks, Ben J. "Lean information management: Understanding and eliminating waste." *International Journal of Information Management* 27, no. 4 (2007): 233–249.

Huang, Chien-Yi, et al. "A lean manufacturing progress model and implementation for SMEs in the metal products industry." *Processes* 10, no. 5 (2022): 835.

Joshi, Rahul R., and G. R. Naik. "Reduction in setup time by SMED a literature review."*International Journal of Modern Engineering Research (IJMER)*, 2, no. 1 (2012): 442–444.

Kanawade, Chaitanya S., Darshak R. Khedekar, and Sunny R. Prasad. "Implementation of Kanban System." (2020). Kumar, Rajesh, Ashwini Kumar, and Rajender Kumar. "Role of lean with green in achieving sustainable manufacturing practices." In AIP Conference Proceedings, vol. 3387, no. 1, p. 070005. AIP Publishing LLC, 2026.

Leksic, I. S. N. V. I., Stefanie, N., and Veza, I. (2020). The impact of using different lean manufacturing tools on waste reduction. Advances in production engineering & management, 15(1), 81–92.

Ondra, Pavel. "The impact of single minute exchange of die and total productive maintenance on overall equipment effectiveness." *Journal of Competitiveness* (2022): 1–6.

Panigrahi, Shrikant, et al. "Lean manufacturing practices for operational and business performance: A PLS-SEM modeling analysis." *International Journal of Engineering Business Management* 15 (2023): 147864.

Patil, Aditya S., Mahesh V. Pisal, and Chandrakant T. Suryavanshi. "Application of value stream mapping to enhance productivity by reducing manufacturing lead time in a manufacturing company: A case study." *Journal of Applied Research and Technology* 19, no. 1 (2021): 11–22.

Pedrinazzi, Maddalena, and Gianmarco Villa. "Enabling successful lean implementation in the fashion industry: A framework for companies' self-assessment of lean maturity." (2020).

Rahardjo, Benedictus, et al. "Lean manufacturing in industry 4.0: A smart and sustainable manufacturing system." *Machines* 11, no. 1 (2023): 72.

Rahman, Nor Azian Abdul, Sariwati Mohd Sharif, and Mashitah Mohamed Esa. "Lean manufacturing case study with Kanban system implementation." *Procedia Economics and Finance* 7 (2013): 174–180.

Ribeiro, Murilo Augusto Silva, et al. "Analysis of the implementation of the single minute exchange of die methodology in an agroindustry through action research." *Machines* 10, no. 5 (2022): 287.

Saad, Sameh M., et al. "Industry 4.0 and lean manufacturing–A systematic review of the state-of-the-art literature and key recommendations for future research." *International Journal of Lean Six Sigma, 15*(5) (2023), 997–1024.

Salwin, Mariusz, et al. "Using value stream mapping to eliminate waste: A case study of a steel pipe manufacturer." *Energies* 14, no. 12 (2021): 3527.

Setiawan, Indra, Ojakma Sihar Panaili Tumanggor, and Humiras Hardi Purba. "Value stream mapping: Literature review and implications for service industry." *Jurnal Sistem Teknik Industri* 23, no. 2 (2021): 155–166.

Singh, Jagdeep, Surjit Kumar Gandhi, and Harwinder Singh. "Assessment of implementation of lean manufacturing in manufacturing unit-a case study." *International Journal of Business Excellence* 21, no. 2 (2020): 274–296.

Soltani, Hossein, and Pawan Bhandari. "Enhancing operational efficiency: A study on lean manufacturing implementation in small and medium-sized enterprises." *Proceedings of the 8th North American International Conference on Industrial Engineering and Operations Management*, Houston, Texas, USA, 2023.

Sujatha, Y., and K. Prahlada Rao. "A study on lean manufacturing tools and techniques implementation in the Andhra Pradesh silk production industry." *International Journal of Mechanical Engineering and Robotics Research* 2, no. 4 (2013): 59–73.

Sundar, Rhitwik, A. N. Balaji, and RM Satheesh Kumar. "A review on lean manufacturing implementation techniques." *Procedia Engineering* 97 (2014): 1875–1885.

Sutrisno, A., et al. "Lean waste classification model to support the sustainable operational practice." *IOP Conference Series: Materials Science and Engineering*, 337, no. 1. IOP Publishing, (2018).

Taher, Md Abu, and M. Al Bashar. "The impact of lean manufacturing concepts on industrial processes' efficiency and waste reduction." *International Journal of Progressive Research in Engineering Management and Science* 4, no. 6 (2024): 338–349.

Thakur, Ashish. "A review on lean manufacturing implementation techniques: A conceptual model of lean manufacturing dimensions." *REST Journal on Emerging Trends in Modelling and Manufacturing* 2, no. 3 (2016): 62–72.

Toki, Gazi Farhan Ishraque, et al. "Single Minute Exchange Die (SMED): A sustainable and well-timed approach for Bangladeshi garments industry." *Cleaner Engineering and Technology* 12 (2023): 100592.

Vergara, Raymond Allan G. "Process and Facility Design and Improvement: A Case Study on the Lean Production Management of the Spanish Fashion Retailer Zara." (2016).

Wahab, Amelia Natasya Abdul, Muriati Mukhtar, and Riza Sulaiman. "A conceptual model of lean manufacturing dimensions." *Procedia Technology* 11 (2013): 1292–1298.

Wu, Peng, Sui Pheng Low, and Xiaohua Jin. "Identification of non-value adding (NVA) activities in precast concrete installation sites to achieve low-carbon installation." *Resources, Conservation and Recycling* 81 (2013): 60–70.

4 The Role of Circular Economy for a Sustainable Future

Rahul Prasad Nandhakumar and Sharvesh Ramesh

4.1 INTRODUCTION TO CIRCULAR ECONOMY

The circular economy (CE), guided by the 3R principles—Reduce, Reuse, and Recycle—serves as a powerful framework that bridges economic growth with environmental sustainability. By minimizing resource extraction, extending product lifespans, and promoting waste recovery, CE fosters a regenerative system where materials retain their value for as long as possible. This approach not only mitigates environmental degradation but also enhances economic resilience by reducing costs, driving innovation, and creating new business opportunities in sustainable industries. China is the third major nation making significant strides toward implementing the CE on a large scale. However, unlike Germany and Japan, which have adopted broader national strategies, China has opted for a gradual approach by first launching pilot projects. This phased implementation allows the government to assess economic feasibility, environmental impact, and policy effectiveness before full-scale adoption. The strategy mirrors China's economic liberalization model, which initially focused on coastal free economic zones before expanding nationwide, ensuring controlled yet progressive development.

The implementation of the CE follows a structured approach involving several key steps. It begins with presenting the CE concept and contrasting it with the traditional linear economy, which is based on the take-make-dispose model. This comparison highlights the urgency of transitioning toward a regenerative and sustainable industrial system that operates in a closed loop. Subsequently, the review introduces current CE practices and discusses the standards used to assess their development and performance. Drawing from existing literature, it then analyzes the core challenges and issues from an entrepreneurial perspective, identifying barriers to effective implementation. Finally, the review concludes by summarizing CE's developmental trajectory and offering policy recommendations aimed at enhancing its future adoption and evolution as part of a broader, innovation-driven national development strategy (Geissdoerfer et al., 2017).

This strategic approach not only enables countries to refine its circular economy framework through real-world testing but also ensures a smoother transition toward

DOI: 10.1201/9781003637677-4

large-scale implementation, balancing economic growth with environmental sustainability.

4.1.1 Circular Economy in India

India, as a rapidly developing and resource-intensive economy, stands at a crucial juncture where adopting CE principles can drive sustainable growth while addressing pressing environmental concerns. With a large and expanding population, the country faces increasing resource demand, making it essential to shift away from the traditional linear economic model of take, make, and dispose toward a more regenerative system. Rooted in its cultural heritage, India has long embraced the ethos of reuse, repair, and minimal waste generation, making CE a natural fit for its economic landscape. By scaling up circular practices in key sectors such as manufacturing, agriculture, and urban development, India can significantly reduce resource dependency, minimize environmental degradation, and foster innovation in sustainable technologies (Singh and Ordoñez, 2016).

Moreover, adopting a CE can enable India to decouple economic growth from excessive resource consumption, ensuring long-term prosperity while reducing carbon emissions and industrial waste. This transformation can create new business opportunities, generate green jobs, and enhance economic resilience, all the while safeguarding natural ecosystems for future generations. By integrating CE strategies into policy frameworks, industrial operations, and consumer behavior, India has the potential to set a global benchmark in sustainability, ensuring a balanced approach to economic progress and environmental stewardship.

According to the Central Pollution Control Board (CPCB), India generates over 62 million tons of waste every year. About 43 million tons (70%) are collected, of

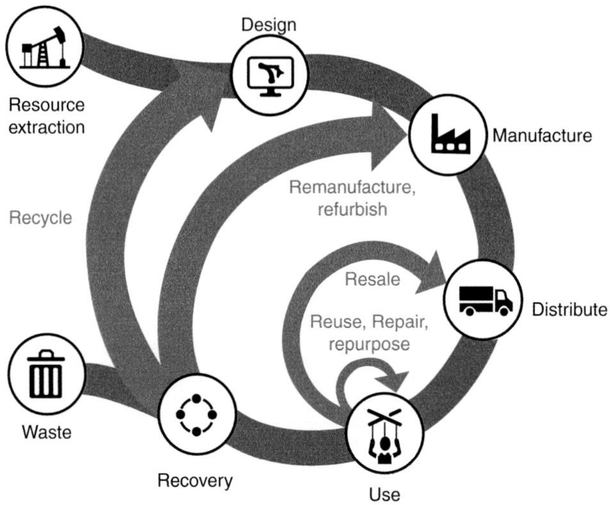

FIGURE 4.1 The different steps in circular economy (Singh and Ghosh, 2021).

which about 12 million tons are treated, and 31 million tons are dumped in landfill sites. This is expected to increase to 165 million tons by 2030. The majority of this waste is disposed of in landfills or dumped in open spaces, leading to environmental degradation and health hazards. India is highly vulnerable to the impacts of climate change. We are already experiencing the adverse effects of climate change, such as more frequent and severe droughts, floods, heat waves, and extreme weather events. More frequent and severe heat waves have led to increased mortality and morbidity (Singh and Ghosh, 2021).

CE principles can contribute to a system that would meet the growing needs of the Indian population, especially in cities. It could help limit negative externalities, such as greenhouse gas emissions, congestion, and pollution and a basis for setting principles to guide business innovation, policy, and education. Applying these principles would help India build on its current CE practices and scale them across sectors and value chains, using existing building blocks to embark on a circular development path and capture its benefits.

4.2 GOVERNMENT INITIATIVES

4.2.1 Circular Economy Opportunities in India

As India continues to invest in long-term infrastructure to enhance the quality of life for its citizens, initiatives like the Smart Cities Mission present a significant opportunity to embed CE principles into urban development. By integrating CE strategies into the design and management of essential services such as water supply, sanitation, and waste management, cities can establish efficient nutrient and material cycles, minimizing resource wastage and environmental impact (Bocken et al., 2016).

A more systemic and holistic approach to urban planning can further drive sustainability by incorporating circular mobility solutions, such as public transport optimization, shared mobility, and green energy-powered transportation. This would not only enhance air quality and reduce traffic congestion but also curb urban sprawl, fostering compact, resource-efficient, and resilient cities. By embedding circular economy practices into its urban infrastructure, India can create smart, sustainable, and self-sufficient cities, ensuring long-term economic and environmental benefits (Zink and Geyer, 2017).

With half of India's workforce are engaged in agriculture, the sector remains a cornerstone of the national economy and a critical pillar of food security. However, traditional farming practices often lead to soil degradation, resource depletion, and nutrient loss, threatening long-term productivity. By adopting a CE approach, the agricultural sector can transition toward a regenerative model that enhances natural capital, improves economic and ecological resilience, and ensures sustainable food production. Additionally, integrating circular bioeconomy solutions—such as converting agricultural waste into bioenergy or biodegradable materials—can create new revenue streams while minimizing environmental impact. A well-designed circular agricultural system will not only support India's growing population with a stable supply of fresh, nutritious, and diverse food but also contribute to climate resilience and sustainable rural development (Mowat and Frost, 2020).

FIGURE 4.2 Circular economy in India for cities and construction.

4.2.2 Government-Based Initiatives

The Indian government is making significant strides in urban infrastructure development through initiatives like the Smart Cities Mission, industrial corridors, the Swachh Bharat Mission, and the 500 AMRUT cities program. These projects are driving large-scale investments aimed at enhancing urban living conditions, boosting economic growth, and improving public services. According to research by the McKinsey Global Institute, India must invest ₹77 lakh crore (US$ 1.2 trillion) in urban infrastructure by 2030 to meet the demands of its rapidly expanding cities (McKinsey and Company, 2016). Government bodies and non-profits in India are aligning with CE principles through initiatives such as vehicle-sharing schemes, investments in renewable energy, and training programs for farmers on regenerative practices (Narayan and Sree, 2019). In 2013, India launched the National Electric Mobility Mission Plan (NEMMP) 2020 to reduce pollution and oil dependence by promoting electric vehicles (EVs) with fiscal and monetary incentives. The plan targets sales of 6–7 million electric or hybrid vehicles by 2020, offering a solution to growing air pollution linked to fossil fuel-burning vehicles (Ranjan and Gupta, 2018). In addition to these policy initiatives, the Indian government has implemented financial incentives, such as tax benefits, subsidies, and low-interest loans, to encourage circular business models and promote sustainable consumption patterns (Sridhar, 2020).

4.2.3 Future Roadmap for Circular Economy in India

India boasts a rich biodiversity, and to preserve it for future generations, the country must focus on creating a sustainable future. Research indicates that a transition to a CE in India could generate significant economic benefits. By 2030, the CE could create an annual value of ₹14 lakh crore (US$ 218 billion), and by 2050, this value could increase to ₹40 lakh crore (US$ 624 billion), offering substantial cost savings compared to the current development trajectory. These savings could account for 11% of India's gross domestic product (GDP) by 2030 and 30% by 2050 (Bansal et al., 2020). Additionally, India's growing middle class presents considerable business opportunities, particularly in industries such as textiles and electronics, where CE practices can drive profitability (Ellen MacArthur Foundation, 2019). The integration of digital technology with CE practices further amplifies value creation, with India's strong IT sector positioning it well to leverage this synergy (Sharma and Gupta, 2021). Furthermore, adopting CE principles in new industries from the outset offers India a competitive edge over developed nations that face challenges in transitioning from a linear economy model. India's unique advantage lies in building circular systems efficiently without the need to overhaul existing infrastructure (Pereira et al., 2021).

4.3 SECONDHAND CIRCULAR ECONOMY PRINCIPLES

A CE aims to minimize waste and maximize the use of resources by keeping products, materials, and components in use for as long as possible. Within this framework, the secondhand market plays a crucial role, embodying core CE principles. This section delves into these principles as they relate to the secondhand economy, particularly within the Indian context.

4.3.1 Core Principles of the Secondhand Circular Economy

The secondhand market inherently supports several core principles of the CE, offering both environmental and socioeconomic benefits, particularly in a country like India. First, it promotes product life extension by enabling goods to be reused by new owners, which delays their entry into the waste stream. This aligns with traditional Indian practices of reuse and repair, where items such as clothing, electronics, and furniture are commonly passed down through families and communities (Borthakur and Govind, 2017). Second, the secondhand market enhances resource efficiency by reducing the demand for new products, thus conserving the raw materials, energy, and water needed for manufacturing. Given India's population growth and consumption patterns, this reduction is crucial for sustainability (Singh and Ordoñez, 2016). Third, it contributes to waste reduction and diversion by preventing functional goods from ending up in landfills. The informal sector plays a significant role in this process, collecting, sorting, and redistributing goods in ways that relieve the pressure on India's still-developing waste infrastructure (Gill, 2021). Fourth, the market contributes to a reduction in environmental impact, helping to lower greenhouse gas emissions, pollution, and resource depletion by reducing demand for virgin production

(Maitre-Ekern and Dalhammar, 2019). Finally, the secondhand market supports economic opportunity and social equity, offering affordable goods to lower-income communities and creating employment in repair, refurbishment, and resale sectors. These economic activities are vital in India, where the informal economy supports millions of livelihoods while also promoting equitable access to consumer goods.

4.3.2 Factors Affecting the Secondhand Market

India's secondhand market is vibrant and diverse, comprising a mix of formal and informal channels including traditional markets, street vendors, specialized shops, and online platforms. This market thrives due to several interlinked factors. Culturally, the reuse and repair of goods are longstanding practices rooted in Indian values, exemplified by the ethos of *jugaad*—a form of innovative frugality that encourages making the most of limited resources. Economically, the secondhand market plays a crucial role in providing affordable access to goods for lower-income populations. Socially, the expanding Indian middle class is driving demand not only for cost-effective alternatives but also for avenues to resell used items as lifestyles evolve. Furthermore, technological advancements have dramatically transformed the landscape of the resale economy. Digital platforms like OLX, Quikr, and Facebook Marketplace have democratized access to used goods, making transactions quicker and more accessible,

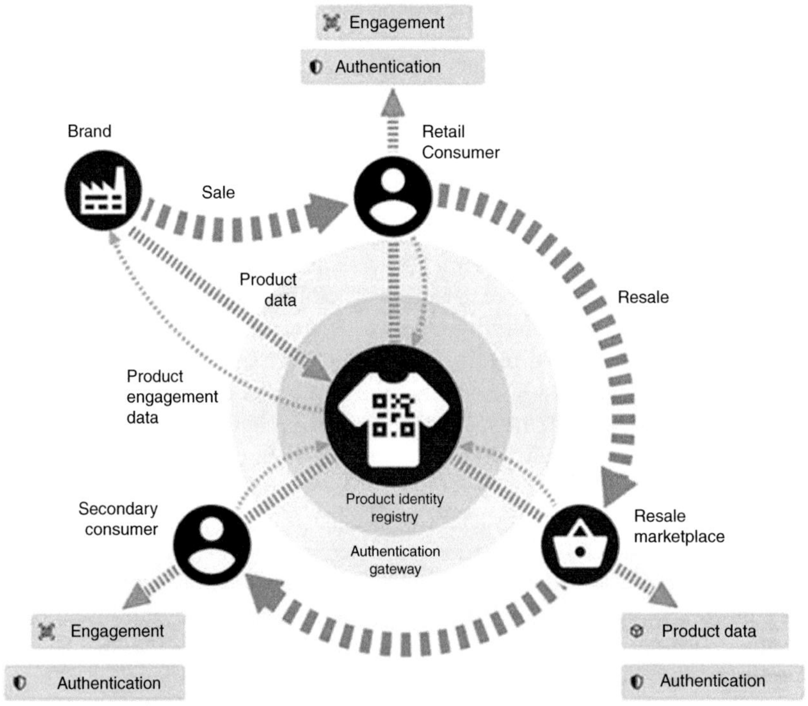

FIGURE 4.3 Creating a digital authentication platform for the secondhand markets.

especially in urban areas. These platforms enhance convenience, widen consumer reach, and increase the circulation of pre-owned products, contributing significantly to the circular economy in India (Borthakur and Govind, 2017).

4.3.3 Challenges and Opportunities

The Indian secondhand market, while offering numerous environmental and economic benefits aligned with circular economy principles, also encounters several challenges that hinder its full potential. Key concerns include the inconsistent quality of goods, lack of standardization and grading systems, and the informal nature of many market participants, which complicates regulation, taxation, and consumer protection. Additionally, hygiene and safety concerns—particularly for categories like clothing and electronics—can deter consumer trust. Despite these hurdles, the market holds vast potential. Formalizing operations through regulatory frameworks, integrating digital technologies to enhance traceability and quality control, and implementing awareness campaigns can address negative perceptions and build consumer confidence. Furthermore, embedding the secondhand market within national circular economy policies can foster a resilient, resource-efficient system that encourages sustainable consumption across socio-economic strata (Borthakur and Singh, 2021).

4.3.4 Linking secondhand firms, ownership structures and circular economy

The relationship between secondhand businesses, their ownership structures, and the circular economy (CE) is a crucial area of study. Scholars like Stahel (2016) suggest that how a company is owned significantly impacts its ability to participate in circular systems. Research has shown a clear link between ownership and governance, and how effectively a company operates (Sánchez-Ballesta and García-Meca, 2007). A company's internal makeup, including its ownership structure, influences its commitment to CE practices. Inefficient organizational and ownership models can actually hinder the adoption of circular economy principles (Liu and Bai, 2014). Furthermore, the motivations behind ownership, influenced by managerial philosophies and broader institutional factors, play a role in shaping a company's sustainability efforts (Wang et al., 2014; Darnall and Edwards, 2006). Companies prioritizing purely economic justifications for sustainability will approach it differently than those with a deeply ingrained sustainability-focused culture.

Secondhand businesses, categorized as "reuse" systems, play a key role in the CE by reintegrating products back into the market after their initial use. These businesses operate on a model of capturing, preserving, and even enhancing value by reselling used goods, often locally. The rise of these businesses and their associated supply chains has shifted market attention toward used products, creating new pathways for reuse and significantly lessening the environmental impact. By keeping products in circulation for longer, secondhand businesses reduce the need for virgin materials, minimize resource consumption, and lower disposal costs. This aligns with increasingly strict regulations aimed at extending product lifespans and reducing waste from incineration and landfills. Beyond environmental benefits, these businesses also have social implications, addressing the needs and aspirations of individuals seeking to improve their quality of life through the local and global redistribution of goods, they contribute to economic activity while offering sustainable solutions. Importantly, at the local level, they preserve the original value of products, making them accessible to consumers who desire high-quality, branded items but cannot afford them new, thus offering a pathway to a better lifestyle.

In essence, secondhand businesses act as facilitators, connecting businesses and communities through the flow of used goods and related information. They operate on a model of value retention and profit generation through the sale of these products. By promoting reuse, these businesses exemplify what is described as "slowing down" and **"dematerializing"** the loops within the CE, making a significant contribution to its effective implementation.

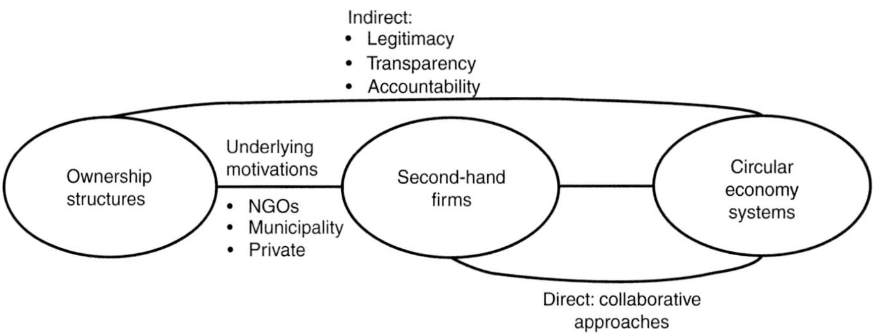

FIGURE 4.4 Pathway of secondhand circular economy.

FIGURE 4.5 The different online and offline secondhand automobile market.

4.3.5 Examples of Secondhand Markets in India

Formal online platforms such as OLX, Quikr, Facebook Marketplace, Refashion, Kiabza, Cashify, and Yaantra have become prominent in facilitating the buying and selling of secondhand goods. These platforms provide a convenient and efficient way to access secondhand products across various categories. Informal markets, including places like Chor Bazaar in Mumbai, Daryaganj Sunday Book Market in Delhi, and Raddi Markets in various cities, continue to play a vital role in India's secondhand economy by providing opportunities for both buying and selling used goods in person. Specialized shops such as used bookstores, antique stores, and vintage clothing shops further support this market by offering curated collections of secondhand items. Additionally, community initiatives, including charitable organizations that collect and distribute used goods, contribute to the circulation of secondhand products, promoting sustainability and social inclusion. These diverse avenues for secondhand trade are essential to India's growing CE, reducing waste and supporting sustainable consumption (Bocken et al., 2016).

4.4 ROLE OF SECONDHAND CIRCULAR ECONOMY PRINCIPLES IN SUSTAINABLE DEVELOPMENT

In India, where resource constraints and economic disparities exist, the secondhand market is not only an environmental necessity but also a socioeconomic opportunity. This chapter explores the principles of the secondhand CE, its role in sustainable development, and how India is leveraging this approach to balance economic growth with environmental responsibility.

4.4.1 Role of the Secondhand Economy in Sustainable Development

The secondhand market in India offers significant environmental, economic, and social benefits. Environmentally, it helps reduce waste by extending product lifecycles, thereby decreasing landfill waste. India generates around 3.5 million tons of e-waste annually, and secondhand markets play a crucial role in alleviating this burden (CPCB Report, 2022). Additionally, reusing and refurbishing products lowers carbon emissions by reducing the need for raw material extraction and manufacturing; for instance, purchasing a secondhand phone saves about 55 kg of CO_2 (GSMA, 2021). In the textile industry, which is the second largest polluter of freshwater resources, secondhand fashion conserves water and energy by minimizing new production. Economically, the secondhand market provides affordable access to goods, allowing low-income populations to obtain durable products such as mobile phones, clothes, and furniture. The refurbishment sector also creates jobs, supporting repair artisans, technicians, and small entrepreneurs in the informal sector. The Indian secondhand market, particularly for electronics and automobiles, is projected to grow at a compound annual growth rate (CAGR) of 14% by 2026 (TechSci Research). Socially, the secondhand economy provides inclusion for informal sector workers, such as scrap dealers and refurbishers, while also bridging the digital divide by making affordable smartphones and laptops available to lower-income communities

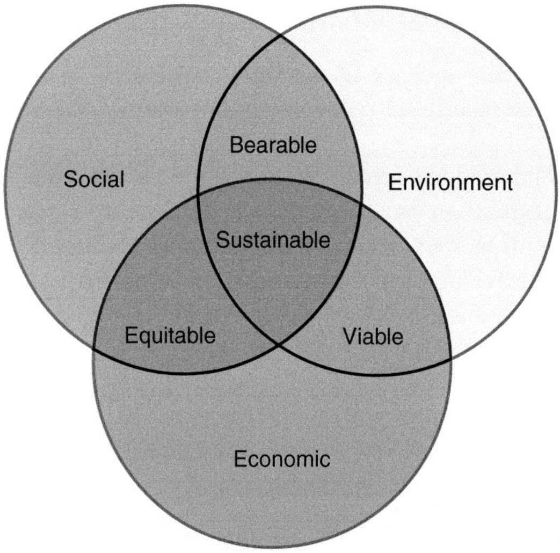

FIGURE 4.6 Venn diagram of sustainability.

for access to education and jobs. This growing market reflects a cultural shift toward sustainability, driven by increasing consumer awareness of environmental impact and a move toward ethical consumption (Niinimäki et al., 2020).

4.4.2 Key Sectors Benefiting from Secondhand Economy

India is one of the largest producers of electronic waste, but the refurbished electronics market plays a crucial role in extending device lifespans, reducing hazardous waste, and lowering consumer costs. Companies like Cashify and Amazon Renewed are leading the way in smartphone refurbishment, helping mitigate the growing e-waste issue. In the automotive industry, India's significant pre-owned vehicle market, supported by initiatives such as Maruti Suzuki True Value and Mahindra First Choice, reduces the need for new vehicle production, thereby minimizing its environmental impact. Similarly, the textile and fashion industry, a major contributor to pollution, benefits from platforms like Kiabza and Relove, which promote sustainable fashion through secondhand clothing and thrift stores. These efforts across multiple sectors support CE practices that reduce waste, conserve resources, and foster environmental sustainability (Geissdoerfer et al., 2017).

4.4.3 Future Prospects and Recommendations

To strengthen the secondhand circular economy in India, several strategies can be implemented. The government can offer tax benefits and subsidies to businesses engaged in refurbishing and reselling products, which would encourage greater participation. Consumer awareness campaigns promoting the environmental and economic advantages of secondhand goods, through social media and educational

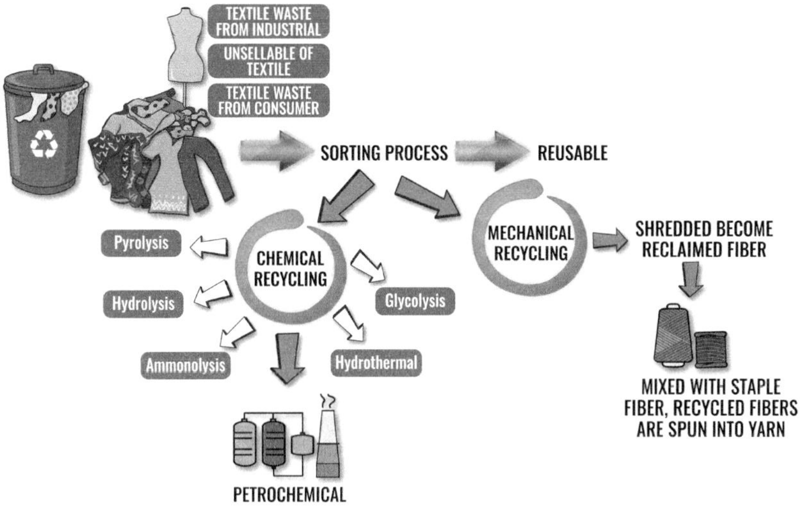

FIGURE 4.7 Mechanisms of possible textile waste recycling routes.

initiatives, can help shift public perceptions. Additionally, integrating technology, such as AI-driven authentication for used goods, improved logistics for collection and distribution, and digital certification for refurbished products, can boost trust and efficiency in the secondhand market. Finally, formalizing the informal sector by incorporating unorganized resale businesses into regulatory frameworks can enhance consumer rights protection and ensure better tax compliance (Niinimäki et al., 2020).

4.5 CASE STUDY EXAMPLE

4.5.1 Brand Introduction

Thaely, an Indian footwear brand founded in 2017 by Ashay Bhave, creates sneakers from recycled plastic waste with the goal of addressing plastic pollution through sustainable innovation. Built entirely on the principles of a CE, Thaely treats waste as a resource by using discarded plastic bags and bottles instead of virgin materials, thereby diverting waste from landfills and reducing environmental degradation. Their proprietary material, ThaelyTex, is crafted entirely from recycled plastic bags, while other components such as laces and canvas are made from recycled PET, and soles from recycled rubber. Emphasizing sustainability throughout its operations, Thaely minimizes energy use, sources materials locally, and employs vegan, eco-friendly production methods. The brand also promotes longevity through durable, timeless sneaker designs and supports a closed-loop system via a product take-back program that enables customers to return worn shoes for recycling or refurbishment. Through these initiatives, Thaely demonstrates how circular economy principles can be effectively applied to fashion, reducing environmental impact and fostering more responsible consumer behavior (Geissdoerfer et al., 2017).

4.5.2 Challenges Faced by the Brand

Plastic waste is a global crisis, with millions of tons ending up in landfills and oceans annually, and the footwear industry—heavily dependent on virgin materials—plays a significant role in exacerbating this issue. Recognizing the urgent need for sustainable alternatives, Thaely set out to develop footwear that is both environmentally responsible and appealing to consumers. Traditional shoes are often made from nonbiodegradable materials like leather, synthetic fabrics, and plastics, all of which have severe environmental impacts. Leather production, for instance, contributes to deforestation, consumes vast amounts of water, and involves toxic chemicals in the tanning process, while synthetic materials derived from petrochemicals further intensify ecological damage. Moreover, the shoe manufacturing process generates substantial carbon emissions throughout sourcing, production, transportation, and packaging stages, especially considering the high frequency of consumer purchases. Additionally, the use of harmful chemicals such as adhesives and dyes poses risks to water sources, ecosystems, and communities. By addressing these interconnected challenges, Thaely presents a model for sustainable innovation in the footwear industry, aligning with broader efforts to mitigate environmental degradation (Niinimäki et al., 2020).

4.5.3 Solution toward the Challenges Faced

Thaely has embedded the mission of upcycling plastic waste into the creation of stylish footwear, addressing both environmental and social challenges in India. In 2020, the country generated 4.1 million metric tons of plastic waste, a significant portion of which was not recycled and ended up polluting waterways and landfills. To combat this, Thaely partnered with Triotap Technologies, a waste-removal company that employs rag pickers, offering them stable pay, protective gear, and safe indoor sorting spaces—ensuring dignified working conditions and promoting social inclusion. Triotap collects and sorts dry waste at its Material Recovery Facility (MRF), and for Thaely's shoe production, workers process plastic bags by soaking, cutting, and fusing them into usable material. Additionally, Triotap has appointed "Green Warriors" in residential areas to raise community awareness about waste segregation at the source, further supporting CE practices. This initiative not only contributes to environmental sustainability but also uplifts marginalized communities by integrating them into a more formal and respectful work environment. Such efforts align with broader goals for sustainable fashion and inclusive development (Niinimäki et al., 2020).

4.5.4 Impact Delivered by the Company

Thaely has made a notable contribution to India's circular economy by transforming plastic waste into value-added products, particularly sustainable footwear. By using recycled plastic bags and bottles to manufacture shoes, the company actively reduces plastic pollution and diverts waste from landfills and waterways (Niinimäki et al., 2020). Thaely's innovative material, Thaely Tex—a fabric created from plastic

FIGURE 4.8 The Brand Thaely's sustainable shoe.

waste—exemplifies sustainable manufacturing by lowering dependence on virgin resources and minimizing environmental impact. In collaboration with Triotap Technologies, Thaely also creates green jobs by employing rag pickers, providing them with dignified, socially inclusive work that supports recycling and contributes to a greener economy. Their initiatives, such as appointing "Green Warriors,' promote waste segregation at the source and raise community awareness about responsible disposal practices. Moreover, Thaely's shoes are designed with an eco-friendly lifecycle in mind, allowing them to be recycled after use, thereby supporting circularity in fashion. This case study underscores how innovative thinking aligned with sustainability can transform industries—Thaely has demonstrated that it is possible to merge profitability with environmental and social responsibility. Their model not only sets a precedent for sustainable fashion but also inspires other businesses to adopt similar practices that benefit both people and the planet. By integrating rag pickers into the formal economy, Thaely also addresses unemployment and poverty, showcasing how circular business models can create inclusive economic opportunities in India.

4.6 CONCLUSION

The CE offers a promising path to a sustainable future, a much better way of doing things than the old "use it and throw it away" system that has put so much pressure on our planet. As this chapter has shown, switching to a CE is not just a nice idea; it's something we really need to do. India, with its fast-growing economy and huge population, can be a real leader in this change. The government's helpful programs, along with clever businesses and people becoming more aware, are starting to build a circular future. The secondhand economy, a key part of this change, shows how important it is to reuse, fix, and share things. This helps a lot by saving resources, reducing waste, and creating jobs. By focusing on these ideas, we can have economic growth without using up all our resources and harming the environment. The example we looked at showed real benefits from circular practices, proving that businesses can do well while being good to the planet. But, moving to a completely CE takes ongoing hard work, teamwork, and new ideas. We need to find ways to make circular solutions bigger, change how people buy and use things, and build the right systems to support it all. In the end, the CE is about much more than just recycling; it's about completely rethinking how our economy works, making a future where everyone can prosper while protecting the Earth. By embracing circularity, we can take care of our planet for future generations and build a better and fairer world.

REFERENCES

Bansal, G., Kumar, P., and Thakur, M. (2020). Circular economy in India: Pathways and implications. *Journal of Cleaner Production, 277*, 123456. https://doi.org/10.1016/j.jclepro.2020.123456

Bocken, N. M. P., De Pauw, I., Bakker, C., and van der Grinten, B. (2016). Product design and business model strategies for a circular economy. *Journal of Industrial and Production Engineering, 33*(5), 308–320. https://doi.org/10.1080/21681015.2016.1172124

Aljaradin, Mohammad, Kenneth M. Persson, and Emad Sood. "The role of informal sector in waste management, a case study; Tafila-Jordan." *Resources and Environment* 5, no. 1 (2015): 9–14. https://doi.org/10.5923/j.re.20150501.02

Borthakur, A., and Singh, P. (2021). Challenges and opportunities of the second-hand market in promoting circular economy: A case study of India. *Journal of Cleaner Production, 297*, 126645. https://doi.org/10.1016/j.jclepro.2021.126645

Central Pollution Control Board. (2022). *India generates about 3.5 million tonnes of e-waste annually*. CPCB.

Darnall, Nicole, and Daniel Edwards Jr. "Predicting the cost of environmental management system adoption: the role of capabilities, resources and ownership structure." *Strategic Management Journal*, 27, no. 4 (2006): 301–320

Ellen MacArthur Foundation. (2019). *The circular economy opportunity in India: From waste to wealth*. www.ellenmacarthurfoundation.org

Geissdoerfer, M., Savaget, P., Bocken, N. M. P., and Hultink, E. J. (2017). The circular economy—A new sustainability paradigm? *Journal of Cleaner Production, 143*, 757–768. https://doi.org/10.1016/j.jclepro.2016.12.048

Gill, K. (2021). Informal recycling networks and circular economy in Indian cities: Exploring the role of waste pickers. *Journal of Cleaner Production, 281*, 124556. https://doi.org/10.1016/j.jclepro.2020.124556

GSMA. (2021). *The impact of second-hand phones on carbon emissions*. www.gsma.com

Liu, Yong, and Yin Bai. "An exploration of firms' awareness and behavior of developing circular economy: An empirical research in China." *Resources, Conservation and Recycling*, 87 (2014): 145–152.

Maitre-Ekern, E., and Dalhammar, C. (2019). Towards a hierarchy of circular economy discourses: A review of the circular economy literature. *Journal of Cleaner Production, 223*, 123–132. https://doi.org/10.1016/j.jclepro.2019.03.294

McKinsey & Company. (2016). *India's urban awakening: Building inclusive cities, sustaining economic growth*. McKinsey Global Institute. https://www.mckinsey.com/~/media/McKinsey/Featured%20Insights/Urbanization/Urban%20awakening%20in%20India/MGI_Indias_urban_awakening_executive_summary.pdf

Mowat, F., and Frost, S. (2020). Circular economy and agriculture: A transformative approach for sustainable development. *Agricultural Systems, 176*, 102694. https://doi.org/10.1016/j.agsy.2019.102694

Narayan, R., and Sree, S. (2019). Circular economy practices in India: Opportunities for growth and sustainability. *Journal of Environmental Management, 240*, 170–182. https://doi.org/10.1016/j.jenvman.2019.04.103

Niinimäki, K., Peters, G., Dahlbo, H., Perry, P., Rissanen, T., and Gwilt, A. (2020). The environmental price of fast fashion. *Nature Reviews Earth & Environment, 1*(4), 189–200. https://doi.org/10.1038/s43017-020-0039-9

Kalauni, Kishor, Ajitanshu Vedrtnam, Sahendra P. Sharma, Abhishek Sharma, and Shashikant Chaturvedi. "A comprehensive review of recycling and reusing methods for plastic waste focusing Indian scenario." *Waste Management & Research* 43, no. 9 (2025): 1378–1399. https://doi.org/10.1007/s44371-025-00125-y

Pereira, M., Sharma, P., and Gupta, R. (2021). Leveraging digital technology for circular economy: The Indian perspective. *Technology in Society, 67*, 101743. https://doi.org/10.1016/j.techsoc.2021.101743

Ranjan, P., and Gupta, S. (2018). National Electric Mobility Mission Plan (NEMMP) 2020: A review of policies and progress. *Energy Policy, 118*, 16–25. https://doi.org/10.1016/j.enpol.2018.03.030

Sánchez-Ballesta, Juan Pedro, and Emma García-Meca. "Ownership structure, discretionary accruals and the informativeness of earnings." *Corporate Governance: An International Review*, 15, no. 4 (2007): 677–691.

Sharma, A., and Gupta, S. (2021). Digitalization and the circular economy: A new path for sustainable business in India. *Business Strategy and the Environment, 30*(4), 1802–1814. https://doi.org/10.1002/bse.2684

Singh, A., and Ghosh, S. (2021). Waste management in India: Challenges, opportunities, and the role of circular economy. *Environmental Science & Policy, 116*, 1–11. https://doi.org/10.1016/j.envsci.2020.10.009

Singh, J., and Ordoñez, I. (2016). Resource recovery from post-consumer waste: Importance of the informal sector in India. *Waste Management, 53*, 17–28. https://doi.org/10.1016/j.wasman.2016.04.015

Singh, Jagdeep, and Isabel Ordoñez. "Resource recovery from post-consumer waste: important lessons for the upcoming circular economy." *Journal of cleaner production* 134 (2016): 342–353.

Sridhar, K. (2020). Financial incentives for circular economy in India: Policy implementation and challenges. *Journal of Cleaner Production, 258*, 120608. https://doi.org/10.1016/j.jclepro.2020.120608

TechSci Research. (2022). *India second-hand market growth and projections*. https://www.credenceresearch.com/report/india-second-hand-apparel-market

Teo, T. C., A. Zhang, Z. Djuraeva, and A. Ettibaev. "Out of Uzbekistan: The Barriers of Adopting Circular Business Practices in Rural SMEs."*Journal of Human Resource and Sustainability Studies*, 11, no. 2 (2023): 356–380.

Wang, Yanjun, Anqian Wang, Jianqing Zhai, Hui Tao, Tong Jiang, Buda Su, Jun Yang et al. "Tens of thousands additional deaths annually in cities of China between 1.5 C and 2.0 C warming." *Nature communications*, 10, no. 1 (2019): 3376.

Zink, Trevor, and Roland Geyer. "Circular economy rebound." *Journal of industrial ecology* 21, no. 3 (2017): 593–602.doi.org/10.1111/jiec.12545Digital Object Identifier

5 The Impact of Circular Economy on Global Trade and Supply Chains

Jishnu Priyan G and Rajyalakshmi G

5.1 INTRODUCTION

The traditional industrial production and consumption system is based on a linear economy model, also known as the cradle-to-grave approach, which involves taking resources, making products, consuming them, and disposing of the waste. However, this linear model has significant limitations that make it unsustainable in the long term. It generates enormous amounts of waste, depletes natural resources, and pollutes the environment (Nasir et al., 2017). The global community currently generates approximately 2.3 billion tons of solid waste annually, and this volume is projected to increase significantly in the coming decades due to rapid urbanization, population growth, and changing consumption patterns. (Lahane et al., 2020). Furthermore, the extraction and processing of natural resources are major contributors to biodiversity loss. By 2060, global materials use is projected to almost double, with greenhouse gas emissions growing significantly (Yamaguchi, 2021). Current production and consumption patterns are unsustainable. They also create social injustice by concentrating wealth in the Global North while placing environmental and social burdens on the Global South (Barrie et al., 2022).

In response to these sustainability challenges, circular economy principles have emerged. The circular economy is defined as an economic system that minimizes resource consumption and waste generation while focusing on sustainable development (Mastos et al., 2021). It operates as a restorative and regenerative system by design, aiming to use resources efficiently across their lifecycle through approaches like closing, extending, and narrowing material loops (Yamaguchi, 2021). This model aligns with the United Nations' Sustainable Development Goals (SDGs) and has gained traction in policy efforts globally (Barrie et al., 2022). Circular Supply Chain Management (CSCM), which integrates circular economy principles, is increasingly recognized for its potential to drive sustainable development (Farooque et al., 2019).

Several drivers are propelling the global shift toward resource efficiency and circularity. These include growing resource scarcity (Zhang et al., 2021), environmental and social concerns, stakeholder demands, policy regulations, international cooperation, market demand for sustainable models, supply chain innovation, and competitive advantages associated with reducing material consumption and waste

DOI: 10.1201/9781003637677-5

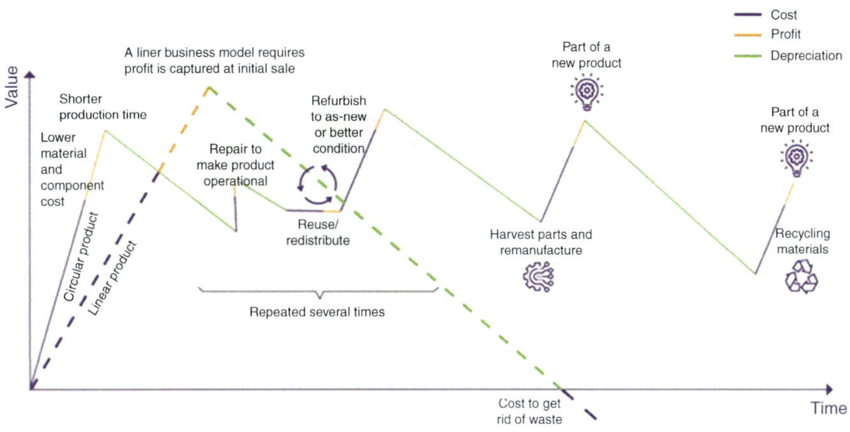

FIGURE 5.1 Lifetime accruements of revenue and profit throughout value chain (Jensen, n.d.).

(Jain et al., 2018). The transition is also supported by over 520 policies aimed at enabling circular economies (Barrie et al., 2022).

Adopting circular economy practices offers numerous benefits. Environmentally, it reduces the extraction of virgin natural resources and minimizes environmental pressures, including contributions to climate objectives (Yamaguchi, 2021). Economically, it creates new opportunities for employment, yields cost savings through waste reduction, enhances supply chain management, and lowers sensitivity to resource price volatility (Bellmann et al., 2021). Socially, it fosters inclusive growth by improving human welfare and creating equitable opportunities (Mangla et al., 2018). Business models aligned with circularity also offer commercial advantages, such as customer relationship strengthening and sustainable growth (Lahane et al., 2020). Overall, transitioning to a circular economy is essential for addressing the triple threat of pollution, climate change, and biodiversity loss while contributing to global sustainable development.

This section provides the foundational context necessary to understand the broader implications of the linear economy's limitations and the emergence of circular economy principles. The subsequent sections will explore how international trade patterns are shifting toward resource efficiency, the challenges posed by linear consumption in global trade, strategies for establishing circular economy principles in supply chains, case studies of successful implementations, and the broader challenges hindering circular economy adoption. By examining these aspects, this chapter aims to provide a comprehensive understanding of the transition from linear to circular economic models in the context of global trade and supply chains.

5.1.1 International Trade and the Transition Toward Resource Efficiency

The global shift toward resource efficiency is reshaping international trade patterns, with growing emphasis on circularity and sustainable practices. This transition is driven by the recognition that traditional linear models are unsustainable, leading to significant environmental degradation and resource depletion. As a result, international trade is increasingly aligning with resource efficiency goals, supported by policy frameworks and market dynamics.

International trade is shifting toward secondary materials, recyclable waste, and second-hand products, now in higher demand than raw materials. This trend highlights a global move toward valuing resources more efficiently. Trade in waste and scrap plays a critical role in supporting the circular economy by enabling cost-effective recycling, remanufacturing, or recovery of materials (Bellmann et al., 2021). Between 2000 and 2019, the value of trade in secondhand goods, secondary raw materials, and waste for recovery increased by more than 230% (Barrie et al., 2022). This trend shows increasing awareness of the economic and environmental advantages of circular trade.

Regional specialization is also emerging as a key feature of this transition. Commodities exporting countries are increasingly specializing in specific commodities to leverage their comparative advantages in resource efficiency (Bellmann et al., 2021). This specialization aligns with global efforts to optimize resource use and reduce waste. Additionally, the creation of efficient reverse supply chains for environmentally sound waste management is essential, particularly in compliance with international agreements like the Basel Convention (Swannell et al., 2021).

Global agreements and policies play a pivotal role in facilitating this transition. Multilateral environmental agreements, such as the Organisation for Economic Co-operation and Development (OECD) Council Decision and the Basel Convention, regulate trade in waste and scrap, ensuring that materials are processed and recovered in an environmentally sound manner (Yamaguchi, 2018). The World Trade Organization (WTO)'s Technical Barriers to Trade Agreement encourages member countries to adopt international standards for national policy-making, fostering harmonization and reducing barriers to trade in sustainable goods and services (Yamaguchi, 2021). The Global Alliance on Circular Economy and Resource Efficiency, launched in 2021, underscores the importance of international cooperation in advancing circularity. Regional alliances, such as the Latin American and Caribbean Circular Economy Coalition and the African Circular Economy Alliance, further highlight the diverse approaches countries are taking to embed circular principles into their trade policies (Swannell et al., 2021). Additionally, initiatives like the WTO's Trade and Environmental Sustainability Structured Discussions (TESSD) have revitalized efforts to integrate environmental sustainability into global trade frameworks (Bellmann et al., 2021).

Certain industries are already gaining a competitive advantage by prioritizing resource efficiency. The recycling industry, for instance, benefits from countries' comparative advantages in sorting and processing activities, which can enhance global recycling rates (Yamaguchi, 2018). Similarly, the remanufacturing sector, particularly in the United States, has grown significantly, with an industry valued at around $43 billion and generating 180,000 full-time jobs (Barrie et al., 2022). Other industries, such as sustainable agriculture and the paper industry, are also leveraging circular practices to reduce waste and enhance resource efficiency (Barrie et al., 2022).

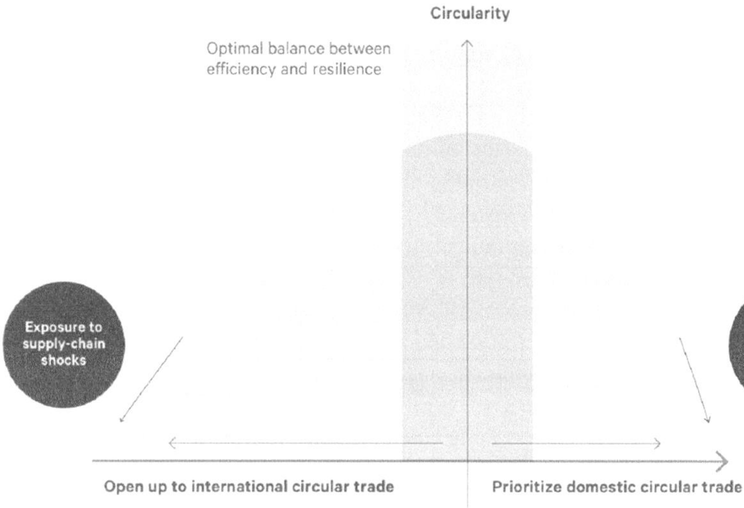

FIGURE 5.2 Achieving optimal circularity requires balancing circular trade flows.

Several countries and regions have emerged as leaders in advancing circular economy principles. China's "Dual Circulation" strategy integrates circularity into its economic framework, although its focus remains largely on domestic development (Barrie et al., 2022). The European Union has taken a leadership role through its Circular Economy Action Plan (CEAP), which aims to increase repair and reuse of used goods, thereby reducing reliance on secondary markets. Finland, France, and the Netherlands have also developed comprehensive circular economy roadmaps, demonstrating their commitment to sustainable resource management (Yamaguchi, 2018).

Other countries, such as Japan, are implementing specific policies like the Fundamental Law for Sound Material-Cycle Society, while the United States has launched initiatives such as the Sustainable Materials Management Action Plan (Yamaguchi, 2018). In India, despite limited domestic steel scrap supplies, the country accounts for 13% of global secondary steel production, underscoring the importance of trade in waste and scrap for material recovery (Yamaguchi, 2018). Similarly, countries like Chile, China, Colombia, Rwanda, South Africa, and Vietnam are embedding circularity into their policy frameworks (Swannell et al., 2021).

In Europe, Finland, France, Spain, and Sweden have made significant strides in establishing national circularity frameworks, while Costa Rica, Morocco, Peru, and South Korea are leveraging bilateral trade agreements to promote remanufacturing and sustainable practices (Barrie et al., 2022). These examples demonstrate the diverse approaches countries are taking to align their trade policies with resource efficiency and circular economy principles.

There is a transformative potential of international trade in driving resource efficiency and circularity. By fostering collaboration, adopting supportive policies, and prioritizing industries that value sustainability, nations can lead the global transition toward a more efficient and sustainable economic model.

5.2 LIMITATIONS OF LINEAR CONSUMPTION IN GLOBAL TRADE

The "take-make-dispose" model of consumption harms both the environment and the economy, slowing sustainable development. This model perpetuates soil and water pollution, greenhouse gas (GHG) emissions, and resource depletion (Lahane et al., 2020). The extraction and processing of natural resources contribute to 50% of all carbon emissions and 95% of terrestrial biodiversity loss, with the toxic effects of key metal extraction and processing expected to double by 2030 (Bellmann et al., 2021). Unsustainable resource consumption exacerbates these issues, particularly as global demand for materials continues to rise.

The environmental consequences of linear consumption are profound. Global waste generation has reached alarming and unsustainable levels, with the world currently producing more than 2.3 billion tonnes of municipal solid waste annually. This volume is projected to rise to nearly 3.8 billion tonnes by 2050 due to rapid urbanization, industrialization, and increasing consumption patterns. (Lahane et al., 2020). The weight of materials consumed worldwide has more than doubled over the past four decades, exceeding 100 billion tons per year since 2017, with over 90% derived from raw materials (Swannell et al., 2021). This surge in material consumption is accompanied by a significant increase in GHG emissions, which are expected to grow by 80% from 41 Gt CO_2 equivalent (Bellmann et al., 2021).

Linear consumption patterns also pose economic drawbacks. Overconsumption, inefficient material use, and poor waste disposal mechanisms contribute to the global burden of waste generation (Mangla et al., 2018). The reliance on resource extraction for linear systems leads to environmental degradation, including pollution and biodiversity loss (Bellmann et al., 2021). Moreover, the material footprint of OECD countries has increased by nearly 70% due to embedded materials in traded goods. These trends highlight the unsustainability of current consumption patterns.

The linear model's alignment with social justice is another critical concern. Wealth accumulation and resource-intensive lifestyles are concentrated primarily in the Global North, while the Global South bears the brunt of environmental impacts and human health threats (Barrie et al., 2022). This disparity undermines efforts to achieve SDGs, as rising living standards and poverty reduction have been accompanied by steep environmental costs (Swannell et al., 2021).

Despite growing awareness of these issues, significant gaps in current systems hinder the transition from linear to circular consumption. Limited awareness among global trade actors about the benefits of circularity remains a barrier (Barrie et al., 2022). Additionally, data scarcity and lack of standardized definitions for circular economy indicators complicate progress tracking (Bressanelli et al., 2019; Circular Economy: From Indicators and Data to Policy-Making, 2024). Policy misalignment also poses challenges, as existing trade policies often conflict with circularity objectives (Bellmann et al., 2021). Furthermore, global inequities in digital trade capabilities, infrastructure, and access to circular finance exacerbate the linear trade divide (Barrie et al., 2022). Regulatory divergence across jurisdictions adds further complexity by creating barriers to investment in circular solutions (Barrie et al., 2022). Finally, limited recycling capacity, particularly for materials like lithium with less than 1% recycling rates, underscores technical and economic challenges in achieving circularity (Yamaguchi, 2021).

These limitations underscore the urgent need for systemic changes to overcome barriers and transition toward a more sustainable, circular economy. To fight against limitations, certain strategies for embedding circular principles in global supply chains can be utilized. There are case studies of successful transitions which can help in identifying broader challenges in circular economy adoption.

5.3 ESTABLISHING CIRCULAR ECONOMY PRINCIPLES IN GLOBAL SUPPLY CHAINS

The circular economy offers a transformative framework for redefining global supply chains by prioritizing resource efficiency, waste reduction, and sustainable practices. Central to this approach are key principles that emphasize the value retention of materials and resources throughout their lifecycle. Circular Supply Chain Management (CSCM) applies these principles to reduce waste, optimize material flows, and lessen environmental impact (Zhang et al., 2021). This involves closing resource loops through reuse, recycling, and remanufacturing while also extending product lifecycles through repair and redesign (Geissdoerfer et al., 2018).

The integration of circularity into international trade agreements is critical for fostering global sustainability. Strategies such as embedding circularity across trade policies, promoting international standards for resource efficiency, and harmonizing regulations can facilitate the transition to circular practices (Barrie et al., 2022). Additionally, public procurement processes and investment in circular economy initiatives are essential for driving market transformation. Addressing potential trade barriers caused by domestic circular policies through stakeholder engagement ensures inclusive growth and avoids unintended negative impacts on global trade.

Technology plays a pivotal role in enabling circular supply chains. Blockchain technology enhances transparency, traceability, and efficiency, particularly in tracking material flows and ensuring compliance with sustainability standards. The Internet of Things (IoT) supports waste reduction by optimizing resource allocation and monitoring supply chain operations in real time (Mastos et al., 2021). Big data analytics further aids in identifying inefficiencies and driving sustainable decision-making. These technologies collectively contribute to reducing costs, emissions, and social impacts while fostering collaboration among stakeholders.

Fostering collaboration among diverse stakeholders is essential for advancing circular economy principles. This includes cross-sector partnerships, extended producer responsibility (EPR) schemes, and public–private initiatives that promote shared responsibility for resource management (Barrie et al., 2022). Capacity-building efforts, particularly for small- and medium-sized enterprises in low-income countries, ensure inclusivity and equitable participation in the circular transition. By aligning policies, sharing knowledge, and engaging stakeholders, global supply chains can be transformed to support SDGs and contribute to a regenerative economy.

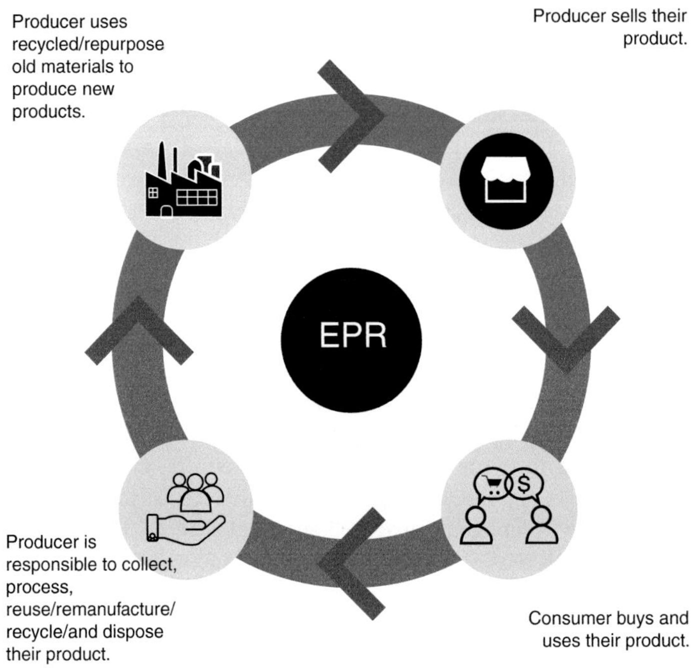

FIGURE 5.3 EPR model.

The role of technology and stakeholder collaboration in embedding circular principles into global trade systems is critical. It is one of the ways in which the limitations of circularity can be tackled.

5.4 CASE STUDIES AND BEST PRACTICES IN CIRCULAR GLOBAL TRADE AND SUPPLY CHAIN

Successful case studies across industries show how circular economy principles benefit global trade and supply chains. These examples highlight the potential of circular business models (CBMs) and circular supply chain management (CSCM) to reduce waste, enhance resource efficiency, and deliver environmental and economic benefits.

i. Case Study: Chemical and Food Industries (Genovese et al., 2017)
   The chemical and food industries were chosen for their significant environmental footprint. These sectors are critical to global trade due to their vast production networks and resource-intensive processes. The study aimed to evaluate how circular economy principles could mitigate their environmental impact while maintaining economic viability. The research employed a Hybrid Lifecycle Assessment (LCA) methodology, combining process LCA with environmentally extended multi-regional input–output (MRIO) models. This

dual approach allowed for a comprehensive analysis of direct and indirect emissions, resource use, and waste recovery across supply chains. Data was sourced from industry-specific databases like Ecoinvent to ensure accuracy.

Chemical Supply Chain (Ferrous Sulphate):

- Circular production utilizing acidic waste from titanium dioxide reduced carbon emissions significantly compared to linear systems.
- This approach not only cut down on resource extraction but also enhanced sustainability by repurposing industrial waste.

Food Supply Chain (Waste Cooking Oil):

- Biodiesel production from waste cooking oil resulted in lower environmental impact than virgin oil-based methods.
- The study highlighted the potential for circular economy practices to reduce reliance on virgin resources and promote sustainable production within global trade networks.

This research shows that circular approaches reduce environmental harm while maintaining economic gains.

ii. Case Study: Medium-Sized Manufacturing Company (Braun et al., 2018)

The selected company produces decorative automotive parts, primarily using aluminum and wooden inlays. Its significant material usage and waste generation made it an ideal candidate for studying circular economy implementation. The study focused on deterministic calculations and simulations to assess material efficiency improvements. It evaluated five main supply chains, including pre-products, auxiliary materials, and operating supplies. The analysis considered material efficiency (η) as the ratio of usable output to input and assessed the impact of reintroducing waste into supply chains.

Material Efficiency Improvements:

- Overall material efficiency increased from 59% to 68% post-optimization.
- The company reduced its reliance on virgin materials by optimizing resource use, thereby decreasing extraction pressures.

Waste Management and Recycling:

- Reintroducing 50% of waste into the supply chain yielded 1,110,433 kg/a of material annually, representing 39% of total material input.
- This underscores the economic benefits of circular practices in reducing waste and enhancing resource recovery.

The study's findings suggest that companies can achieve significant cost savings and operational efficiency by integrating circular economy measures. These improvements not only bolster a company's competitive edge in global trade but also contribute to broader sustainability goals, aligning with global trends toward eco-friendly production practices.

iii. Case Study: Circular Supply Chain Framework in OEMs (Amir et al., 2023)

This case study focuses on two Original Equipment Manufacturers (OEMs) in the white goods and automotive spare parts industries. Each company applied circular economy principles in unique ways, creating different value models in their industries.

White Goods Manufacturer:

- A leading European home appliance manufacturer piloted a pay-per-wash scheme for washing machines. This innovative business model involves rethinking product design, manufacturing, usage, and recovery, with the company retaining full ownership of the products throughout their lifecycle.
- The ASKO85 washing machine model is designed for multiple lifecycles (three cycles of 5 years each) and includes connectivity features for monitoring and predictive maintenance. This approach aims to reduce costs and environmental impact by combining sustainable business models with regenerative practices, such as reuse, remanufacture, and recycle.
- The pay-per-wash model successfully reduced environmental impact by extending product lifecycles and promoting resource efficiency.
- Better connectivity and monitoring improved customer service and cut operational costs.

Automotive Spare Parts Manufacturer:

- A long-standing player in the automotive aftermarket implemented a product sales model with after-sales service incentives for spare parts return. Driven by the benefits of remanufacturing—reduced raw material extraction and waste generation—the company designed its products for re-manufacturability.
- Challenges included integrating forward and reverse supply chains, which were addressed through an ICT platform to enhance transparency and efficiency in value chains. This platform facilitates core returns and streamlines financial and information operations, improving cross-organizational cooperation.
- Remanufacturing initiatives significantly reduced waste generation and raw material extraction.
- The ICT platform proved essential in overcoming reverse logistics challenges, enhancing the overall effectiveness of circular practices.

These case studies demonstrate how OEMs can lead the way in adopting circular economy principles to enhance global trade sustainability. By integrating innovative business models and advanced technologies, companies can achieve both economic efficiency and environmental stewardship. The findings underscore the importance of systemic approaches and technological solutions in driving the circular transition across industries.

The three case studies collectively illustrate the diverse ways industries are embracing circular economy principles to transform their supply chains and contribute to sustainable global trade. From chemical and food production to automotive manufacturing, each sector offers unique insights into overcoming challenges and capitalizing on opportunities for circularity. By learning from these examples, businesses can adopt tailored strategies that enhance competitiveness while promoting environmental sustainability.

5.5 CHALLENGES IN CIRCULAR ECONOMY ADOPTION

The transition toward circular economy principles faces numerous challenges that hinder its widespread adoption across industries and geographies. These challenges can be categorized into technological barriers, financial constraints, regulatory and policy hurdles, and behavioral and cultural challenges.

5.5.1 Technological Barriers

The integration of advanced technologies such as Industry 4.0 innovations—leveraging big data, IoT, and blockchain—remains a significant challenge in CSCM. These technologies improve traceability, visibility, and efficiency, but their adoption remains limited. For instance, the lack of an automated and trusted waste tracking system creates inefficiencies in material flows (Mastos et al., 2021). Additionally, technical standards and regulations often fail to align with circularity goals, leading to fragmented and costly compliance processes (Barrie et al., 2022).

5.5.2 Financial Constraints

High costs make it hard for businesses to switch to circular models. High upfront investments in circular strategies, such as take-back systems and product remanufacturing, coupled with limited access to capital, deter many companies from adopting these practices (Ayati et al., 2022). In developing economies like India, small and medium-sized enterprises (SMEs) face additional challenges due to insufficient government subsidies and a lack of long-term financial incentives (Ayati et al., 2022; Mangla et al., 2018).

5.5.3 Regulatory and Policy Hurdles

Regulatory divergence across jurisdictions is a major obstacle to circularity. Misaligned policies, conflicting standards, and trade-restrictive measures such as import bans on plastic waste create barriers for companies operating in multiple markets (Barrie et al., 2022). Furthermore, the absence of harmonized methodologies for measuring circular economy progress complicates international collaboration. Weak government policies often block 3R (reduce, reuse, recycle) efforts, especially in recycling (Ayati et al., 2022).

5.5.4 Behavioral and Cultural Challenges

Behavioral shifts are essential for embedding circularity into global trade. However, consumer resistance to purchasing recovered or remanufactured products due to perceived quality issues remains a significant barrier (Geissdoerfer et al., 2018). Cultural attitudes toward sustainability also play a role, with societal norms often prioritizing convenience and disposable lifestyles over resource efficiency (Bressanelli et al., 2019). Awareness programs are crucial to fostering informed decision-making among consumers, yet many regions still lack adequate initiatives to educate the public on circular practices (Circular Economy: From Indicators and Data to Policy-Making, 2024).

These challenges underscore the complexity of transitioning from linear to circular consumption patterns. Governments, businesses, and society must work together to align policies, invest in innovation, and encourage change. The next section will conclude the chapter by summarizing key insights and offering recommendations for accelerating the global adoption of circular economy principles.

5.6 CONCLUSION

The transition toward a circular economy represents a critical shift in how we approach resource management, production, and consumption in the context of global trade and supply chains. By addressing the limitations of linear models, leveraging technological advancements, overcoming financial and regulatory barriers, and fostering behavioral change, stakeholders across industries can drive the adoption of circularity principles.

This chapter has explored the foundational context of circular economy principles, examined the transformative potential of international trade in advancing resource efficiency, highlighted successful case studies and best practices, and identified key challenges hindering widespread adoption. The insights gained from these discussions underscore the importance of collaborative efforts between governments, businesses, and civil society to accelerate the transition toward a more sustainable and regenerative economic model. As global trade continues to evolve, embedding circularity into its core principles will be essential for achieving the SDGs and addressing pressing environmental and social challenges. By learning from successful examples and overcoming existing barriers, the global community can pave the way for a future where resources are valued, reused, and recycled, ensuring a sustainable path forward for generations to come.

REFERENCES

Amir, S., Salehi, N., Roci, M., Sweet, S., and Rashid, A. (2023). Towards circular economy: A guiding framework for circular supply chain implementation. *Business Strategy and the Environment*, *32*(6), 2684–2701. https://doi.org/10.1002/bse.3264

Ayati, S. M., Shekarian, E., Majava, J., and Wæhrens, B. V. (2022). Toward a circular supply chain: Understanding barriers from the perspective of recovery approaches. *Journal of Cleaner Production*, *359*. Elsevier Ltd. 131775. https://doi.org/10.1016/j.jclepro.2022.131775

Barrie, Jack, Patrick Schröder, Marianne Schneider-Petsinger, Richard King, and Tim Benton. "The role of international trade in realizing an inclusive circular economy." *Chatham House, Environment and Society Programme, London* (2022), Chatham House.

Bellmann, C., Keitanen, M., Lizano, A., Vanderloo, C., Ming, Z., Steinfatt, K., Yamaguchi, S., Aila, M., Bauer, K., Dessi, R., John, U., Larshans, P., Schroeter, L., Wallstrom, S., Botwright, K., Deere, C., Sell, M., and Tipping, A. (2021). *The circular economy and international trade options for the World Trade Organization (WTO).*

Braun, A. T., Kleine-Moellhoff, P., Reichenberger, V., and Seiter, S. (2018). Case study analysing potentials to improve material efficiency in manufacturing supply chains, considering circular economy aspects. *Sustainability (Switzerland)*, *10*(3). 880. https://doi.org/10.3390/su10030880

Bressanelli, G., Perona, M., and Saccani, N. (2019). Challenges in supply chain redesign for the circular economy: A literature review and a multiple case study. *International Journal of Production Research*, *57*(23), 7395–7422. https://doi.org/10.1080/00207543.2018.1542176

Farooque, M., Zhang, A., Thürer, M., Qu, T., and Huisingh, D. (2019). Circular supply chain management: A definition and structured literature review. *Journal of Cleaner Production*, *228*, 882–900. Elsevier Ltd. https://doi.org/10.1016/j.jclepro.2019.04.303

Geissdoerfer, M., Morioka, S. N., de Carvalho, M. M., and Evans, S. (2018). Business models and supply chains for the circular economy. *Journal of Cleaner Production*, *190*, 712–721. https://doi.org/10.1016/j.jclepro.2018.04.159

Genovese, A., Acquaye, A. A., Figueroa, A., and Koh, S. C. L. (2017). Sustainable supply chain management and the transition towards a circular economy: Evidence and some applications. *Omega (United Kingdom)*, *66*, 344–357. https://doi.org/10.1016/j.omega.2015.05.015

Jain, S., Jain, N. K., and Metri, B. (2018). Strategic framework towards measuring a circular supply chain management. *Benchmarking*, *25*(8), 3238–3252. https://doi.org/10.1108/BIJ-11-2017-0304

Jensen, H. H. (n.d.). *8 Ways the Circular Economy Outperforms Traditional Business Models*. Retrieved February 10, 2025, 15, from https://dxc.com/uk/en/cp/intelligence_delivered/perspectives/blogs/8-ways-the-Circular-Economy-outperforms-traditional-business-models

Lahane, S., Kant, R., and Shankar, R. (2020). Circular supply chain management: A state-of-art review and future opportunities. *Journal of Cleaner Production*, *258*: 9, Elsevier Ltd. https://doi.org/10.1016/j.jclepro.2020.120859

Mangla, S. K., Luthra, S., Mishra, N., Singh, A., Rana, N. P., Dora, M., and Dwivedi, Y. (2018). Barriers to effective circular supply chain management in a developing country context. *Production Planning and Control*, *29*(6), 551–569. https://doi.org/10.1080/09537287.2018.1449265

Mastos, T. D., Nizamis, A., Terzi, S., Gkortzis, D., Papadopoulos, A., Tsagkalidis, N., Ioannidis, D., Votis, K., and Tzovaras, D. (2021). Introducing an application of an industry 4.0 solution for circular supply chain management. *Journal of Cleaner Production*, *300*, 2. https://doi.org/10.1016/j.jclepro.2021.126886

Nasir, M. H. A., Genovese, A., Acquaye, A. A., Koh, S. C. L., and Yamoah, F. (2017). Comparing linear and circular supply chains: A case study from the construction industry. *International Journal of Production Economics*, *183*, 443–457. https://doi.org/10.1016/j.ijpe.2016.06.008

Pauwels, Michiel, René Reich, An Vercalsteren, Maarten Christis, Luc Alaerts, and Karel Van Acker. "Understanding the Problem Space for Effective Use of a Circular Economy Monitor in Policy Making." In International Conference on Electronic Participation, pp. 163–178. Cham: Springer Nature Switzerland, 2024. DOI: https://doi.org/10.1007/978-3-031-70804-6_11

Swannell, Richard. "The role of business in moving from linear to circular economies." (2021). https://wedocs.unep.org/20.500.11822/36830

Yamaguchi, S. (2018). *International Trade and the Transition to a More Resource Efficient and Circular Economy: Vol. 2018/03* (OECD Trade and Environment Working Papers). https://doi.org/10.1787/847feb24-en

Yamaguchi, S. (2021). *International trade and circular Economy – Policy alignment: Vol. 2021/02* (OECD Trade and Environment Working Papers). https://doi.org/10.1787/ae4a2176-en

Zhang, A., Wang, J. X., Farooque, M., Wang, Y., and Choi, T. M. (2021). Multi-dimensional circular supply chain management: A comparative review of the state-of-the-art practices and research. *Transportation Research Part E: Logistics and Transportation Review, 155*: 11. https://doi.org/10.1016/j.tre.2021.102509

6 Circular Strategies for Waste Management in Manufacturing

Neelesh Mishra, Nishkarsh Singhal, Pratyush Pandey, and Shiven Shanmugananthan

6.1 INTRODUCTION

Manufacturing waste comes in various forms including hazardous materials, scrap metals, and chemical byproducts, each of which pose specific challenges. However, this waste also presents an opportunity for resource recovery. By considering waste as a valuable resource manufacturing processes can reduce their reliance on virgin materials, innovate sustainably, and decrease the environmental impact from their operations. To transition to circular waste management practices, financial investments and incentives play a key role, as current waste management practices are limited by outdated infrastructure. Implementing circular strategies that focus on reusing, recycling, and remanufacturing also faces obstacles due to technological constraints, and complexity of existing supply chains. In today's world, environmental challenges and resource scarcity are major concerns, and traditionally used linear model based on "take-make-dispose" mindset is being reconsidered. In modern manufacturing, waste management and circular strategies play a crucial role in achieving sustainable and efficient production. Many industries face environmental and economic challenges, and new methods to eliminating waste and utilizing resources wisely are becoming key drivers of change (Zeller et al., 2023).

6.2 MAJOR TYPES OF WASTE IN MANUFACTURING

Manufacturing processes generate various types of waste, which are generally categorized into solid, liquid, and gaseous forms. Solid waste commonly includes metal scraps, packaging materials, defective products, and other discarded manufacturing residues. Liquid waste comprises chemical effluents, cooling water waste, and waste from chemical byproducts. Additionally, most of the gaseous waste is generated via combustion and chemical processing (Figure 6.1).

6.2.1 Solid Waste

Various production processes in manufacturing generate a significant amount of scrap metal waste, which is unavoidable. The various production processes include

DOI: 10.1201/9781003637677-6

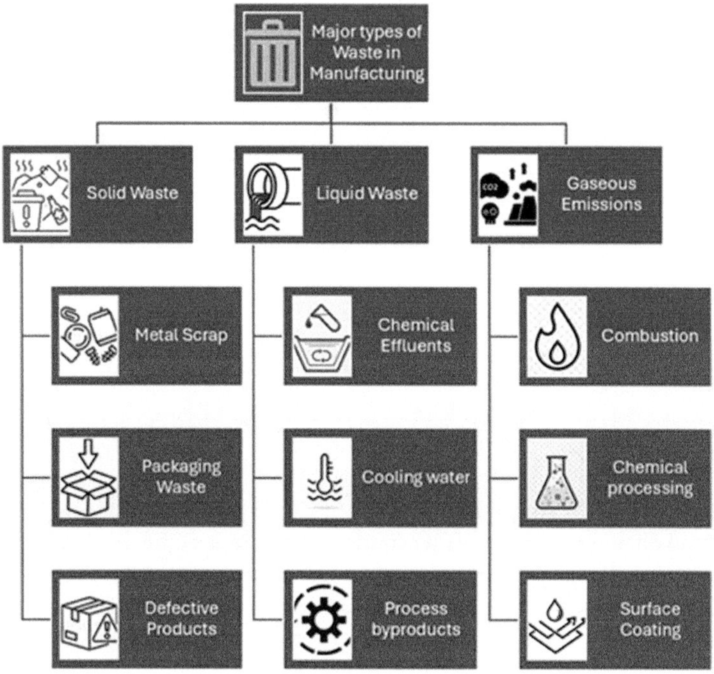

FIGURE 6.1 Major types of waste in manufacturing.

operations like machining, cutting, drilling, milling, and turning. These operations produce large amounts of metal shavings, chips, and turnings. Quality control also plays an important role in waste generation, as components that fail to meet required standards due to dimensional inaccuracies, surface defects, or material inconsistencies are discarded and considered as metal waste. Additionally, the manufacturing process generates a surplus amount of material at the end of the process, which includes setup components and test runs that are no longer needed (Shahbazi et al., 2013). Proper equipment maintenance may also lead to waste generation; this waste can be worn-out metal parts, tools, or replacement of machine parts. Errors, whether caused by machines due to wrong input settings or operator mistakes, can also lead to the generation of unwanted metal waste of defective products.

On the other hand, packaging waste is also an important challenge in the manufacturing process. This waste is being generated almost at every stage of operations. The major sources of packaging waste generation may include protective materials like cardboard boxes, wooden pallets, and foam to safeguard goods from getting damaged, creating a significant amount of waste. Work-in-progress packaging, which safeguards goods during their movements, also leads to packaging waste. In high-volume manufacturing, items like containers and boxes are used for bundling several units together, which leads to waste generation. Other sources may include the large amount of packaging waste generated by the transport packaging; it is necessary to secure products during transportation and marketing or display packaging, including retail boxes and promotional materials, but at last it leads to waste (Demko-Rihter et al., 2023).

In addition to scrap and packaging waste, waste due to defective goods also represents a significant waste stream in manufacturing. There are various factors that contribute to the defective goods generation, which are material quality, environmental conditions, human error, equipment malfunction, process variations, and quality control failures. The use of poor-quality raw materials also results in faulty products. Equipment malfunctions, including mechanical failure, worn-out machine parts, and calibration mistakes, can cause a large amount of defective goods. Process variations may include changes in temperatures, pressure, or timing, which can contribute to the generation of defective goods. Errors made by humans, such as poor assembly, improper design, or unskilled labor, further increase defect rates. By proper training, we can solve these issues. Environmental factors like humidity, dust, and extreme temperatures can also impact production quality. Additionally, quality control failure, such as improper inspection or maintenance, allows defective goods to reach later stages of production (Fernando et al., 2016).

To mitigate these challenges, manufacturers are now moving toward sustainable management strategies, which include lean production practices, circular economy principles, and reverse logistics. Lean production mainly focuses on the detection of defects due to inefficiencies, quality, or other sources at the early stage of production, preventing them from turning into waste. Circular economy promotes reuse and recycling of materials, which reduces dependence on single-use packaging. Strategies like reverse logistics play a crucial role by allowing manufacturers to remake and recover materials from waste. In the case of scrap metal, effective strategies could lead manufacturers to get valuable materials from waste, which reduces the dependence on virgin materials. Reverse logistics in scrap metal management has shown lower inventory levels and decreased scrap generation rates, ultimately conserving resources and lowering production costs.

Similarly, now companies are exploring alternative packaging options such as reusable materials to reduce negative impact on the environment. By integrating these waste management strategies, manufacturers can enhance sustainability while optimizing operational efficiency.

6.2.2 Liquid Waste

There are multiple sources for liquid waste generation, such as chemical effluents, cooling water discharge, and process byproducts. Effective management of this waste is critical to reducing environmental impact and ensuring regulatory compliance.

Chemical effluents are an important contributor to liquid waste generation, which includes major sources like chemical processing, equipment maintenance, and surface treatment. Surface treatment includes processes like metal plating and anodizing, which generate acidic and alkaline effluents containing heavy metals (Narayanan et al., 2021). Cooling systems, including industrial cooling towers and heat exchangers, discharge water containing chemicals and minerals. Paint and coating also release harmful compounds, solvent waste, and pigments. Moreover, chemical reactions, laboratory operations, and research and development (R&D) works generate hazardous effluents. Techniques like neutralization, precipitation, and filtration help to reduce the environmental impact (Robin et al., 2024).

Cooling water discharge represents another major waste stream in manufacturing. The systems, like heat exchanger systems, use large amounts of water to cool equipment, which leads to evaporation losses and thermal pollution, which can cause harm to aquatic life. Machining operations like CNC milling use water, which gets mixed with metal particles and lubricants. On-site power generation uses cooling towers, which results in heated water discharge containing concentrated chemicals. Cleaning operations further contaminate water with oils and residues, while system leaks and inefficiencies contribute to unnecessary water waste. Implementing closed-loop systems, recycling technologies, and regular maintenance can significantly reduce cooling water losses (Al Falah et al., 2023).

Another liquid waste can be byproducts from industrial operations. Food and beverage industries produce organic waste such as fruit pulp and fermentation residues. Oil waste, solvents, and reaction byproducts are generated by various petrochemical and pharmaceutical industries. Metal processing and textile manufacturing also discharge harmful, chemical-laden waste streams, affecting water quality. To overcome these challenges, manufacturers are adopting circular economy principles, utilizing byproduct recovery systems, and implementing closed-loop processing to repurpose waste streams (Constance et al., 2024).

6.2.3 Gaseous Emissions

Manufacturing processes promote environmental and health concerns due to the release of harmful products into the environment. There are some important sources of gaseous emissions, which play an important role in waste generation into the atmosphere.

One of the major sources is combustion process. The industrial boilers, furnaces, and power systems burn fuel and emit carbon dioxide (CO_2), oxides of sulfur, and nitrogen. These pollutants all contribute to the poor air quality and global warming. Similarly, chemical processes in manufacturing generate volatile organic compounds (VOCs), which degrade air quality and can cause serious health risks to workers (Mulindwa et al., 2024).

Surface operations like coating, painting, and finishing also release harmful solvents and VOCs, further increasing pollution levels. Moreover, processes like cutting, welding, and heat treatment produce harmful fumes and pollutants. Waste treatment is another major source, as organic waste decomposes and releases greenhouse gases like methane (CH_4) and CO_2. Even material handling, mainly powder and granules, can lead to dust and fine particulates entering the air and causing respiratory risks.

To mitigate these challenges, modern manufacturing facilities are investing in advanced emission control technologies. Like scrubbers and filters, trap harmful particulates and gases before they leave into the atmosphere (Schiavon et al., 2024).

By controlled combustion, we can convert VOCs and organic pollutants into less harmful substances by using thermal oxidizers. Additionally, air quality monitoring systems help to track pollutant levels and their compliance with environmental regulations.

With growing concerns about climate change, industries are now shifting toward green manufacturing practices. By following cleaner production technologies, alternative fuels, and better emission controls, manufacturers can significantly reduce the negative impact on the environment and can make processes more sustainable.

6.3 RESOURCES FROM WASTE: OPPORTUNITIES FOR RECOVERY

6.3.1 Material Recovery

Material recovery helps to transform the waste into valuable resources through proper collection, processing, and reuse of materials in manufacturing. This technique reduces the dependence of manufacturers on virgin materials, which further reduces raw material costs, minimizes landfill waste, and generates revenues. The recovered materials can be again used in the production process or sold to other industries. This whole system contributes to maintaining a circular economy in the manufacturing process.

Recovery of metal and alloys plays a crucial role in circular manufacturing processes. The recovered scrap metals are sorted, shredded, and separated to further recover precious metal components from it. The ferrous metals are separated through magnetic force, and non-ferrous metals are separated through eddy currents. This process can attain metal recovery rates of 95% (Ishfaq et al., 2024). Additionally, techniques like solid-state crushing and room temperature ball milling transform waste chips from the machining process into metal powders. This method is very effective for aluminum alloys, where we can control the size distribution of particles by regulating milling parameters (Baramate et al., 2024). Metals can also be recovered through chemical recovery; there are some techniques that are used to remove metals, such as through leaching, electrowinning, hydrometallurgical techniques, and precipitation. This is particularly very useful in extracting rare earth metals from electronic trash. With some treatments like pyrometallurgical treatment, we use elevated temperatures to isolate and purify metals. This process includes melting and refining. This technique is widely useful to recover bulk metals but needs proper management and control over the release of pollutants (Bankole et al, 2013).

Material recovery also includes plastic and polymer recovery. There are many ways to optimize material quality and their reclamation. The various recovery methods include mechanical recycling, in which operations like sorting, washing, and reprocessing are performed on the plastic garbage to get new goods. This operation is very effective on thermoplastics, which can be repeatedly melted and reformed into new shapes and designs without affecting their properties. Chemical recycling is also used for recovery, where some complex methods like depolymerization methods decompose plastic waste into its fundamental chemical constituents, which leads to the production of virgin quality products. This method has demonstrated significant potential for polydiketoenamine elastomers, as acidolysis may efficiently recover valuable monomers (Fricke et al, 2024). Use of additive manufacturing integration also contributes to the circular economy. Recovered plastics are effectively integrated into 3D printing materials, generating new value streams for recycled polymers. This method has demonstrated efficacy in the automotive sector for waste plastics, where Acrylonitrile Butadiene Styrene (ABS) and Thermoplastic Polyolefin (TPO can be amalgamated with other materials for novel applications (Laoutid et al., 2021).

In case of recovering materials like glass and ceramics, there are some specialized techniques. The recovery processes include crushing and sorting of glass waste. The glass waste is mechanically crushed and sorted to produce cullets of different diameters that can further be melted and transformed into new products. Advanced sorting technologies guarantee excellent recovered material by separating glass by color and removing pollutants. Processes like vitrification refer to controlled melting and cooling techniques that allow industrial waste materials into glass ceramic goods. It is an excellent technique to transform waste into practical building materials. Recovered waste may also include broken or faulty ceramic goods, which are ground into powder form for use as additions in building materials or as a raw material. It further includes screening, quality control, and grinding to have consistent particle size distribution. The recovered glass and ceramic can be further used in innovative applications, such as to make glass-reinforced composites or ceramic matrix material (Turan et al., 2023).

6.3.2 Energy Recovery

With the use of cutting-edge technologies, the energy recovery processes from manufacturing have gained worldwide importance. By means of devices such as heat exchangers and steam recovery units, studies reveal that industrial enterprises may recover up to 45% of waste heat.

The various thermal and biological processes can convert manufacturing wastes into usable power through waste-to-energy technologies. Important techniques are anaerobic digestion of organic waste to create biogas, gasification to produce synthetic gas, and incineration with heat recovery. By having these systems, we can reduce volume waste by up to 90% and can achieve energy recovery rates up to 45% (Gangon et al., 2022). Advanced emissions control systems are included in modern buildings to satisfy strict legal standards and reduce environmental effects.

Heat recovery systems in manufacturing capture and reuse thermal energy that would otherwise be lost to the environment. These systems include heat exchangers, recovery of steam, thermal storage, and use of combined heat and power (CHP). The heat exchangers facilitate the transfer of thermal energy between process streams, preheating incoming materials, and cooling existing products. Steam recovery recovers leftover steam from industrial operation and uses it for power or heat generation. Thermal storage includes storing recovered heat in specialist materials and using them when there is maximum demand. CHP generates both electricity and useful thermal energy from a single fuel source, thus enhancing the overall energy efficiency.

6.3.3 Chemical Recovery

It involves extracting and reprocessing valuable chemical compounds from waste in manufacturing. There are many advanced separation techniques including crystallization, distillation, membrane filtration, etc., that are used to recover solvents, acids, and other chemical compounds. These recovered materials can be further used for production processes as a raw material to other manufacturers. Focusing on purification and reuse of organic solvents from industrial operations, reclamation of solvents is a critical component of chemical recovery in manufacturing.

This process involves a series of stages of separation and purification, which are as follows:

- Depending on boiling point variations, we can separate mixed solvents; this distillation system recovers up to 95% of industrial solvents.
- Membrane filtration technologies are used to remove pollutants and particles. This technique can generate very pure recovered solvents, which can be further reused.
- Crystallization techniques is useful to extract high-value specialized solvents, which are used in pharmaceutical industries. This method removes dissolved particles and contaminants from waste solvents.

In manufacturing, chemical recovery also depends critically on the regeneration of acids and bases. Modern facilities recover and clean waste acids and bases from industrial operations using ion exchange and membrane separation methods. Manufacturers can recover 95% of sulfuric acid from discarded pickle liquor and other acidic waste streams by using electrodialysis and electrochemical techniques. Multiple phases of purification including filtration to eliminate suspended particles and metals, followed by concentration processes to restore the acids and bases to their original strength, are included into advanced treatment systems. Once recovered, the chemicals can be put back into use in manufacturing operations, therefore lowering raw material prices and lessening hazardous waste disposal needs.

Modern facilities guarantee product quality and maximize reclamation efficiency by means of automated solvent recovery systems with real-time monitoring (Bankole et al., 2013). These technologies minimize environmental effect and hazardous waste disposal needs as well as raw material prices.

As businesses try to lower running costs and environmental effect, water recovery and reuse in industrial plants has grown ever more important. Modern water recovery systems use several treatment phases to turn wastewater into useable process water, some of which include the use of advanced technologies such as nanofiltration and ultrafiltration which can greatly reduce disposal costs and recover premium water fit for reuse again for manufacturing purposes (Alfano et al., 2021). Chemical treatment methods such as coagulation, flocculation, and precipitation are employed to remove dissolved metals and suspended solids from wastewater. In addition, biological treatment processes, including aerobic and anaerobic digestion, are used to decompose organic contaminants present in the wastewater. These treatments produce purified water that can be reused in cooling systems and various other industrial applications.

Manufacturing plants can achieve reuse rates of up to 90% by implementing comprehensive water recovery systems. This approach not only reduces the consumption of fresh water but also minimizes wastewater generation and the costs associated with it. For companies operating in water-scarce regions, transitioning from a traditional linear water system to a more sustainable circular model is increasingly essential (Fernandes et al., 2023).

For combining multiple treatment technologies, rigorous monitoring procedures are needed to satisfy the quality criteria for uses. This methodical approach to water management promotes operational effectiveness in industrial plants as well as contributes to sustainability environmental objectives.

6.4 FINANCING AND SYNERGIES IN CIRCULAR WASTE MANAGEMENT

Funding circular waste management necessitates creative business ideas and strategies. Various plans such as build-operate-transfer (BOT) and design-build-finance-operate (DBFO) plans can distribute expenses while preserving operational control flow manufacturers to obtain funds (Kaza et al., 2020). Industrial symbiosis, resource sharing, and infrastructure can help to minimize individual investment loads and maximize material recovery. Integration of sustainable waste management with green supply chain practices creates better resource efficiency and generates extra value streams (Zahran et al., 2024).

6.4.1 Investment Models for Circular Initiatives

To make waste management more sustainable and cost-effective, manufacturing companies are turning to innovative investment models that help them implement circular practices without putting too much strain on their finances. These strategies allow businesses to reduce waste, recover resources, and minimize environmental impact while ensuring long-term economic benefits.

One effective approach is the BOT model, where a private company builds a waste disposal facility, operates it for a set period of time to recover its investment, and then hands it over to the manufacturer. This not only reduces the upfront cost for businesses but also ensures that waste is managed by experts during the initial years. Similarly, the DBFO model allows manufacturers to integrate facility design, construction, financing, and operation into one contract. By spreading out the cost over time, companies can implement advanced waste management systems without worrying about heavy initial investments.

Another smart strategy is industrial symbiosis partnerships, where multiple companies come together to share waste management costs and infrastructure. In this setup, one company's waste can become another company's raw material, creating an efficient cycle that reduces waste, lowers costs, and optimizes resource use.

For companies looking to fund sustainability projects, Green Bonds provide a great solution. These bonds are designed specifically to finance eco-friendly waste management initiatives, attracting investors who are interested in supporting environmental causes. At the same time, they offer manufacturers better financial terms to implement circular waste practices (Climate Bonds Initiative, 2020).

By embracing these investment models, manufacturers can reduce waste, cut costs, and contribute to a cleaner environment—all the while ensuring their businesses remain financially strong. These strategies make sustainable waste management a practical and profitable choice for industries worldwide.

6.4.2 Government Incentives and Support Programs

Circular waste management techniques are promoted by government incentives and assistance schemes. These initiatives comprise incentives for using recycling technologies, tax credits for waste reduction efforts, and subsidies for embracing circular economy solutions. Many governments provide accelerated depreciation allowances for waste management equipment and lower permission fees for establishments proving exceptional environmental performance. Extended Producer Responsibility (EPR) policies developed by the European Union offer manufacturers financial incentives to enhance trash recovery and product design (Kaza et al., 2020). To enable manufacturers switch to circular waste management systems, local authorities sometimes augment these initiatives with technical support and training expertise.

6.4.3 Industrial Symbiosis Networks

This includes a cooperative agreement connecting industrial plants, allowing the flow of byproducts and resources. Through well-coordinated networks, they convert trash from one facility into another's raw materials, generating value. Companies in the network trade waste products, excess heat, water, and byproducts, lowering disposal costs while providing low-cost inputs to other manufacturers. For example, plastic waste from automotive manufacturing can be transformed into 3D printing materials for other industries. Network partners commonly share waste treatment facilities, energy recovery systems, and transport infrastructure, which optimize the use and sharing of capital costs. In underdeveloped areas where individual businesses can lack funds for thorough waste management systems (Adalla et al., 2024), this cooperative method has shown great success. Technical knowledge and operational best practices are exchanged by symbiotic network partners, therefore hastening the acceptance of circular economy solutions throughout sectors. This kind of knowledge exchange helps to maximize waste reduction and resource recovery plans. Studies indicate that industrial symbiosis networks can lower waste disposal costs and raw material expenditures by 20–30%. Through the sales of shared services and recovered materials, the networks also create fresh income sources.

For successful industrial symbiosis networks, proper and effective coordination is required between partners. They often benefit from tracking material flows and facilitate exchanges. These arrangements show how collaborative approaches to waste management can deliver both environmental and economic benefits.

6.4.4 Collaborative Research and Development

In circular waste management, collaborative R&D gathers all research institutes, manufacturers, and technology companies to create innovative and creative ideas for waste reduction and resource recovery. These partnerships accelerate the development of new technologies while sharing risks and costs across multiple stakeholders. Important elements of effective collaborative R&D consist of:

- Cross-Industry Innovation—Here, the manufacturers collaborate with technological companies to create creative systems or techniques. For example, manufacturers of cars collaborate with material scientists to create 3D printing materials from plastic trash (Garwacki et al., 2024).
- Digital integration companies work together to apply blockchain technology, IoT sensors, and artificial intelligence (AI) to maximize resource tracking and waste control practices.
- Academic Partnership Manufacturing companies collaborate with universities to create unique waste processing methods, such as powdering aluminum machining chips for additive manufacturing.
- Supply Chain Integration—Companies work across supply chains to design closed-loop systems that increase material recovery and reduce waste creation.
- By means of these cooperative efforts, technological difficulties are overcome, and individual research expenses are lowered, therefore hastening the shift to circular manufacturing techniques.

6.5 LIMITATIONS OF CURRENT WASTE MANAGEMENT PRACTICES

Current waste management practices face several limitations, which include a general linear economic mindset of the masses that leads to reliance on landfills which cause environmental degradation. Technological constraints that involve ineffective sorting and collecting systems, along with regulatory and economic challenges, impede the adoption of sustainable solutions.

6.5.1 Linear Economy Mindset

The "take-make-dispose" mindset or the linear economy mindset considers resources to be unlimited and waste to be an acceptable byproduct. This thinking leads to severe environmental repercussions. Manufacturing practices and consumer behavior still mostly reflect this mindset and create waste streams that damage both the environment and economic sustainability.

The linear system produces waste through several mechanisms. Many products have single use design characteristics, they are intended for short-term use and have limited lifespans, and this leads to unnecessary waste creation when the product reaches its end-of-life. Many manufacturers treat waste as an acceptable byproduct of their operations and hence do not try to reduce it. Additionally, there is a disposal mindset where people assume that products don't have value after initial usage and it leads to potentially valuable resources that could've been recovered, reprocessed, and reused to end up in landfills (De Carvalho et al., 2019). Also, virgin materials being extracted at unsustainable rates to fulfil demand lead to resource depletion.

This linear mindset creates environmental harm as it leads to increased landfill usage, increased greenhouse gas emissions from waste decomposition, and the energy-intensive extraction of virgin materials. It also highlights untapped economic opportunities, as valuable materials are discarded rather than being recovered and used again (Pham et al., 2019).

6.5.2 Technological Constraints

Current waste management practices face several technological challenges. One is the limited sorting abilities of existing automated waste sorting technologies. They usually struggle to separate mixed and contaminated materials. This reduces the material recovery rate and increases processing costs. Another issue is the insufficient integration of waste management processes in manufacturing operations. Many systems do not have real-time monitoring and control features, hence making it difficult to track and manage waste streams efficiently and prevent waste generation at its source. Quality control also poses a significant issue as current recycling technologies often produce lower quality materials compared to materials produced from virgin materials, especially for plastics and composites where degradation is relatively common. Additionally, many waste processing technologies have high energy requirements which can outweigh their environmental benefits. Also, many innovative waste management solutions that work well in laboratory environments face difficulties in industrial settings due to scaling issues (Lisboa et al., 2024). These technological constraints force the manufacturers to rely on traditional waste disposal methods rather than implement sustainable and circular solutions.

6.5.3 Regulatory Challenges

Effective waste management in different regions is greatly affected by regulatory hurdles. Manufacturers find it hard to apply uniform waste management techniques due to rules changing between nations and states.

A major challenge is due to inconsistent definition and classification of waste materials. Waste that might be considered as hazardous in one jurisdiction may be considered non-hazardous in another, hence forcing manufacturers to maintain multiple waste handling protocols. This regulatory split across regions increases compliance costs and inefficiencies.

Different countries have different requirements for waste transportation and processing, which complicates international material recovery efforts. Some nations restrict the import of certain waste types entirely, thus limiting opportunities for efficient resource recovery and recycling. Hence, cross border movement of waste is also an issue (Awino et al., 2023).

The lack of a standard and consistent Extended Producer Responsibility (EPR) scheme across regions further adds to the challenge. While some jurisdictions direct companies to have a comprehensive producer responsibility for end-of-life product management, others have minimal requirements which lead to inconsistent approaches to waste reduction and recovery (Global Waste Management Outlook, UNEP, 2024).

The approval processes for new waste treatment facilities often involve lengthy approval times and varying requirements across jurisdictions. This regulatory burden can discourage investment in innovative waste management technologies and infrastructure. Furthermore, newer recycling technologies or circular economy approaches are not accounted for in outdated laws, making it hard to implement. Hence, there is a need for consistent reform (Schulte et al., 2021).

6.5.4 Economic Barriers

Economic factors play a crucial role in implementing effective waste management practices. The high investments required to set up advanced waste management infrastructure make it difficult for companies, especially smaller ones, to invest in such systems. Market volatility, such as fluctuating oil prices, complicates this issue by increasing transportation and energy costs, which makes the consistency of the recycling initiatives difficult. Additionally, limited financial incentives from the government, such as tax breaks and subsidies, make traditional disposal methods seem much more cost-effective in the short term. Moreover, competitive pressures force manufacturers to cut operational costs. Hence, there is little investment in waste reduction and management technologies. This focus on short-term financial performance prevents the much-needed long-term investments for sustainable waste management systems, even if it could lead to potential future savings (Ezeudu et al., 2024).

6.6 CHALLENGES IN IMPLEMENTING CIRCULAR WASTE MANAGEMENT PRACTICES

Implementing circular waste management practices faces several challenges, including the need for advanced infrastructure and technology to efficiently recycling. Many regions struggle with insufficient facilities and logistics systems to collect, sort, and process waste effectively. Economic issue, such as high upfront investment costs and uncertain market returns, complicates the transition to circular models. Also, consumer behavior and awareness remains a potential challenge.

6.6.1 Technological Barriers

Transitioning to circular practices in manufacturing operations faces several key issues. One of the main challenges is data integration; many manufacturing facilities lack effective tracking mechanisms for materials throughout their production and recovery processes. There are also technological gaps, as current manufacturing methods often lack the capability to handle recovered materials properly. This leads to suboptimal recovery rates, especially when dealing with complex materials and composites. Quality control issues may arise since the use of recycled materials can sometimes result in inconsistent product quality, which is a significant concern for companies that rely on high-precision parts.

Furthermore, many manufacturing facilities do not have the necessary infrastructure to implement circular practices effectively. They lack advanced sorting systems, material recovery equipment, and waste processing technology, which are essential for closing material loops. Scaling up solutions from laboratory settings to industrial levels is complex, as large-scale material recovery and reprocessing often exceed current technological capabilities. The absence of real-time monitoring and control systems prevents effective tracing of material flows and resource recovery optimization (Islam et al., 2024).

6.6.2 Consumer Awareness and Behavioral Challenges

Circular waste management practices in manufacturing face some potential challenges due to the behavior of the consumers. The already established consumer patterns and the process of waste disposal are way too far from being called sustainable and environmentally friendly. There has been a reluctance on the consumer's side to give up on these practices. And this has become a major problem for circular economy adoption.

The more challenging aspect of this is the unawareness of the consumers. They lack the understanding of the proper segregation of wastes and disposal practices. Many individuals have exactly no idea about the different types of materials, which is one of the reasons for hazardous waste practices. One such material is plastic, which poses a significant threat to the environment. This can also sometimes lead to contamination of recyclable waste streams (Jongsuksomsakul, 2024). This lack of knowledge makes it tough on the manufacturer's part to procure recyclable materials of superior quality for reprocessing.

People often tend to adopt disposal practices that are convenient and easier for them. Most of the time, the concern is their own comfort rather than adopting environmentally friendly or sustainable means. For instance, research points out that consumers are unwilling to give up on their practice of using single-use plastic mainly because they find the reusable alternatives inconvenient for themselves to carry.

The behavior of the consumers is also largely dependent on the economic incentives. Their waste management practices are largely influenced by financial rewards and monetary inducements. Plastic bottles, for instance, having low redemption value in comparison to metal waste or paper waste, result in little or no attention for their proper disposal. Moreover, due to price variation, the recycling rates, especially in the urban areas, turn out to be uneven (Jongsuksomsakul, 2024).

A new challenge with regard to the young generation of consumers is that they tend to hoard a large number of items. They are tempted to buy more, even more than they actually require for their needs, and store them. This prevents them from following adequate disposal practices. This behavior keeps the items stored and stagnant, preventing them from entering recycling streams.

6.6.3 Supply Chain Coordination

Supply chain coordination issues pose a significant challenge in the implementation of circular waste management in manufacturing operations. The transition from linear to circular model requires complex coordination across multiple stakeholders. For instance, many manufacturing facilities face problems related to material flow tracking as their data systems do not work well. This makes it hard to track materials through production and recovery processes and a lack of clear data exchange can reduce trust among partners in the value chain. Additionally, reverse logistics is also a challenge as the collection and transportation of waste materials back through the supply chain involves extra costs and difficulties. Many companies currently do not have the infrastructure required to efficiently gather, sort, and process for reuse (Solomon et al., 2024).

Maintaining consistent quality is also a challenge as recovered materials might vary in quality, and this forces manufacturers to keep larger stocks of virgin materials, leading to excess resource use (Zhang et al., 2020). Also, the irregular availability of recycled materials often leads to supply demand mismatches, so manufacturers often prefer to rely on virgin materials. Stakeholder alignment is also an issue as manufacturers, recyclers, and material processors may have conflicting goals that may hinder optimal resource recovery (Sami et al., 2023). Lastly, the cost of implementing circular strategies falls short across the supply chain. Smaller suppliers might lack resources to invest in necessary infrastructure and technology, thereby creating bottlenecks in the proposed circular system (Zahran et al., 2024).

6.7 CURRENT INNOVATIONS AND FUTURE TRENDS

As companies look for ways to reduce waste, conserve resources, and improve their overall sustainability, circular strategies are gaining significant traction. Transition from linear to circular model is becoming crucial in addressing environmental challenges and resource scarcity. This is driving innovation and creating new opportunities across various manufacturing sectors.

6.7.1 Current Innovative Circular Strategies

Manufacturers are increasingly adopting a design approach that prioritizes circularity from the very beginning, that is, from the design phase. This involves creating modular, repairable, and recyclable products. By designing products that can be easily disassembled, repaired, and recycled, manufacturers can increase the product's life and enable efficient material recovery when the product is no longer in use. Many electronics manufacturers such as Fairphone and Framework, now build smartphones and laptops with components that are easy to replace, allowing for upgrades and repairs without needing to replace the entire device.

Remanufacturing and refurbishment is another key circular strategy which is common in the automotive and heavy machinery sectors. Remanufacturing involves restoring used products to a condition that is as good as, or even better than, new. This process helps extend the life of products, reduces waste, and conserves resources by reusing existing components. In the automotive industry, many companies are successfully remanufacturing engines, transmissions, and other critical parts. This method reduces the demand for new raw materials while offering consumers more affordable replacement options.

Industrial symbiosis is yet another innovative approach where companies collaborate by using each other's waste streams as raw materials. In such arrangements, waste heat from one facility might be used to power another, or byproducts from one manufacturing process become the inputs for a different process. A well-known example is found in Kalundborg, Denmark, where various industries form an eco-industrial park and exchange materials and energy. This closed-loop system not only reduces overall resource consumption and waste generation but also delivers significant economic benefits to the companies involved (Danielsson et al., 2018).

6.7.2 Future Trends

Future trends in circular manufacturing are opening up exciting opportunities for a more sustainable industry. One key development is the use of bio-based materials and processes. These materials are made from renewable biological resources and are often biodegradable or easy to recycle, which helps in reducing environmental impact and dependence on fossil fuels. For example, companies are exploring the use of mycelium—fungal roots—as an eco-friendly alternative to traditional packaging, and they are developing bio-based plastics that can be composted or recycled more efficiently. Additionally, additive manufacturing (or 3D printing) is gaining ground by enabling on-demand production, reducing inventory waste, and making it easier to produce specific replacement parts when needed. This technology also allows for innovative product designs that use materials more efficiently and enhance overall performance.

Digital technologies are also transforming how products are designed and managed throughout their life cycle. AI is being used to analyze material choices, predict the environmental impacts over a product's life, and suggest improvements to enhance recyclability and durability. Such AI-powered tools help manufacturers to create designs that are easier to disassemble and repair, thus extending product lifecycles. Meanwhile, blockchain technology offers a promising solution for improving transparency and traceability in supply chains by creating secure, unchangeable records of material flows from extraction to recovery. Also, the rise of product-as-a-service models means that companies are now offering products as services rather than one-time sales. This approach encourages manufacturers to design products for longevity and continuous improvement, while consumers benefit from high-quality products without large initial investments. Together, these future trends are set to transform manufacturing by turning waste into a valuable resource and supporting a resilient, circular economy (GlobeNewswire, 2024).

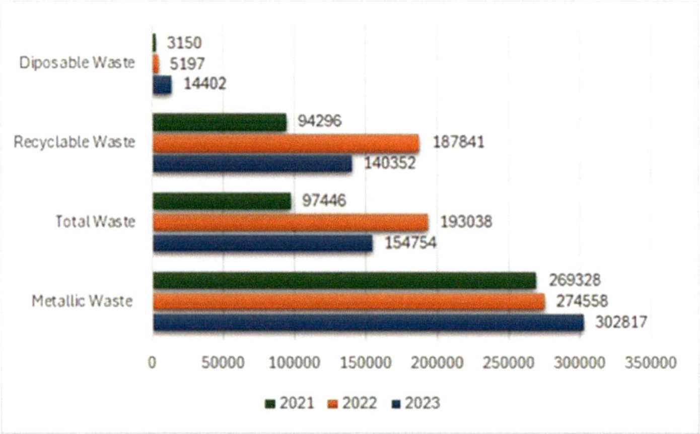

FIGURE 6.2 Total weight of waste generated (metric tons) and segregated based on recyclable and disposable waste by the Audi Group (Audi Report, 2023).

6.8 CASE STUDIES IN CIRCULAR WASTE MANAGEMENT

6.8.1 Automotive Industry

The automotive industry is a significant contributor to the GDP and resource consumption and is undergoing a transformative shift toward sustainability by adopting circular economic practices. The driving forces behind this transition are environmental concerns, resource scarcity, regulatory obligations, consumer demand, and tapping into new business opportunities.

The Audi Group is a key player in the premium automotive segment and has been at the forefront of implementing circular economy initiatives across its manufacturing operations. The Audi Group has implemented various circular strategies which range from preventative measures to recycling and recovery. These strategies span across multiple aspects of their operations from vehicle production, material usage, waste management to energy consumption.

The 9R model provides a framework for categorizing and implementing circular strategies, including Refuse, Rethink, Reduce, Reuse, Repair, Refurbish, Remanufacture, Repurpose, Recycle,. Based on this, the Audi Group's circular economic practices can be classified as:

- **Preventative Measures (Level 0 Circularity):** This includes Refuse, Rethink, and Reduce. The Audi Group has focused on minimizing unnecessary fuel consumption by not producing gasoline and diesel vehicles. In 2022, they produced 127,927 fully electric vehicles, with plans to make all new models fully electric from 2026 onward. They are also rethinking their manufacturing process by adopting digital technologies to optimize resource consumption. The Audi Group has also made it their priority to reduce material consumption through lightweight design and increasing the production of fully electric vehicles, thereby achieving reductions in CO_2 emissions and resource usage.

- **Reuse Activities (Level 1):** This includes Reuse, Repair, and Refurbish. The Audi Group has adopted practices such as reusing water multiple times which significantly reduced the consumption of freshwater and generation of wastewater. They incorporate general maintenance strategies to prolong the life of their products. They are also involved in refurbishing their older models by upgrading them technologically, to ensure the older models stay competitive in the market.
- **Remanufacturing and Repurposing Activities (Level 2):** The Audi Group remanufactures discarded parts to produce new parts that perform the same function as damaged automotive glass to produce new windshields. They also repurpose materials such as Li batteries that have reached their end in the electric vehicle but can be utilized elsewhere, use recycled PET bottles and textiles to make material for seat upholstery.
- **Recycling (Level 3):** The Audi recycles materials like aluminum to produce new aluminum coils of same quality and reintroduces them into production. They also dismantle old batteries from their vehicles and recycle them to extract valuable materials like cobalt, nickel, and lithium which are then used to produce new batteries.
- **Recover (Level 4):** The Audi Group has started a pilot project called MaterialLoop where they aim to recycle 100 end-of-life vehicles to analyze how and to what extent recovered parts can be reused in production of new vehicles. They also incinerate the waste that cannot be recycled for energy recovery in the form of heat or electricity (Imre et al., 2023).

6.8.2 Textile Manufacturing

The textile industry generates a large amount of waste of discarded garments which leads to landfills. Fast fashion worsens the situation by promoting short-lived trends, excessive resource use, and low recycling rates. Here, the "take-make-dispose" model is unsustainable, leading to degradation of environment. A circular economy offers a solution to this problem by keeping textiles in use longer through recycling, sustainable design, and reuse. Companies like H&M Group focus on circular transition of waste textile products, responsible supply chains, and by engaging the consumer in repair and resale programs. This case study explores the H&M Group's efforts in promoting circularity in fashion (Ellen McArthur Foundation, December 14, 2021). The H&M Group's main aim is to become climate positive by 2040 by transitioning their linear fashion model to a circular economy model. They ensure that every stage of the value chain supports sustainability. It involves designing of durable products, circular supply chains, and enabling customers to engage in fashion reuse, repair, and rental.

The H&M Group is committed to designing garments that can be used for longer periods, thus reducing waste. They ensure that the products are created using safe, recycled, and renewable materials that minimize environmental impacts. The H&M

Group introduced the circular tool, which guides the design teams in selecting materials and methods that enhance product longevity. It ensures that the products are made with high durability, recyclability, and sustainability.

Monki launched the circular collection of garments containing timeless designs in January 2021. These designs are made to be recycled by ensuring that waste is minimized while maintaining style and quality. ARKET's main focus is on the production of timeless and high-quality designs that last longer. The company prioritizes and explores different innovative methods to remake old clothes into new and innovative ones. For example, ARKET introduced limited collections of patchwork denim made from repurposed materials.

The H&M Group categorizes materials based on their environmental impact. The company prioritizes products that are renewable, recyclable, and long-lasting. By ensuring ethical sourcing and focusing on selective materials, the Group contributes to biodiversity conservation and lowers ecological footprints. The H&M Group collaborates with innovators like Renewcell, Worn Again, Amercycle, and Infinited Fiber to explore textile-to-textile recycling. This collaboration helps transform old garments into high-quality raw materials, reducing the use of virgin materials and promoting circularity in fashion.

By 2026 and beyond, the Group aims to further increase the share of recycled materials in its overall production processes, building upon its sustainability targets through the adoption of advanced recycling technologies and circular manufacturing practices that significantly reduce dependence on virgin raw materials. The company is also committed to sustainable production by ensuring the use of only nontoxic inputs, renewable sources, and sustainable water management in manufacturing operations. This approach reduces harm to the environment and aligns production with the principles of the circular economy (Ellen McArthur Foundation, 2021).

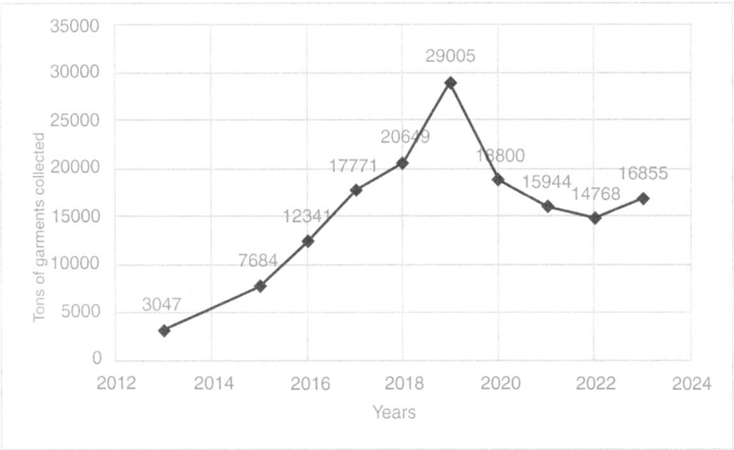

FIGURE 6.3 Quantity of apparel collected by the H&M Group's reuse and recycling scheme worldwide from 2013 to 2023 (Statista, September 25, 2024).

The H&M Group has expanded its circularity by entering the secondhand market through partnerships with resale platforms like Sellpy and Reflaunt. Customers in over 20 markets can now buy and sell their clothing through these platforms. The Group has also launched rental services in select cities, allowing customers to rent garments instead of purchasing new ones. This initiative not only reduces waste generation but also promotes a shared fashion economy. In Sweden, Weekday and &OtherStories offer rental services through the Gemme Collective platform.

To extend the lifespan of clothing, the Group has piloted in-store repair services. The Mintcare and "Take-care" initiatives aim to help customers mend damaged garments, ensuring that fewer clothes are discarded (H&M Group, 2024).

6.9 CONCLUSION

Managing manufacturing waste presents not only substantial challenges but also significant opportunities for innovation and transformation. The diversity of industrial waste—ranging from solid residues and liquid effluents to gaseous emissions—underscores the complexity of byproduct streams generated during production. Yet, within this complexity lies an immense, largely untapped potential to recover valuable materials, regenerate energy, and minimize environmental impact.

Despite advancements in waste management, current systems often fall short due to a range of limitations, including inefficient segregation, inadequate recycling infrastructure, lack of advanced processing technologies, and insufficient policy support. These barriers inhibit the full realization of a circular economy, where waste is reintegrated as a resource within the production cycle.

Transitioning to a circular strategy is further complicated by a host of technical, economic, and regulatory challenges. These include high implementation costs, limited market readiness for secondary materials, gaps in policy frameworks, and the need for cross-sector collaboration. Overcoming these barriers requires systemic change and bold, forward-looking approaches.

Leading companies like the Audi Group and the H&M Group offer compelling examples of how industries can begin this transition. By adopting innovative technologies and sustainable design principles, these organizations are pioneering methods to transform waste into valuable assets—such as alternative raw materials, bioenergy, or recycled components—thus contributing to a more resilient and regenerative industrial ecosystem.

Their initiatives serve as important case studies, demonstrating that with the right investment in innovation, infrastructure, and policy alignment, it is possible for manufacturers across sectors to shift from a linear model of take-make-dispose to a circular paradigm that prioritizes sustainability, efficiency, and long-term value creation.

REFERENCES

Adalla MK, Kapiyo RA, Oindo BO, Obong'o BO. Influence of Industrial Symbiosis on Solid Waste Reuse in Manufacturing Industries in Kisumu County, Kenya. *European Journal of Environment and Earth Sciences*. 2024; 5(1):1–6. doi:10.24018/ejgeo.2024.5.1.422

Al Falah KA, Ghouri AM. Integrating Lean Management, Sustainability, Circular Capabilities, and Waste Minimization to Optimise Operational Performance in Manufacturing. *SSRN Electronic Journal*. Published online, 2023; 6(2): 99–122. doi:10.2139/ssrn.4605337

Alfano A, D'ambrosio S, D'Agostino A, Finamore R, Schiraldi C, Cimini D. Concentrated Buffalo Whey as Substrate for Probiotic Cultures and as Source of Bioactive Ingredients: A Local Circular Economy Approach towards Reuse of Wastewaters. *Fermentation*. 2021; 7(4):281. doi:10.3390/fermentation7040281

Almendro-Candel MB, Jordán Vidal MM. Glasses, Frits and Glass-Ceramics: Processes and Uses in the Context of Circular Economy and Waste Vitrification. *Coatings*. 2024; 14(3):346. doi:10.3390/coatings14030346

Audi Report. 2023. https://www.audi.com/en/publications/corporate/2023-combined-annual-and-sustainability-report-1520. Accessed on 12/06/2025

Awino FB, Apitz SE. Solid Waste Management in the Context of the Waste Hierarchy and Circular Economy Frameworks: An International Critical Review. *Integrated Environmental Assessment and Management*. 2023; 20(1):9–35. doi:10.1002/ieam.4774

Bankole OE, Gong C, Lei L. Battery Recycling Technologies: Recycling Waste Lithium Ion Batteries with the Impact on the Environment In-View. *Journal of Environment and Ecology*. 2013; 4(1):14. www.macrothink.org/journal/index.php/jee/article/view/3257/0

Baramate KA, Lacan F, Ryan M, Gangopadhyay S, Badhuri D. Development of a Net Zero Route for the Circular Production of Additive Manufacturing Powders. *Proceedings of the Cardiff University Engineering Research Conference 2023*. Published online May 1, 2024:123–127. doi:10.18573/conf1.ac

Bjørnbet MM, Skaar C, Fet AM, Schulte KØ. Circular Economy in Manufacturing Companies: A Review of Case Study Literature. *Journal of Cleaner Production*. 2021; 294(4):126268. doi:10.1016/j.jclepro.2021.126268

Climate Bonds Initiative. *Financing waste management, resource efficiency and circular economy: Green bond market briefing*. ETC ST 2025/5, 2020.

Constance Obiuto N, Ninduwezuor-Ehiobu N, Chigozie Ani E, Andrew Olu-lawal K, David Ugwuanyi E. Implementing Circular Economy Principles to Enhance Safety and Environmental Sustainability in Manufacturing. *International Journal of Advanced Multidisciplinary Research and Studies*. 2024; 4(2):22–29. doi:10.62225/2583049x.2024.4.2.2432

Danielsson, Stefan E, Møller, Per, Randers, Lisbeth. (2018). *Modelling CO_2 savings and economic benefits for the Kalundborg Symbiosis*. Symbiosis Center Denmark, Denmark, 10p.

De Carvalho Araújo CK, Salvador R, Moro Piekarski C, Sokulski CC, de Francisco AC, de Carvalho ACSK. Circular Economy Practices on Wood Panels: A Bibliographic Analysis. *Sustainability*. 2019; 11(4):1057. doi:10.3390/su11041057

Demko-Rihter J, Sassanelli C, Pantelic M, Anisic Z. A Framework to Assess Manufacturers' Circular Economy Readiness Level in Developing Countries: An Application Case in a Serbian Packaging Company. *Sustainability*. 2023; 15(8):6982. doi:10.3390/su15086982

Ezeudu OB, and Ugochukwu CN. Financing Methods for Solid Waste Management: A Review of Public and Private Sector Financing Options. *Sustainable Development*. 2024; 32(1):45–56. https://doi.org/10.1002/sd.3256:contentReference

Fernandes E, Cunha Marques R. Review of Water Reuse from a Circular Economy Perspective. *Water*. 2023; 15(5):848. doi:10.3390/w15050848

Fernando Y, Tew MM. Reverse Logistics in Manufacturing Waste Management: The Missing Link between Environmental Commitment and Operational Performance. *International Journal of Integrated Supply Management*. 2016; 10(3/4):264. doi:10.1504/ijism.2016.10002256

Fricke SN, Haber S, Hua M, Salgado M, Helms BA, Reimer JA. Magnetic Resonance Insights into the Heterogeneous, Fractal-Like Kinetics of Chemically Recyclable Polymers. *Science Advances*. 2024; 10(14). doi:10.1126/sciadv.adl0568

Gagnon B, Tanguay X, Amor B, Imbrogno AF. Forest Products and Circular Economy Strategies: A Canadian Perspective. *Energies*. 2022; 15(3):673. doi:10.3390/en15030673

Garwacki M, Cudnik I, Dziadowiec D, Szymczak P, Andrzejewski J. The Development of Sustainable Polyethylene Terephthalate Glycol-Based (PETG) Blends for Additive Manufacturing Processing—The Use of Multilayered Foil Waste as the Blend Component. *Materials*. 2024; 17(5):1083. doi:10.3390/ma17051083

Global Waste Management Outlook 2024. Unknown. 2024.

GlobeNewswire. *Global market for bio-based and sustainable materials 2024–2035: Insights into challenges and opportunities in the sustainable materials market.* 2024, July 15.

González-Sánchez, R, Alonso-Munoz S, and Medina-Salgado MS. Circularity in Waste Management: A Research Proposal to Achieve the 2030 Agenda. *Operations Management Research*. 2023; 16(3): 1520–1540. Retrieved 25 December 2023, from https://pmc.ncbi.nlm.nih.gov/articles/PMC10120487

Hamraoui L, Bergani A, Ettoumi M, et al. Towards a Circular Economy in the Mining Industry: Possible Solutions for Water Recovery through Advanced Mineral Tailings Dewatering. *Minerals*. 2024; 14(3):319. doi:10.3390/min14030319

H&M Group. *H&M Group's commitment to circular fashion and sustainability: Transforming the fashion industry through innovation and collaboration.* H&M Group. 2024, February 21.

How H&M Group is progressing towards a circular economy in fashion. 2021. December 14, 2021. www.ellenmacarthurfoundation.org/circular-examples/hm-group

Imre G, Remsei S. Towards Sustainability in the Automotive Industry – Circular Economic Practices of the Audi Group. *Chemical Engineering Transactions*. 2023; 107:211–216. doi:10.3303/CET23107036

Ishfaq K, Sana M, Waseem MU, Anwar S, Zia AW. Circular Usage of Waste Cooking Oil towards Green Electrical Discharge Machining Process with Lower Carbon Emissions. *The International Journal of Advanced Manufacturing Technology*. 2024; 131(9–10):5133–5153. doi:10.1007/s00170-024-13322-3

Islam, T, and Jabber, A. Realization of Circular Economy Principles in Manufacturing. *Circular Economy and Sustainability*. 2024; https://doi.org/10.1007/s43621-024-00689-2

Jongsuksomsakul P. "Waste Is Wealth": Circular Economy Strategies from Media Persuasion to Utilization. *Sustainability*. 2024; 16(12):5205. doi:10.3390/su16125205

Kaza S, Yao L, Bhada-Tata P, Van Woerden F. (2020). *What a waste 2.0: A global snapshot of solid waste management in 2050*. World Bank Group.

Kouhizadeh M., and Sarkis J. Blockchain Practices, Potentials, and Perspectives in Greening Supply Chains. *Sustainability.* 2018; 10(10):3652. https://doi.org/10.3390/su10103652

Laoutid F, Lafqir S, Toncheva A, Dubois P. Valorization of Recycled Tire Rubber for 3D Printing of ABS- and TPO-Based Composites. *Materials*. 2021; 14(19):5889. doi:10.3390/ma14195889

Lisboa HM, Pasquali MB, dos Anjos AI, Sarinho AM, de Melo ED, Andrade R. et al. Innovative and Sustainable Food Preservation Techniques: Enhancing Food Quality, Safety, and Environmental Sustainability. *Sustainability.* 2024; 16(18):8223.

Mulindwa MS, Akor J, Auta M, Nijman-Ross E, Ogutu MO. Assessing the Circularity Status of Waste Management among Manufacturing, Waste Management, and Recycling Companies in Kigali, Rwanda. *Frontiers in Sustainability*. 2024; 4:1215554. doi:10.3389/frsus.2023.1215554

Narayanan J, Manjith Kumar B, Jishnu M. An assessment on the Effect of Titanium Dioxide & Iron Oxide Nano-Particles in Industrial Waste Water Decontamination. *IOP Conference Series: Materials Science and Engineering*. 2021; 1114(1):012076. doi:10.1088/1757-899x/1114/1/012076

Pham TT, Kuo TC, Tseng ML, et al. Industry 4.0 to Accelerate the Circular Economy: A Case Study of Electric Scooter Sharing. *Sustainability*. 2019; 11(23):6661. doi:10.3390/su11236661

Robin M, Rivonkar A, Suzuki-Muresan T, Abdelouas A, Mokili M. Optimized Precipitation Process for the Treatment of Radioactive Effluents from Ni-Alloy Decontamination Using a Chemical Oxidation Reduction Process. *Frontiers in Nuclear Engineering*. 2024; 3:1396821. doi:10.3389/fnuen.2024.1396821

Schiavon M, Ravina M, Zanetti M, Panepinto D. State-of-the-Art and Recent Advances in the Abatement of Gaseous Pollutants from Waste-to-Energy. *Energies*. 2024; 17(3):552. doi:10.3390/en17030552

Shahbazi S, Kurdve M, Bjelkemyr M, Jönsson C. Industrial Waste Management within Manufacturing: A Comparative Study of Tools, Policies, Visions and Concepts. In: *Proceedings of the Proceedings of the 11th International Conference on Manufacturing Research (ICMR2013), Cranfield University, UK*, 2013.

Solomon NO, Simpa P, Adenekan OA, Obasi SC. Circular Economy Principles and Their Integration into Global Supply Chain Strategies. *Finance & Accounting Research Journal*. 2024; 6(5):747–762. doi:10.51594/farj.v6i5.1133

Statista. *Amount of clothing reused and recycled by H&M worldwide from 2013 to 2023*. 2024, September 25. www.statista.com/statistics/961998/quantity-of-apparel-collected-by-handm-s-reuse-and-recycling-scheme-worldwide

Tang X, He Y. Supply Chain Coordination in Coordination in Industrial Symbiosis Networks with Uncertain Waste Demand and Supply. Mohammadzadeh A, ed. *Mathematical problems in engineering*. 2022; 2022:1–14. doi:10.1155/2022/3579378

Trollman H, Colwill J, Brejnholt A. Ecologically Embedded Design in Manufacturing: Legitimation within Circular Economy. *Sustainability*. 2020; 12(10):4261. doi:10.3390/su12104261

Turan, S, and Yılmaz, E. Recycled Waste Glass Usage for Construction Materials. *Journal of Physics: Conference Series*. 2023; 2423(1):012002. https://doi.org/10.1088/1742-6596/2423/1/012002

Zahran S. Investigating the Nexus between Green Supply Chain Practices and Sustainable Waste Management in Advancing Circular Economy. *Sustainability*. 2024; 16(9):3566. doi:10.3390/su16093566

Zhang N, Li B. Pricing and Coordination of Green Closed-Loop Supply Chain with Fairness Concerns. *IEEE Access*. 2020; 8:224178–224189. doi:10.1109/access.2020.3045152

7 Circular Practices to Regulate Agro-Waste

Joyston Jacob S, Jayakarthikeyan, and Suresh Babu C

7.1 INTRODUCTION

Agricultural industry has fundamentally been at the heart of living, and most ironically it still is among those sectors with gigantic contributions toward producing wastes. Crop residues, by-products of food processing, and organic waste are some examples of agro-waste commonly discarded or not properly managed. This results in large-scale environmental problems such as soil degradation, pollution of waterways, and emissions of greenhouse gases into the atmosphere. More specifically, resources are extracted, used, and finally disposed of through a linear approach that contributes to inefficiencies and accumulation of wastes in traditional agriculture systems. With concerns for sustainability, climate change, and food security growing, it is high time to shift toward a more circular and regenerative agricultural system. The circular economy presents a groundbreaking way to tackle these issues by reimagining how we produce, use, and reuse resources. Rather than viewing agricultural waste as a problem, circular methods emphasize extending the lifespan of farm materials, transforming organic waste into useful assets, and cutting down on waste across the entire supply chain. The transition cuts down on the environmental footprint and, at the same time, ensures a higher efficiency in resource utilization, generates new opportunities for economic gain, and supports the sustainability of an agricultural system in the long term.

This chapter treats the key circular strategies to be developed for regulating agro-waste and six important aspects of this: introducing a circular economy into agriculture, reconsidering production and consumption, reducing waste, lengthening the life of agricultural inputs, lengthening the life of food, and transforming waste into resources. Implementing these principles and aspects could propel the agricultural sector ahead toward a much more resilient, efficient, and environmentally responsible system through the optimal use of resources while minimizing adverse planetary impacts (Dey, S., Pal, A., Nandy, A., 2022).

7.2 CIRCULAR ECONOMY APPROACH IN AGRICULTURE

The circular economy, as it emerges in agriculture, would create an almost completely regenerative system in which waste is constantly minimized and resources are

DOI: 10.1201/9781003637677-7

repeatedly reused, to contrast with the old "take-make-dispose" linear standard. Such an approach would promote sustainability and reduce environmental degradation and will further economic resilience in the agricultural production sector.

7.2.1 Linear versus Circular Agriculture

In a nutshell, linear agriculture follows the cycle of taking, making, and disposing. Farmers make use of the natural resources—water, soil nutrients, and energy—to grow crops and livestock to maximize their production (Adewale, O., and Kazeem, M. (2020)). The entire process of crop production puts an emphasis on near-term yields, therefore leading to overapplication of fertilizers, monocropping, and inefficient disposal of waste. Hence, massive quantities of agro-waste, including plant waste such as surplus crop residues, animal excreta, and processing by-products, are often disposed of by burning or dumping into open areas or just left to decay, creating a great negative impact on the environment. The same is the case for a classic example: the burning of paddy straw spread all over north India. Every year, millions of tons of this produce are burned to clear the fields for the next growing season, thus adding to the terrible air pollution and smog in cities like Delhi and other neighboring states. It leads to the loss of such valuable organic materials that could otherwise have been repurposed for soil enrichment, bioenergy, or industrial applications. Circular agriculture adopts a complete paradigm shift from the end-of-life paradigm associated with farming toward systems that regenerate resources while producing value from waste streams. Instead of this waste being aggregated, it finds its way back into a production cycle in a circular system.

7.2.2 Achieving Sustainable Agriculture through Resource Optimization

A truly sustainable agricultural system minimizes waste, energy consumption, and soil degradation for long-term productivity and low environmental impact. This can be achieved by:

- Animal manure, crop residues and food waste can be composted and put back into the soil as organic fertilizers reducing the need for chemical fertilizers. This boosts soil microbial activity and water retention and reduces greenhouse gas emissions from synthetic inputs. Denmark and the Netherlands have circular farming models where organic waste is recycled into agricultural production.
- Diverse farming systems that combine trees, crops, and livestock restore soil fertility and biodiversity and create climate-resilient farms. Agroforestry systems like alley cropping and silvopasture reduce soil erosion, increase carbon sequestration, and provide multiple income sources for farmers. In Brazil and Kenya, agroforestry has increased crop yields and forest cover, proving that nature-based agricultural solutions work.
- Organic agricultural waste like manure, crop husks, and food scraps can be converted into biogas, bioethanol, and biomass energy, a renewable alternative

to fossil fuels. Biogas plants in India, China, and Germany provide farmers with affordable energy for cooking, heating, and electricity generation, reducing their dependence on firewood, Liquefied Petroleum Gas (LPG), and coal. Large-scale biomass power plants convert crop residues into electricity and make rural areas energy self-sufficient.

- A notable real-world example of circular farming is dairy farms in the Netherlands. They have started using tech to recycle poop which cuts methane gas by about 30%. They turn animal waste into energy and fertilizers. Also, their smart vertical farms grow crops better while saving water and land usage.

By transitioning from a linear to a circular model, agriculture can become more resilient, sustainable, and economically viable. Circular practices not only reduce environmental harm but also create new opportunities for farmers through waste valorization, improved resource efficiency, and climate-friendly innovations.

7.2.3 Benefits of Circular Agriculture

Focusing on resource efficiency, waste minimization, and ecological balance, circular agriculture represents a transformative shift in farming. This model offers significant environmental, economic, and social benefits, making it a cornerstone of sustainable food production. Sustainable agricultural practices are crucial for mitigating climate change by reducing greenhouse gas emissions and enhancing carbon sequestration. A key strategy is minimizing dependence on chemical inputs, as excessive use of synthetic fertilizers and pesticides significantly contributes to nitrous oxide (N_2O) emissions, a potent greenhouse gas. Transitioning to organic fertilizers, composting, and precision farming not only lowers emissions but also improves soil health and reduces water contamination. Agroforestry—integrating trees, crops, and livestock—serves as a natural carbon sink, storing carbon in both biomass and soil (Smith, J., and Brown, T. (2023)). This method offsets CO_2 emissions while enhancing biodiversity, preventing soil erosion, and improving farm resilience. Countries like Brazil and Kenya have successfully implemented agroforestry, demonstrating how tree-based agriculture can restore degraded land and combat climate change. As global food demand rises, adopting low-emission farming techniques will be essential in building a climate-resilient and sustainable food system that benefits both farmers and the environment. Circular agriculture offers significant cost savings by reducing reliance on external fertilizers and pesticides. Utilizing farm-generated inputs, such as compost from organic waste, improves soil fertility and lowers input costs (Adewale and Kazeem, 2020).

Additionally, converting agricultural waste into bioenergy through anaerobic digestion creates new revenue streams for farmers. The global potential of this process was highlighted, demonstrating that how organic waste can be transformed into valuable energy resources (Patel and Kumar, 2024). Circular agricultural practices strengthen food security by enhancing soil fertility and promoting diversified farming systems. It was explored how integrated organic farming within a circular economy framework supports sustainable development and stable food supplies (Selvan et al., 2023).

7.2.4 Environmental Benefits

Implementing circular economy principles in agro-food systems reduces environmental impact. It was reviewed by various models demonstrating how recycling organic waste through composting and other techniques fosters sustainable agricultural practices (Velasco-Muñoz et al., 2021) and also further detailed the efficiency and application of compost in farming, underscoring its role in environmental conservation (Martínez and Ramírez, 2020).

7.3 RETHINK PRODUCTION AND CONSUMPTION OF RESOURCES

Agriculture heavily depends on natural resources such as soil, water, and energy. However, conventional farming often results in overuse and wastage, leading to environmental degradation and inefficiencies in production. A circular approach to agriculture requires rethinking how resources are produced and consumed—from optimizing farming techniques to reducing input dependency and improving overall sustainability. By adopting sustainable farming practices and efficient resource utilization strategies, farmers can produce more with fewer resources while minimizing environmental harm. This shift not only conserves valuable inputs but also reduces costs and increases resilience against climate change and resource scarcity.

7.3.1 Sustainable Farming Practices

Sustainable farming practices aim to optimize agricultural productivity while preserving natural resources, ensuring long-term viability. Conventional farming methods, reliant on excessive chemical inputs and intensive land use, have led to soil degradation, water scarcity, and declining biodiversity. In contrast, sustainable approaches integrate ecological balance with technological advancements to improve efficiency and resilience. These practices include regenerative agriculture, precision farming, and integrated pest management, all of which contribute to reducing environmental impact while enhancing yields and economic returns.

Regenerative agriculture focuses on restoring soil health through minimal disturbance and organic enrichment. Practices such as crop rotation, cover cropping, and no-till farming enhance soil structure, increase water retention, and boost carbon sequestration. In Australia, farmers implementing regenerative techniques have observed up to a 30% increase in soil organic matter, leading to improved drought resistance and reduced dependency on synthetic fertilizers. Additionally, large-scale regenerative farming projects in the United States have reported a 15–20% reduction in input costs, making it both an economically and environmentally viable solution.

Precision agriculture leverages advanced technologies such as satellite imagery, IoT sensors, and artificial intelligence to optimize resource application. By using real-time data, farmers can apply water, fertilizers, and pesticides precisely where needed, minimizing waste and maximizing efficiency. In the United States, studies have shown that precision agriculture can reduce fertilizer use by 20–40% while maintaining or even increasing crop yields. In India, startups like Fasal and CropIn provide sensor-based irrigation and climate monitoring systems, allowing smallholder farmers to increase efficiency and cut costs. The widespread adoption of precision farming has also led to a 50% reduction in pesticide use in certain regions, mitigating environmental contamination and improving food safety. Another key sustainable approach is Integrated Pest Management (IPM), which reduces reliance on synthetic pesticides by utilizing natural predators, crop diversification, and targeted biological control methods. This technique not only minimizes chemical exposure but also helps prevent pesticide-resistant pest populations. In India, rice farmers adopting IPM techniques, such as planting pest-repelling companion crops, have successfully cut pesticide usage by over 50%, leading to healthier ecosystems and improved soil biodiversity. Similarly, in Kenya, maize farmers have introduced *Desmodium* plants between crops to repel pests naturally, reducing chemical pesticide dependence and increasing yields by 20–30% (Pretty et al., 2014). These sustainable agriculture practices are significant in solving current challenges in farming. The implementation of regenerative practices, precision farming, and ecological pest management will help the farmers shift to resource-efficient, climate-resilient agriculture. These benefits can go beyond productivity for the long-term sustainability and economic stability, ensuring environmental conservation.

7.3.2 Efficient Resource Utilization

Efficient resource utilization is at the core of circular agriculture, ensuring that every input—whether water, nutrients, or energy—is used optimally to minimize waste and maximize productivity. Traditional farming practices often lead to overuse of resources, resulting in soil depletion, excessive water consumption, and environmental pollution. By adopting sustainable methods such as water-efficient irrigation, circular nutrient management, and renewable energy integration, farmers can reduce costs, improve yields, and contribute to long-term agricultural resilience. One of the most critical aspects of resource efficiency is water conservation. Agriculture accounts for nearly 70% of global freshwater withdrawals, with much of it lost due to inefficient irrigation techniques. Drip irrigation has emerged as a game-changing solution, delivering water directly to plant roots, reducing evaporation, and cutting water use by up to 60% compared to flood irrigation. Israel, a leader in water-efficient agriculture, has widely implemented drip irrigation, increasing crop yields while using 50% less water than conventional methods. In India, the adoption of micro-irrigation techniques has helped smallholder farmers reduce water use by 30–50%, ensuring better productivity in drought-prone regions. Furthermore, rainwater harvesting systems, such as those used in parts of Africa, enable farmers to store and utilize rainwater, reducing dependence on groundwater (Postel et al., 2001).

Another essential component of resource efficiency is circular nutrient management, where organic materials such as crop residues, manure, and food waste are repurposed to enrich soil. Instead of relying on synthetic fertilizers, which contribute to soil degradation and water pollution, circular farming emphasizes composting, biofertilizers, and closed-loop nutrient cycles. The Netherlands has pioneered manure recycling in dairy farming, using anaerobic digesters to convert animal waste into biofertilizers and biogas, reducing methane emissions by 30% while cutting fertilizer costs. Similarly, rice farmers in Vietnam have adopted the "Straw to Gold" initiative, converting rice husks into organic compost, reducing chemical fertilizer use by 40% while improving soil quality. By reintegrating organic waste back into the farming system, these practices enhance soil fertility, prevent nutrient runoff, and create sustainable input sources for future harvests (Jurgilevich et al., 2016).

Integrating renewable energy into agricultural operations is another crucial step toward resource efficiency. Traditional farms rely heavily on fossil fuels for irrigation, processing, and transportation, increasing carbon emissions and operational costs. Renewable energy solutions such as solar-powered irrigation, wind turbines, and biogas plants not only reduce dependence on fossil fuels but also offer long-term economic savings. In Kenya, the adoption of solar-powered irrigation pumps has increased farm productivity by 150%, allowing smallholder farmers to expand cultivation without increasing energy expenses. Similarly, Germany has integrated biogas plants into dairy farms, converting manure into energy that powers farm operations while reducing greenhouse gas emissions. By transitioning to sustainable energy sources, farms can lower operational costs and contribute to climate change mitigation. (Mukumba et al., 2022). Efficient resource utilization in agriculture is not just about conservation; it is about creating smarter, self-sustaining systems that reduce waste while increasing productivity. By optimizing water use, recycling nutrients, and integrating renewable energy, farmers can build resilient agricultural models that thrive despite climate challenges, resource limitations, and market fluctuations. These strategies not only enhance profitability but also ensure that future generations inherit a farming system that is both productive and sustainable.

7.4 RETHINK AND REDUCE AGRO-WASTE

Agricultural waste is a growing concern worldwide, with millions of tons of organic and inorganic waste generated annually. Crop residues, spoiled food, livestock manure, and agro-industrial by-products often end up in landfills, burned in fields, or improperly disposed of, contributing to greenhouse gas emissions, soil degradation, and water pollution. A circular approach to agriculture requires a fundamental shift in how waste is perceived—not as a by-product to be discarded, but as a valuable resource to be repurposed. By rethinking and reducing agri-waste through improved harvesting, storage, processing, and distribution techniques, farmers and agricultural industries can minimize losses, increase efficiency, and create new economic opportunities.

One of the most pressing challenges is post-harvest losses, which account for a significant portion of agricultural waste. According to the Food and Agriculture Organization (FAO), nearly 14% of the world's food production is lost between harvest and retail, often due to improper handling, inadequate storage, and inefficient supply chains. In India, post-harvest losses of fruits and vegetables exceed 40%, largely due to a lack of cold storage facilities and efficient transportation. To combat this, innovative solutions such as solar-powered cold storage units and smart supply chain management have been introduced. In Nigeria, for example, the company ColdHubs provides solar-powered refrigeration units, extending the shelf life of perishable produce by up to 21 days and reducing food waste for smallholder farmers (Gustavsson et al., 2011).

In addition to better storage and handling, sustainable harvesting techniques play a crucial role in minimizing waste. Many farmers harvest crops prematurely or inefficiently, leading to quality deterioration and spoilage. Mechanized harvesting methods, precision agriculture tools, and improved drying techniques can help ensure that more of the harvested produce reaches markets in optimal condition. In Thailand, rice farmers using mechanized harvesters have reduced grain losses by 10–15%, ensuring higher yields and reducing waste at the farm level. Beyond primary production, processing and packaging innovations can further reduce agricultural waste. Many food processing industries discard imperfect or surplus produce, even though it remains nutritionally viable. Creative solutions, such as transforming "ugly" fruits and vegetables into processed products like juices, dried snacks, and purees, have gained traction globally. In the European Union, initiatives such as "Too Good To Go" and "Wonky Veg" campaigns promote the sale of imperfect produce at lower prices, reducing food waste while making fresh produce more affordable. Similarly, in the United States, companies like Misfits Market and Imperfect Foods have built entire business models around redistributing surplus or cosmetically flawed food, preventing thousands of tons of produce from being wasted each year.

Biodegradable and sustainable packaging also plays a critical role in reducing waste in agriculture. Conventional plastic packaging contributes to significant environmental pollution, as it often ends up in landfills or waterways. Alternatives such as banana leaf packaging (used in Thailand and India), biodegradable cassava starch bags (developed in Indonesia), and compostable food wraps made from beeswax are increasingly being adopted as eco-friendly packaging solutions. These innovations help extend the shelf life of agricultural products while minimizing plastic waste in supply chains (Siracusa et al., 2020). Another important aspect of reducing agriwaste is waste valorization, where agricultural by-products are repurposed into useful materials instead of being discarded. In Vietnam, for instance, rice husks—a major by-product of rice milling—are converted into biochar and biodegradable packaging, creating additional revenue streams for farmers. Similarly, sugarcane bagasse, traditionally seen as a waste product, is now widely used in paper production, biofuels, and biodegradable food containers, replacing plastic-based alternatives (Pfaltzgraff et al., 2013).

By rethinking how agricultural waste is managed, farmers and agribusinesses can transition toward a more sustainable and profitable system. Reducing agri-waste not only benefits the environment by cutting down pollution and emissions but also contributes to food security, economic growth, and resource efficiency. Through better post-harvest management, innovative processing techniques, sustainable packaging, and waste valorization, agriculture can move toward a truly circular model where waste is minimized and resources are fully utilized.

7.5 EXTENDING THE LIFE OF RESOURCES

The concept of extending the life of agricultural resources is central to a circular economy, where every material is seen as having potential beyond its primary use. Agriculture is an industry that generates vast amounts of organic and inorganic waste, from crop residues and food processing by-products to packaging materials and animal manure. In traditional farming and agro-industrial practices, much of this material is discarded or underutilized, leading to environmental pollution and economic losses. However, by redefining waste as a resource, the agricultural sector can drastically reduce inefficiencies, lower production costs, and generate additional income streams while contributing to environmental sustainability. Extending the life of agricultural resources can be achieved through two primary approaches: upcycling agricultural by-products and implementing circular design principles in agro-industries. These strategies focus on maximizing the usefulness of materials beyond their conventional lifespan, turning them into new, valuable products, energy sources, or industrial inputs (Mohan et al., 2016).

7.5.1 Upcycling Agricultural By-Products

Upcycling is the process of transforming agricultural waste into higher-value products rather than discarding them. Many parts of crops, such as husks, stalks, leaves, and peels, are rich in fibers, nutrients, or organic compounds that can be repurposed into sustainable materials, food additives, bioenergy, or fertilizers. One major success story in upcycling is found in India's wheat and rice production. India is one of the largest producers of wheat and rice, generating over 500 million tons of crop residues annually. Traditionally, much of this agricultural waste is burned in open fields, leading to severe air pollution, particularly in northern states like Punjab and Haryana. However, innovative companies have started turning wheat straw and rice husks into biodegradable packaging, tableware, and paper products, providing an eco-friendly alternative to single-use plastics. Similarly, banana stems and pineapple leaves—once discarded as useless agricultural waste—are now being upcycled into natural fibers for textiles. In countries like the Philippines and Thailand, companies are producing banana fiber-based fabrics, which are stronger and more sustainable than conventional cotton. This innovation not only reduces textile industry pollution but also provides farmers with additional revenue from by-products that were previously discarded.

Upcycling also plays a significant role in food waste reduction. Fruits and vegetables that do not meet size or appearance standards are often rejected by retailers, leading to enormous food losses. However, companies like Misfits Market and Imperfect Foods in the United States are purchasing these "ugly" fruits and vegetables and selling them at discounted prices, reducing food waste while making fresh produce more affordable. Meanwhile, in the European Union, the "Wonky Veg" initiative encourages supermarkets to sell imperfect produce at lower prices, successfully reducing food waste by 30% in participating stores. By repurposing agricultural by-products into new products and industries, upcycling extends the usefulness of resources, enhances sustainability, and creates new economic opportunities for farmers and agribusinesses (Mohan et al., 2016).

7.5.2 Circular Design in Agro-Industries

While upcycling focuses on repurposing waste into new products, circular design in agro-industries ensures that waste is prevented in the first place through closed-loop systems, where materials are continuously reintegrated into production cycles instead of being discarded. This requires re-engineering supply chains, redesigning manufacturing processes, and finding innovative ways to use waste as raw materials. One of the best examples of circular design is found in Brazil's sugarcane industry. Sugarcane processing produces vast amounts of bagasse, the fibrous residue left after juice extraction. In traditional industries, bagasse was either discarded or burned as waste. However, Brazilian ethanol producers have developed a system to repurpose bagasse into biofuels, biodegradable plastics, and even electricity. Today, nearly 8% of Brazil's electricity is generated using sugarcane waste, drastically reducing fossil fuel dependency and turning a waste product into an energy solution.

Circular design principles are also transforming packaging materials in agriculture. Traditional plastic packaging used for fruits, vegetables, and processed foods contributes to massive amounts of waste that often end up in landfills. Agro-industries are now developing biodegradable and compostable packaging solutions made from corn starch, cassava, and even seaweed. In Indonesia, researchers have developed cassava-based biodegradable bags, which fully decompose within months and offer a sustainable alternative to plastic grocery bags. Similarly, in Thailand, banana leaves are being used as natural food wrappers, reducing reliance on synthetic materials while offering a traditional, eco-friendly packaging solution. The success of circular design in agro-industries highlights the potential to eliminate waste through better system planning, smart material choices, and innovative repurposing. By redesigning agricultural supply chains and integrating circular processes, companies can improve efficiency, cut production costs, and drastically lower their environmental footprint (Leong et al., 2021).

7.5.3 The Future of Extending Resource Life in Agriculture

The push toward circularity in agriculture is not just an environmental necessity—it is an economic opportunity that allows farmers, industries, and supply chains to operate more efficiently while reducing waste. Upcycling agricultural by-products creates high-value secondary industries, from sustainable textiles to biodegradable packaging, while circular design eliminates inefficiencies by reintegrating waste into production processes. By extending the life of agricultural resources, the sector can transition from a waste-heavy, resource-intensive model to one that is regenerative, resilient, and economically viable for future generations. Moving forward, it is crucial to support innovation, encourage investments in circular solutions, and educate farmers and businesses on the economic and environmental benefits of rethinking waste as a valuable resource (Esposito et al., 2020).

7.6 INCREASE THE LIFESPAN OF FOOD

Food loss and waste are among the most pressing challenges in modern agriculture, affecting both food security and environmental sustainability. According to the

FAO, approximately 1.3 billion tons of food—about one-third of all food produced globally—is wasted every year. This waste is not only an economic loss but also a major environmental concern, contributing to 8–10% of global greenhouse gas emissions. The food supply chain is highly vulnerable to inefficiencies, with losses occurring at every stage—from production and processing to distribution and consumption. Food spoilage due to inadequate storage, transportation mishandling, and inefficient market systems leads to substantial resource wastage, including water, energy, and labor. At the consumer level, improper storage and over-purchasing result in significant food disposal. Addressing these issues requires a multi-faceted approach that includes advanced food preservation techniques and improved redistribution of surplus food, both of which extend the usability of food products, reduce waste, and contribute to a more circular economy.

7.6.1 Advanced Food Preservation Techniques

Food preservation has been an essential part of agriculture and human civilization for centuries. Traditional methods such as drying, salting, and fermentation have long been used to extend the shelf life of food, but modern innovations now offer highly effective, scalable, and sustainable solutions. Advances in refrigeration, controlled atmosphere storage, and edible coatings have revolutionized food preservation, significantly reducing spoilage and waste. Refrigeration is one of the most effective methods for extending food lifespan, yet an estimated 45% of global food losses occur due to lack of proper cold storage, particularly in developing countries. Fruits, vegetables, dairy, and meat require precise temperature control to prevent microbial growth and spoilage. However, many smallholder farmers in rural regions lack access to reliable electricity, making traditional refrigeration impractical. To address this issue, solar-powered cold storage units are being deployed in regions like India and Africa. Companies such as Ecozen and ColdHubs provide off-grid, solar-powered refrigeration systems, allowing farmers to store perishable produce for up to 21 days longer, reducing post-harvest losses by up to 80%. In Kenya, the adoption of solar refrigeration systems has enabled dairy farmers to keep milk fresh for over 48 hours, significantly reducing waste and increasing income. In large-scale food storage and distribution, Controlled Atmosphere (CA) Storage plays a crucial role in extending shelf life. CA storage modifies oxygen and carbon dioxide levels within a storage unit to slow down ripening and microbial activity. This technique is widely used for apples, pears, and berries, keeping them fresh for up to 12 months without the use of chemical preservatives.

Edible food coatings offer a chemical-free and sustainable solution to preserving perishable foods. These coatings create a thin, invisible layer around fruits, vegetables, and other fresh produce, reducing moisture loss and oxidation. Apeel Sciences, a company based in the United States, has developed plant-based coatings that double the shelf life of avocados, citrus fruits, and cucumbers.

Research has shown that edible coatings can reduce food waste by up to 50% in supermarkets, as treated fruits remain fresh longer, preventing premature disposal. Some startups in India and China are now exploring rice-starch-based coatings for dairy products, aiming to extend the shelf life of milk and cheese in markets with limited refrigeration (Lipinski et al., 2013). In developing countries where cold storage is not readily available, preservation in the traditional low-tech method is still important. Sun-drying, fermentation, and root cellaring are effective methods of preservation in rural areas. South Korea and Japan have used fermented foods such as kimchi and miso for months or even years without spoilage and serve as examples of natural, long-lasting food preservation. By integrating modern and traditional food preservation techniques, food loss can be drastically reduced at the production, distribution, and consumer levels, ensuring that more food reaches those who need it (Gustavsson et al., 2011).

7.6.2 Redistribution of Surplus Food

Even with effective preservation techniques, food surplus remains a challenge. Every year, millions of tons of edible food are discarded due to market oversupply, strict retail standards, or inefficient distribution systems. Much of this food waste occurs in developed nations, where supermarkets discard "imperfect" fruits and vegetables simply because they do not meet aesthetic standards. Redistributing this surplus food is an essential step toward reducing waste while addressing food insecurity. Technology-Driven Solutions for Surplus Food: Digital platforms and mobile apps are playing an increasingly significant role in redistributing surplus food. Apps like Too Good To Go (Europe) and Olio (global) connect consumers with restaurants, bakeries, and grocery stores selling surplus food at discounted prices. These initiatives not only prevent food waste but also provide affordable meals for consumers.

Upcycling Surplus Food into New Products, it was beyond redistribution, many businesses are now upcycling surplus food into new products, ensuring that food waste is transformed into valuable resources. Breweries in the UK are using leftover bread to make beer, while in the United States, companies are converting ugly fruits into fruit bars, jams, and dried snacks. In Denmark, the restaurant Rub & Stub prepares gourmet meals entirely from surplus ingredients, turning food waste into an opportunity for culinary creativity (Witzel et al., 2015). The Future of Food Waste Reduction is to build a truly circular food system, both businesses and consumers must take responsibility for reducing food waste. Governments worldwide are introducing stricter policies on food waste, encouraging donation programs and supporting innovative technologies to track and manage food supply chains. Supermarkets are experimenting with dynamic pricing, where food nearing its expiration date is sold at lower prices rather than being discarded. By implementing advanced preservation techniques, improving redistribution systems, and upcycling surplus food, the global food industry can significantly reduce waste, ensure better food security, and lessen agriculture's environmental impact. Through these strategies, food can be utilized more efficiently and sustainably, benefiting both people and the planet.

7.7 LIFE TO RESOURCES

In traditional agricultural systems, waste materials such as crop residues, livestock manure, and food processing by-products are often considered low-value or unusable. This leads to widespread burning, dumping, or inefficient disposal, contributing to environmental pollution, soil degradation, and greenhouse gas emissions. However, in a circular agricultural economy, these so-called waste materials are valuable resources that can be repurposed into compost, organic fertilizers, and renewable energy sources. By integrating biological recycling, energy recovery, and soil regeneration techniques, farmers and agro-industries can reduce reliance on synthetic inputs, enhance productivity, and create a self-sustaining, waste-free agricultural system. Two of the most effective strategies to achieve this are composting and soil enrichment and bioenergy and biogas production (Jurgilevich, 2016).

7.7.1 Bioenergy and Biogas Production: Transforming Agricultural Waste into Renewable Energy

Agricultural waste is more than just a soil-enhancing resource—it is also a largely untapped source of renewable energy that can drive sustainability and energy security. Organic waste materials such as crop husks, livestock manure, and food scraps contain carbon-rich compounds that, when processed efficiently, can be transformed into biogas, biofuels, and biomass energy. These technologies provide environmentally friendly alternatives to fossil fuels, reducing greenhouse gas emissions while fostering energy self-sufficiency in both rural and urban settings. As governments, researchers, and private industries recognize the vast potential of bioenergy, investments in anaerobic digestion, biofuel refining, and biomass power plants are expanding worldwide (Patel and Kumar, 2024).

Example 1 Biogas Production from Agricultural Waste

Biogas is a methane-rich gas generated through anaerobic digestion, where bacteria decompose organic matter in the absence of oxygen. This process yields **biogas**, which can be used for cooking, heating, and electricity generation, offering an affordable and sustainable energy source for farming communities. In addition to energy production, anaerobic digestion reduces agricultural waste accumulation, mitigates indoor air pollution, and lowers reliance on firewood, coal, and liquefied petroleum gas (LPG) (Sharma et al., 2021). Countries such as India, China, and Germany have successfully implemented biogas digesters to convert livestock manure and crop residues into renewable energy. In India, over 5 million households operate small-scale biogas plants, significantly cutting down fuel expenses while improving air quality in rural homes (Patel and Kumar, 2024). Meanwhile, Germany has pioneered the development of large-scale biogas plants, processing over 8 million tons of agricultural waste annually to supply electricity to millions of households (Muñoz et al., 2021). The integration of biogas into national energy systems demonstrates its potential as a scalable and efficient waste-to-energy solution.

Example 2 Agricultural Biomass Power Plants: Generating Electricity from Waste

Beyond liquid and gaseous biofuels, biomass power plants play a crucial role in renewable energy production by utilizing crop residues, wood chips, and organic agricultural waste as fuel. Biomass combustion produces heat and electricity, offering a sustainable alternative to coal-fired power plants while reducing waste disposal challenges (Calzolari et al., 2024). In Denmark, biomass accounts for over 30% of the country's renewable energy supply, significantly decreasing reliance on imported fossil fuels. The success of Denmark's biomass industry highlights the importance of government policies and investment in sustainable energy infrastructure (Mulya et al., 2024). Similarly, the "Rice Husk Power Project" in Cambodia demonstrates the impact of small-scale biomass initiatives, where discarded rice husks are used to generate off-grid electricity for rural communities. This project not only provides energy access to underserved populations but also reduces agricultural waste accumulation, promoting environmental sustainability (Selvan et al., 2023).

7.7.2 The Future of Agricultural Waste Repurposing

As global food demand continues to surge and natural resources become increasingly scarce, the transformation of agricultural waste into renewable resources is becoming a necessity rather than an option. The transition to a circular agricultural economy requires innovative strategies that optimize waste utilization, reduce environmental impact, and enhance sustainability. Governments, private enterprises, and research institutions are actively investing in waste-to-energy projects, sustainable soil management programs, and bio-based innovations to ensure that every component of the agricultural cycle is efficiently repurposed. These efforts are not only improving farm productivity but also contributing to environmental conservation and long-term food security. Another significant avenue for waste repurposing is government support for biogas and biofuel development. Agricultural waste, such as crop residues and livestock manure, serves as a valuable feedstock for biogas production through anaerobic digestion. Biogas technology provides a renewable energy source that can be utilized for electricity, heating, and transportation fuels, reducing dependence on fossil fuels. Many countries are already implementing policies to subsidize biogas infrastructure and incentivize biofuel production from agricultural byproducts (Patel and Kumar, 2024). Expanding these initiatives can enhance energy self-sufficiency in rural agricultural regions, reducing waste accumulation while simultaneously providing a stable energy supply. Moreover, advancements in biochemical and thermochemical conversion technologies have enabled the production of second-generation biofuels, which are derived from non-food biomass, further promoting sustainable waste utilization (Zhao et al., 2024). A circular agricultural economy also relies on collaboration between farmers and agro-industries to optimize waste collection, processing, and redistribution. Many agro-industrial operations generate substantial amounts of organic waste that, if managed properly, can be converted into valuable inputs for other agricultural activities. For instance, food processing industries produce byproducts that can be utilized as livestock feed, compost, or raw materials for bioenergy production. Supply chain integration between farmers and industries ensures that waste is diverted from landfills and repurposed into productive applications, promoting efficiency and reducing environmental impact (Muñoz et al., 2021). Creating waste exchange platforms and enhancing logistical support can further facilitate the effective redistribution of agricultural residues, maximizing their value within the food production system (Moreno et al., 2021).

Furthermore, public awareness and education play a crucial role in fostering the adoption of waste repurposing strategies. Many farmers are still unaware of the economic and environmental benefits associated with circular agricultural practices. Training programs, government-led initiatives, and academic collaborations can equip farmers with the knowledge and technical skills necessary to implement sustainable waste management solutions on their farms (Sharma et al., 2021). Increasing public engagement through awareness campaigns, workshops, and knowledge-sharing networks can help accelerate the transition toward a **zero-**waste agricultural system, ensuring that waste management innovations reach smallholder and commercial farmers alike (Selvan et al., 2023). By giving "life to resources," agriculture can evolve from a high-waste industry into a self-sustaining, zero-waste system that benefits farmers, economies, and the environment. Through composting, bioenergy production, biochar application, and innovative material reuse, agricultural waste can become a cornerstone of sustainability and economic resilience. The integration of these solutions into mainstream agriculture will be critical for achieving global food security, climate change mitigation, and resource efficiency in the coming decades. Governments, research institutions, and private stakeholders must work together to scale up investments, implement policy support, and drive innovative collaborations that accelerate the transformation of agricultural waste into valuable resources.

7.8 CONCLUSION

Transitioning towards a circular approach in agriculture is no longer just an option—it is an essential strategy for sustainability, resource efficiency, and long-term food security. Traditional agricultural models, driven by a linear "take-make-dispose" mentality, have led to excessive waste, resource depletion, and environmental degradation. However, by rethinking the way we produce, consume, and repurpose agricultural resources, we can create a system that is both economically viable and environmentally responsible. This chapter has explored several key circular strategies to regulate agro-waste, starting with a circular economy approach in agriculture, where waste is seen not as a by-product but as a valuable input for new cycles of production. By rethinking production and consumption patterns, farmers can adopt sustainable farming practices and efficient resource utilization methods that optimize the use of water, soil nutrients, and energy. Reducing agricultural waste at the source—through improved harvesting, storage, and processing—helps minimize losses while unlocking new economic opportunities.

Beyond waste reduction, extending the lifespan of resources is critical. Upcycling agricultural by-products into biodegradable materials, sustainable textiles, and eco-friendly packaging demonstrates how waste can become a profitable and environmentally friendly industry. Similarly, adopting circular design in agro-industries ensures that waste materials are continuously reintegrated into production cycles, reducing reliance on non-renewable resources. Another significant challenge in agriculture is food waste, which can be drastically reduced through advanced food preservation techniques and the redistribution of surplus food. The use of solar-powered cold storage, edible coatings, and controlled atmosphere storage has already proven successful in extending the shelf life of perishable food, reducing losses, and improving food security. Likewise, initiatives like food banks, digital surplus food redistribution platforms, and upcycled food products demonstrate how surplus food can be redirected to feed people rather than landfills.

Finally, this chapter emphasized the importance of giving life to resources, where agricultural waste is transformed into organic fertilizers, biochar, biogas, and renewable energy. Composting and soil enrichment programs have shown remarkable benefits in improving soil health, crop yields, and carbon sequestration, while bioenergy solutions such as biogas, bioethanol, and agricultural biomass power plants are creating new avenues for sustainable rural electrification and energy independence. By embracing circular agricultural practices, we can move toward a system that not only reduces waste but also enhances economic resilience, restores ecosystems, and ensures food security for future generations. Governments, farmers, researchers, and agribusinesses must work together to scale up these circular solutions, integrating them into mainstream agricultural policies and practices. The future of agriculture must be one that closes the loop on waste, maximizes resource efficiency, and fosters a regenerative, waste-free food system—one that is not just sustainable, but truly circular and resilient.

REFERENCES

Adewale, O., and Kazeem, M. (2020). Impact of compost on soil properties and crop productivity. *Journal of Natural Sciences Research, 10*(4), 56–68. Retrieved from www.iiste.org/Journals/index.php/JNSR/article/download/56409/58250

Ali, J., and Ali, T. (2024). Circular economy and agriculture: Mapping scientific productivity, research pattern, and future research direction. *Environment, Development and Sustainability, 26*, 30063–30108. https://doi.org/10.1007/s10668-023-03963-x

Aschemann-Witzel, J., de Hooge, I., Amani, P., Bech-Larsen, T., and Oostindjer, M. (2015). Consumer-related food waste: Causes and potential for action. *Sustainability, 7*(6), 6457–6477. https://doi.org/10.3390/su7066457

Calzolari, T., Magrini, C., and Menconi, M. E. (2024). Advances and challenges of a circular economy in agriculture in the European Union. *Sustainability, 16*(24), 11266. https://doi.org/10.3390/su162411266

Dey, S., Pal, A., Nandy, A., Baishnab, K., and Singh, P. K. (2022). Role of circular economy in achieving sustainable growth in agriculture and food sector. In Walter Leal Filho, Anabela Marisa Azul, Federica Doni, Amanda Lange Salvi. *Handbook of Sustainability Science in the Future* (pp. 1–19). Springer. https://doi.org/10.1007/978-3-030-68074-9_75-1

Esposito, B., Sessa, M. R., Sica, D., and Malandrino, O. (2020). Towards circular economy in the agri-food sector: A systematic literature review. *Sustainability, 12*(18), 7401. https://doi.org/10.3390/su12187401

Gustavsson, J., Cederberg, C., Sonesson, U., Van Otterdijk, R., and Meybeck, A. (2011). *Global Food Losses and Food Waste: Extent, Causes and Prevention*. Food and Agriculture Organization of the United Nations (FAO). www.fao.org/3/mb060e/mb060e.pdf

Jurgilevich, A., (2016). Transition to circular economy in food systems. *Sustainability, 8*(1), 69. https://doi.org/10.3390/su8010069

Kumar, S., and Singh, J. (2023). Circular economy perspective and implications for livestock farming in India. *The Open Agriculture Journal, 18*(1), e18743315312132. https://doi.org/10.2174/0118743315312132240611074625

Leong, H. Y., Chang, C.-K., Khoo, K. S., Chew, K. W., Chia, S. R., Lim, J. W., and Show, P. L. (2021). Waste biorefinery development toward circular bioeconomy with a focus on sugar- and starch-based waste streams. *Bioresource Technology, 344*(Pt B), 126297. https://doi.org/10.1016/j.biortech.2021.126297

Lipinski, B., Hanson, C., Lomax, J., Kitinoja, L., Waite, R., and Searchinger, T. (2013). Reducing food loss and waste. *World Resources Institute Working Paper*, 1–40. https://doi.org/10.46830/wriwp.13.00080

Martínez, J., and Ramírez, L. (2020). Recycling organic wastes through composting: A sustainable approach. *Agronomy, 10*(11), 1838. https://doi.org/10.3390/agronomy10111838

Mukumba, P., et al. (2022). Solar irrigation for smallholder farmers. *Renewable & Sustainable Energy Reviews, 158*, 112196. https://doi.org/10.1016/j.rser.2022.112196

Mulya, S. P., Hudalah, D., Salim, W., and Prilandita, N. (2024). Review of the circular economy implementation in the agriculture sector: A regional development approach. *Circular Economy and Sustainability*. 1509–1533. https://doi.org/10.1007/s43615-024-00463-6

Patel, S., and Kumar, R. (2024). Anaerobic digestion of agricultural waste: A global perspective. *Environmental Chemistry Letters*. 22(6). https://doi.org/10.1007/s10311-024-01789-1

Pfaltzgraff, L. A., De bruyn, M., Cooper, E. C., Budarin, V., and Clark, J. H. (2013). Food waste biomass: A resource for high-value chemicals. *Green Chemistry, 15*(2), 307–314. https://doi.org/10.1039/C2GC36978H

Postel, S., Polak, P., Gonzales, F., and Keller, J. (2001). Drip irrigation for small farmers. *Water International, 26*(1), 3–13. https://doi.org/10.1080/02508060108686882

Pretty, J., and Bharucha, Z. P. (2014). Sustainable intensification in agricultural systems. *Annals of Botany*, 114(8), 1571–1596. https://doi.org/10.1093/aob/mcu205

Rincón-Moreno, J., Velasco-Muñoz, J. F., and Aznar-Sánchez, J. A. (2021). An evolutionary approach on the framework of circular economy applied to agriculture. *Sustainability, 13*(4), 1881. https://doi.org/10.3390/su13041881

Selvan, T., Panmei, L., Murasing, K. K., Guleria, V., and Ramesh, K. R. (2023). Circular economy in agriculture: Unleashing the potential of integrated organic farming for food security and sustainable development. *Frontiers in Sustainable Food Systems, 7*, 1170380. https://doi.org/10.3389/fsufs.2023.1170380

Sharma, R., Gupta, A., and Singh, P. (2021). Role of composting in sustainable agriculture: A review. *International Journal of Creative Research Thoughts (IJCRT), 9*(5), 1234–1245. Retrieved from https://ijcrt.org/papers/IJCRT2105548.pdf

Siracusa, V., and Blanco, I. (2020). Bio-packaging from agri-food waste: A sustainable circular economy approach. *Polymers, 12*(4), 815. https://doi.org/10.3390/polym12040815

Smith, J., and Brown, T. (2023). Agricultural waste biomass production for bioenergy and bioproducts: Challenges and opportunities. *Agriculture, Special Issue*, 12(3), 101–119. https://doi.org/10.1016/j.biortech.2016.03.130

Velasco-Muñoz, J. F., Mendoza, J. M. F., Aznar-Sánchez, J. A., and Gallego-Schmid, A. (2021). Circular economy models in agro-food systems: A review. *Sustainability, 13*(6), 3453. https://doi.org/10.3390/su13063453

Venkata Mohan, S., Nikhil, G. N., Chiranjeevi, P., Nagendranatha Reddy, C., Rohit, M. V., Kumar, A. N., and Sarkar, O. (2016). Waste biorefinery models towards sustainable circular bioeconomy: Critical review and future perspectives. *Bioresource Technology, 215*, 2–12. https://doi.org/10.1016/j.biortech.2016.03.130

Zhao, L., Wang, Y., and Chen, H. (2024). Thermochemical and biochemical conversion of agricultural waste into bioenergy: A comparative study. *Biomass Conversion and Biorefinery*. 2(1):1–18. https://doi.org/10.1007/s44274-024-00171-w

8 Circular Strategies for Hazardous Solid Waste Management

Sri Charan P, Jonathan Jacob, Akhil Gokulan, and Sahik Naseem Ahmed

8.1 TYPES OF HAZARDOUS WASTE

Hazardous waste is defined as any solid, semisolid, or liquid waste that poses a significant threat to human health and the environment due to its toxicity, ignitability, reactivity, or corrosivity. According to the United Nations Environmental Programme, waste is classified as hazardous if it possesses characteristics that may lead to fires, corrosion, toxic gas emissions, or long-term environmental damage. Understanding the different types of hazardous waste is essential for developing effective management strategies.

Industrial hazardous waste is generated from various industrial activities such as manufacturing processes, mining, and public services. Examples include chemicals, heavy metals, solvents, spent acid etchant, spent pickling liquid, and spent dimethyl formamide (DMF)-contained liquid. Additionally, industries involved in oil and gas extraction produce hazardous waste like waste oils and drilling fluids. To effectively manage industrial hazardous waste, it is crucial to implement robust waste management practices, including proper segregation, treatment, and disposal methods to minimize environmental impact. Moreover, regulatory compliance is essential to ensure that industries adhere to safety and environmental standards.

Household hazardous waste is another significant category, which includes common household items such as batteries, pharmaceuticals, insecticides, cleaning products, pesticides, and electronics. The US EPA defines household hazardous waste (HHW) as "leftover household products that can catch fire, react, or explode under certain circumstances, or that are corrosive or toxic." These items, though seemingly innocuous, can contain hazardous materials that require special disposal methods. Proper management of household hazardous waste involves educating the public about the risks and providing accessible disposal options. Furthermore, local governments and waste management organizations play a vital role in collecting and safely disposing of these materials to prevent environmental contamination. In addition to industrial and household waste, medical waste is a critical category of hazardous waste. This type of waste is generated from healthcare facilities, including hospitals, clinics, and laboratories. Medical waste can include infectious materials, sharps, pharmaceuticals, and chemical waste. The management of medical waste requires stringent regulations to prevent the spread of infections and protect public

 DOI: 10.1201/9781003637677-8

health. Healthcare facilities must follow specific guidelines for the segregation, storage, and disposal of medical waste to ensure safety.

Classification systems are crucial for identifying and managing hazardous waste. The U.S. Environmental Protection Agency (EPA) classifies hazardous waste into four categories: listed, characteristic, universal, and mixed wastes. Listed wastes are specific hazardous wastes from industrial and commercial activities, while characteristic wastes exhibit hazardous properties such as ignitability, corrosivity, reactivity, or toxicity. Universal wastes include common items like batteries and electronics, and mixed wastes contain both hazardous and radioactive components. Other international classification systems, such as the European Waste Catalogue, also provide frameworks for managing hazardous waste. F-list wastes are nonspecific source wastes from common industrial processes, such as solvents and wastewater treatment sludges. K-list wastes are source-specific wastes from particular industries, including waste from petroleum refining and pesticide manufacturing. P-list wastes are acutely hazardous wastes, and U-list wastes are toxic wastes, both of which include specific chemicals and compounds.

Industries that generate hazardous waste include chemical manufacturing, petroleum refining, metal production, electronics manufacturing, and healthcare facilities. Each industry produces unique waste streams that require specialized management practices (Zhou et al., 2024). For example, chemical manufacturing can produce solvents and heavy metals, while petroleum refining generates waste oils and sludge. Effective management of hazardous waste, whether industrial, household, or medical, is crucial for protecting the environment and public health. By implementing robust waste management practices and adhering to regulatory standards, we can minimize the risks associated with hazardous waste and promote sustainable waste management solutions. Environmental and health impacts of hazardous waste include short-term and long-term environmental damage, as well as health risks to humans and ecosystems. Case studies and real-world examples highlight the importance of proper hazardous waste management, and the challenges faced in different regions (Zhao et al., 2008).

The EPA maintains a Toxics Release Inventory (TRI) database, which lists more than 600 potentially hazardous chemicals, providing valuable information for regulatory compliance and risk assessment. In Taiwan, industrial waste is divided into "general industrial waste" and "industrial hazardous waste" (IHW) (Yu et al., 2016). IHW is determined by a regulated list and hazardous characteristics, including manufactured hazardous industrial waste, scrap mixed metal, and biomedical and infectious waste (Zhao et al., 2008). IHW categories are further grouped into A-type (manufactured hazardous industrial waste), B-type (industrial waste containing toxic chemical substances), C-type (biomedical, infectious, and industrial waste identified by Toxicity Characteristic Leaching Procedure (TCLP), E-type (scrap mixed metal), and R-type (including spent acid etchant, spent pickling liquid, spent N,N-Dimethylformamide (DMF)-contained liquid, and spent developer/fixer liquid). These classifications help streamline waste management practices and ensure compliance with environmental regulations. Different countries employ different standards for identifying and managing hazardous waste. For instance, India uses the Hazardous and Other Waste Rules—2016, while China relies on the GB 5085.7 identification standard for

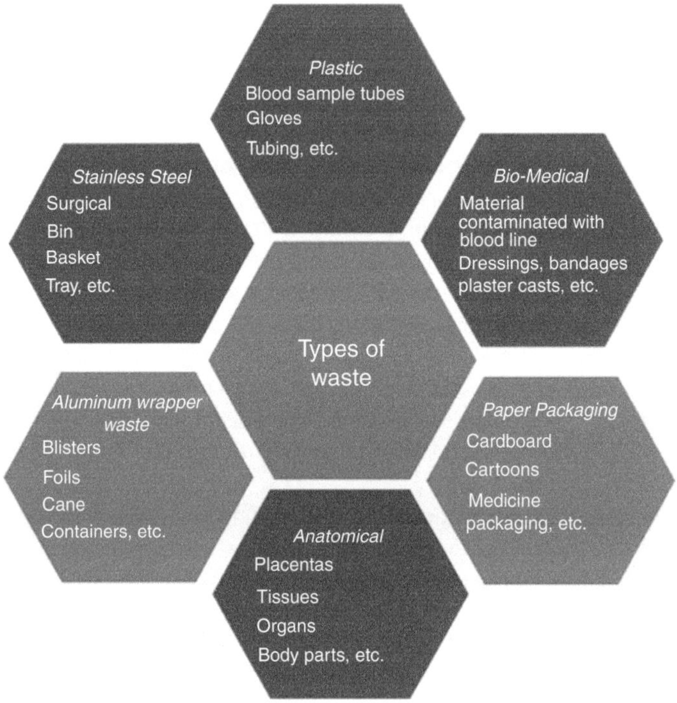

FIGURE 8.1 Classification of wastes.

hazardous waste. These frameworks reflect regional priorities and challenges, underscoring the importance of tailoring waste management strategies to local contexts. Several industries are significant contributors to hazardous waste generation (Yang et al., 2020). These include manufacturing sectors such as chemical industries, metal formation, and automobile industries, which produce a wide range of hazardous by-products. Healthcare facilities, including hospitals, medical service organizations, and laboratories, generate medical waste that requires specialized treatment (Wei and Huang, 2001). Mining operations and oil and gas extraction also produce substantial volumes of hazardous waste, including heavy metals and chemical residues. Other notable contributors include thermal power plants, integrated steel and iron mills, cement industries, paper industries, fertilizers, textiles, waste incineration plants, and waste recycling plants (Unnikrishnan and Singh, 2009). Each of these industries plays a crucial role in the economy but must adopt sustainable practices to minimize their environmental footprint (United Nations Economic Commission for Europe, 2022).

Mihai FC identifies several types of hazardous waste relevant to Eastern Europe. It specifically mentions expired pesticides and agrochemicals as a significant problem in Ukraine. It also notes that hazardous components of HHW require special attention (United Nations Industrial Development Organization, 2024). Additionally, the authors discuss waste oils, waste tires, and waste electrical and electronic equipment

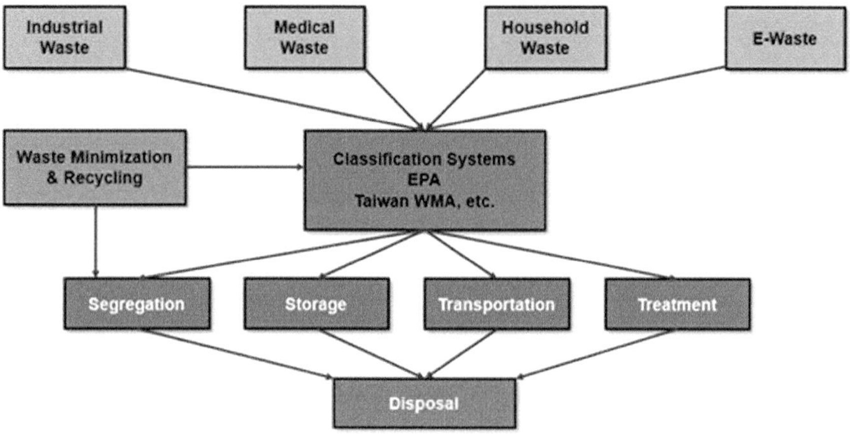

FIGURE 8.2 Hazardous waste classification and management flow.

(WEEE) as categories that need specific regulations and also mention medical infectious waste as a category that has seen increased recycling rates in Ukraine.

As we delve into the specific characteristics and management strategies for each type of hazardous waste in the following sections, it is important to recognize the interconnected nature of these waste streams and the need for comprehensive management approaches.

8.2 RESPONSIBLE STORAGE AND HANDLING OF HAZARDOUS WASTE

Effective hazardous waste management requires responsible storage and handling practices to minimize risks before treatment or disposal. Key aspects include segregation, proper labeling and containment, safe storage conditions, transportation guidelines, personnel training, and emergency preparedness. Segregation is crucial for preventing dangerous reactions. Incompatible waste must be separated; for example, mixing acids and bases can cause violent chemical reactions, and combining incompatible chemicals can lead to the formation of toxic gases. In a hospital setting, non-toxic wastes should be separated from hazardous or cytotoxic wastes at the point of generation. Proper Labeling and Containment involves storing hazardous waste in leak-proof and durable containers appropriate for the type of waste to prevent leaks, spills, or accidents. Containers must be properly labeled to identify the waste type, its hazards, and any special handling requirements. For instance, hospitals may use blue bins for designated hazardous drugs to ensure easy recognition by staff. Safe Storage Conditions are essential, including temperature control, proper ventilation, and fire safety measures. Some hazardous materials may become more volatile or reactive at higher temperatures, making temperature control critical. Proper ventilation prevents the build-up of toxic fumes, and fire safety measures such as sprinklers and fire-resistant storage cabinets are necessary to prevent fires and explosions.

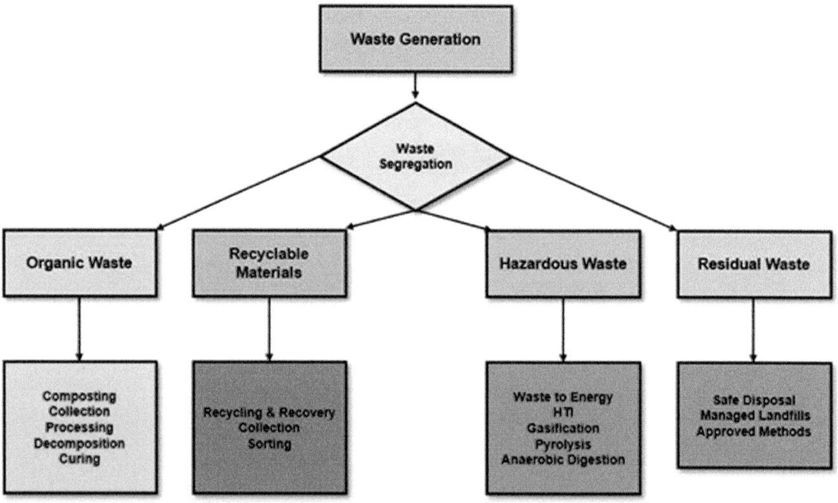

FIGURE 8.3 Sustainable waste management practices.

Storage facilities should be designed to prevent the mixing of incompatible materials and should have secondary containment systems to capture any leaks or spills.

Transportation guidelines must comply with local and international regulations, such as the European Agreement concerning the International Carriage of Dangerous Goods by Road (ADR) or the International Maritime Dangerous Goods (IMDG) code. In Taiwan, a tracking system follows the life cycle of industrial waste from generation to final disposal. In India, the Hazardous Waste (HW) Transboundary Movement, Handling, and Management Rules of 2008 require authorization for the collection, treatment, processing, and disposal of hazardous waste. Enterprises should establish record-keeping systems for hazardous waste transport vehicles, strengthen risk control for long-distance transportation, and address environmental violations strictly. Sustainable Waste Management Practices are critical for meeting the requirements of Sustainable Development Goal (SDG) 12. These practices follow a systematic approach, beginning with waste generation and moving through various treatment and disposal methods based on waste characteristics. Waste generation, the starting point in the management process, involves the production of waste materials from various sources, including residential, commercial, industrial, and institutional activities. The first crucial step after waste generation is waste segregation, which involves separating waste into distinct categories based on their characteristics and potential for treatment or recovery. This segregation typically results in four main streams: organic waste, recyclable materials, hazardous waste, and residual waste. Each of these streams requires specific handling and treatment approaches.

Organic waste, comprising biodegradable materials such as food waste and garden waste, is treated through composting. Composting diverts organic waste from landfills and converts it into valuable soil amendments. The composting process involves collecting organic materials, processing them through methods like shredding, decomposing under controlled conditions, and finally curing to produce stable compost. Recyclable materials, including paper, plastics, glass, and metals, are collected, sorted, and processed to reduce the demand for virgin resources and minimize waste sent to landfills. This material recovery is essential for sustainable waste management. Hazardous waste requires specialized handling due to its potential to harm human health or the environment. This category includes chemicals, medical waste, and certain electronic components. For hazardous waste treatment, various technologies are employed, including waste-to-energy (WtE) processes such as High-Temperature Incineration (HTI), gasification for converting waste into synthetic gas, pyrolysis for producing fuel products, and anaerobic digestion (AD) for generating biogas. Bioremediation is another promising technology for treating hazardous waste, which employs biological processes to break down harmful substances. Additionally, adsorption technology is used to treat wastewater and remove pollutants and dyes. Residual waste, which cannot be recycled or treated through other means, requires safe disposal through properly managed landfills and other approved methods (Tsai et al., 2020). These disposal facilities are engineered to minimize environmental impact and must comply with strict regulatory requirements.

Personnel training and emergency preparedness are vital for effective hazardous waste management. Personnel involved in handling hazardous waste must receive thorough training covering waste identification, safe handling practices, emergency procedures, and the use of personal protective equipment (PPE). In a hospital, staff handling or administering drugs must be able to recognize medications that require special handling. Training should also emphasize proper documentation and record-keeping practices. Emergency Preparedness involves having spill response plans and protective equipment in place. Facilities should have procedures to address accidental releases, including spill containment, clean-up, and decontamination procedures. Emergency plans should be developed for potential industrial disasters caused by hazardous waste, including emergency response plans, responders, human resources capacities, infrastructure, response facilities, coordination, and communication systems. Personnel should be trained on the proper use of PPE, such as respirators, gloves, and protective suits. Regular drills and exercises ensure that personnel are prepared to respond effectively in real emergencies (Su et al., 2010). Technologies and tools for monitoring hazardous waste are essential for effective management. Real-time monitoring systems enable tracking and management of hazardous waste, promoting information sharing and collaboration. Hazardous waste information control platforms facilitate whole process tracking management. In China, an Internet-based hazardous waste intelligent environmental protection construction management pattern is emerging (Reddy et al., 2019). Enterprises should implement systems that connect with national hazardous waste management frameworks to ensure compliance and effective tracking.

Several studies and case studies highlight critical issues in hazardous waste management. A study by Li et al. (2024) focused on extracting and recovering critical metals using molten salt electrolysis, a green process widely applied in advanced technology fields (Li et al., 2023). Such methods contribute to the circular economy (CE) by reclaiming valuable materials from waste streams (Tzou et al., 2019).

Another study examined the health impacts of hazardous waste, emphasizing the need for proper management strategies to prevent health crises. A case study on hazardous waste management in Nigeria by highlighted the challenges faced by African countries, including limited resources and informal waste disposal practices (Lv et al., 2021). This underscores the importance of developing sustainable waste management systems that align with CE principles, especially in developing nations. By implementing these responsible storage and handling practices, along with sustainable waste management strategies, facilities can reduce the risks associated with hazardous waste and ensure compliance with environmental regulations.

8.3 DUTIES OF THE ENTITY THAT HANDLES HAZARDOUS WASTE

Mihai FC explore the complexities of waste management in Eastern Europe, highlighting several key challenges and areas for improvement. Extended Producer Responsibility (EPR) is emphasized as a crucial system that shifts part of the waste management burden, especially for hazardous waste, from governments to producers. This approach leverages manufacturers' knowledge of their products' risks to ensure safer handling, better product design, and improved resource recovery, promoting sustainability by making producers financially and logistically responsible for end-of-life management.

In Ukraine, fragmented authority and lack of coordination among multiple government bodies, including the Ministry of Environmental Protection and Natural Resources and the Ministry of Health, lead to inefficiencies. Better coordination and clearer role distribution among these authorities are crucial for effective hazardous waste management, as the absence of a systematic approach at the local level complicates the situation. Local governments often struggle due to the lack of well-defined waste management plans, which are supposed to follow national and regional strategies but are delayed, leaving communities without the necessary guidance and resources to manage hazardous waste efficiently. Additionally, there is a lack of expertise in implementing joint waste management projects between municipalities and local budget constraints. Specific waste streams, such as expired pesticides in Ukraine, are often stored in unsafe, deteriorating facilities, highlighting the urgent need for proper storage and handling infrastructure. Ukraine also lacks the technology and treatment facilities for certain hazardous waste types, necessitating their export for processing. Economic and regulatory barriers, including the lack of financial incentives and outdated legislation, make it unprofitable for sanitation operators to deliver waste to treatment and recovery centers, discouraging investment in the sector. The Basel Convention on the Control of Transboundary Movements of Hazardous Wastes and Their Disposal plays a crucial role in regulating the international trade of hazardous and solid waste to prevent environmental harm. Effective implementation requires strict enforcement by global and national agencies, with the World Customs Organization (WCO) and national customs authorities serving as key players in monitoring and controlling hazardous waste movements across borders. The WCO develops standardized Harmonized System (HS) codes for categorizing hazardous waste, provides capacity-building programs for customs officials, and collaborates

TABLE 8.1
Key Practices of Hazardous Waste

Duty	Key Practices	Reference
Identifying and Classifying Hazardous Waste	Conducting chemical analysis and toxicity tests Categorizing waste based on regulatory standards (e.g., EPA's RCRA, EU Waste Framework Directive) Segregating hazardous and nonhazardous materials	
Implementing Waste Minimization Strategies	Optimizing processes to reduce waste generation Using less toxic or alternative materials Recycling solvents, oils, and heavy metals Adopting lean manufacturing to minimize waste	
Safely Collecting and Storing Hazardous Materials	Using leak-proof, corrosion-resistant containers (e.g., UN-certified drums) Separating incompatible waste (e.g., acids from bases) Labeling for traceability Implementing temperature-controlled storage for volatile waste	
Arranging for Proper Transportation and Treatment	Using licensed waste transporters with hazard labels and manifests Ensuring proper containment and packaging during transit Selecting suitable treatment methods (e.g., incineration, chemical neutralization, or stabilization)	
Monitoring Disposal Sites and Maintaining Records	Regular groundwater and air quality monitoring near landfills Keeping detailed waste logs for compliance audits Conducting periodic site inspections	

with UNEP, INTERPOL, and the Basel Convention Secretariat to strengthen global efforts against hazardous waste trafficking. Operation DEMETER, a notable WCO initiative, aims to detect and prevent illegal shipments of hazardous waste. At the national level, customs authorities enforce Basel Convention regulations by implementing the Prior Informed Consent (PIC) Mechanism, conducting inspections, and verifying shipments to detect fraudulent declarations. They also seize and return illegal waste shipments, working closely with environmental agencies to ensure compliance with international agreements and domestic laws.

The WCO uses various tools to manage hazardous solid waste (HSW), focusing on controlling transboundary movements and combating illegal trade. These tools include the Harmonized System (HS) for identifying waste types, the Customs Enforcement Network (CEN) for intelligence sharing, and Operation DEMETER for joint international operations. The WCO Environment Programme combats environmental crimes, while the Green Customs Initiative (GCI) raises awareness about trade-related environmental agreements. Single Window (SW) systems facilitate electronic transmission of documents for transboundary waste movements, and the

Authorized Economic Operator (AEO) concept helps identify compliant traders. The WCO Data Model provides standardized data definitions, and the CENcomm System facilitates secure intelligence exchange. The Environmental Network for Optimizing Regulatory Compliance on Illegal Traffic (ENFORCE) enhances cooperation among relevant entities.

In India, the management of hazardous waste involves a system of shared responsibility among various authorities, including the Ministry of Environment and Forests (MoEF), State Governments (SGs), State Pollution Control Boards (SPCBs), and the Central Pollution Control Board (CPCB). The MoEF sets environmental policy and regulations, monitors state-level implementation, and grants permits for importing and exporting hazardous waste. SGs implement environmental regulations at the state level and share responsibility with the MoEF for identifying sites for hazardous waste treatment and disposal facilities. SPCBs are responsible for granting authorizations for handling hazardous waste, inspecting facilities, and establishing systems for reporting accidents. The CPCB sets standards, provides technical support, and participates in surveying and inventorizing hazardous waste generators and processors. Some responsibilities, such as surveying and creating inventories, are shared among the MoEF, SG, and SPCBs, while import record inspections are shared between SPCBs and the CPCB. Both the MoEF and SGs play roles in identifying and notifying sites for hazardous waste treatment and disposal facilities, and they inform port authorities to ensure safe handling at ports (P. Khanna et al.). The European Union (EU) and the Organisation for Economic Co-operation and Development (OECD) have established their own HSW management policies, which align with and complement the Basel Convention. These frameworks aim to regulate the transboundary movement, treatment, and disposal of hazardous waste, ensuring environmental sustainability and preventing illegal waste trafficking. The OECD, comprising 38 developed economies, has established policies for waste management and transboundary movements of hazardous waste, with rules that are more flexible for waste trade between its members compared to the broader application of the Basel Convention. The OECD Decision C(2001)107/Final introduced a Control System for Transboundary Movements of Wastes Destined for Recovery within OECD countries, utilizing a three-tiered waste classification system (Green, Amber, and Red) based on risk levels. This system allows simplified procedures for waste shipments within OECD members, reflecting a more economically driven approach that facilitates the trade of certain hazardous wastes for recycling or treatment. However, when exporting waste to non-OECD countries, members must comply with Basel Convention rules, especially after the Basel Ban Amendment of 2019, which prohibits developed countries from exporting hazardous waste to developing nations.

The European Union (EU) has some of the strictest waste management laws globally, many of which exceed Basel Convention requirements. The EU Waste Shipment Regulation (Regulation (EC) No 1013/2006) directly implements Basel Convention rules for hazardous waste trade, banning exports of hazardous waste to non-OECD countries and requiring strict prior informed consent (PIC) for waste movements, even within the EU. The EU Circular Economy Policy focuses on waste minimization, recycling, and recovery rather than disposal, encouraging EPR to ensure that manufacturers take responsibility for the waste generated by their products. The EU's and OECD's policies are interconnected through their shared commitment to the Basel Convention, but while the EU takes a more stringent approach, the OECD allows controlled flexibility for its members. Both, however, prohibit the dumping of hazardous waste in developing countries, ensuring environmental protection on a global scale.

8.4 LINKING SOLID WASTE MANAGEMENT AND SDGS

SDG refers to Sustainable Development Goals. The United Nations devised 17 Sustainable Development Goals based on five pillars: people, planet, prosperity, peace, and partnership. The SDGs ultimately influence the triple bottom line (economic, environmental, and social), thus transitioning the world into a resilient and sustainable place. Improved solid waste management (SWM) can contribute to the achievement of many SDGs. Effective hazardous waste management significantly contributes to achieving several United Nations Sustainable Development Goals (SDGs). While not explicitly mentioned in the 17 SDGs, SWM, including hazardous waste, is a cross-cutting issue that impacts various aspects of sustainable development. Improved waste management is linked to key development priorities including health, employment, and the environment.

SDG 3 (Good Health and Well-being) emphasizes reducing exposure to toxic substances. Proper hazardous waste management is essential in reducing exposure to toxic substances, thereby preventing various illnesses and fatalities. Improper disposal contaminates soil and water, leading to severe health issues. Exposure to hazardous waste can cause respiratory problems, neurological damage, and cancer. Effective management of this waste significantly lowers these risks. Contaminated soil and water act as vectors for various diseases. Effective waste management and disposal can break the chain of transmission of such diseases. Dumpsite leachates can penetrate the soil and pollute groundwater, causing diseases to spread. Proper hazardous waste management helps prevent this, safeguarding water sources. In healthcare settings, effective waste management reduces risks associated with healthcare waste, protecting medical personnel, waste handlers, and the public from infections and other health problems. Proper segregation, treatment, and disposal methods are crucial in minimizing these risks. Additionally, eliminating breeding sites for disease vectors through regular collection and proper disposal prevents the spread of vector-borne diseases. Effective waste management practices, including regular collection and proper disposal, eliminate these breeding sites, preventing the spread of vector-borne diseases. By reducing exposure to hazardous materials and eliminating disease vectors, proper waste management contributes directly to the SDG 3 goal of ending preventable deaths and reducing illnesses from pollution. This includes targets related to decreasing mortality rates from specific diseases, reducing pollution-related illnesses, and promoting overall well-being. Proper e-waste management is important for achieving SDG targets related to health. E-waste contains toxic substances that can harm human health if not managed correctly. Proper handling and recycling of e-waste are crucial for preventing health issues.

SDG 6 advocates for the implementation of preventing water pollution. Hazardous waste, if not properly managed, can significantly pollute water sources. Leachate from landfills, containing various toxic substances, can seep into the ground and contaminate groundwater, which is a vital source of drinking water. This contamination can introduce harmful chemicals and pathogens into the water supply, making it unsafe for consumption and other uses. Proper hazardous waste management, including secure containment and treatment, helps prevent this form of pollution. Improper disposal of hazardous waste can also directly affect surface water bodies such as rivers and lakes. Direct dumping or runoff from waste sites can carry pollutants into these water systems, harming aquatic life and potentially impacting human health through contaminated drinking water or food sources. Effective waste management practices ensure that hazardous materials are not discharged into these water bodies. Hazardous substances in wastewater can overwhelm treatment facilities, reducing their effectiveness and leading to the discharge of contaminated water into the environment. Proper management involves the pre-treatment of hazardous waste before it enters wastewater systems, ensuring that treatment plants can operate effectively and remove harmful substances. By preventing water pollution, effective hazardous waste management directly contributes to several targets of SDG 6, which include SDG 6.3; "improve water quality by reducing pollution, eliminating dumping and minimizing release of hazardous chemicals and materials." SDG 6 includes targets related to the sustainable use and protection of water resources. Hazardous waste management helps safeguard these resources from pollution. Biological methods can be used for the decomposition or degradation of waste material and can simultaneously address environmental remediation as well as generate value-added products in the form of energy and chemicals.

Effective hazardous waste management is crucial for achieving the United Nations Sustainable Development Goal 11, which focuses on creating sustainable cities and communities. This goal emphasizes the importance of making cities inclusive, safe, resilient, and sustainable. Proper management of hazardous waste contributes significantly to this objective by ensuring that urban environments are protected from the adverse effects of improper waste handling. Sustainable cities prioritize minimizing their environmental footprint. Improper handling and disposal of hazardous waste can lead to soil and water contamination, air pollution, and other environmental problems that undermine sustainability. By implementing effective hazardous waste management strategies, cities can significantly reduce their adverse impact on the environment, thereby enhancing their overall sustainability.

SDG 11 also emphasizes the importance of ensuring access to basic services and promoting public health in urban areas. Hazardous waste poses significant health risks, with exposure to toxic substances causing various illnesses and diseases. Effective management practices reduce these risks and ensure that residents have a healthier environment.

SDG 11.6 specifically aims to reduce the adverse environmental impact of cities by paying special attention to waste management. Effective hazardous waste management is a key part of this. This includes proper collection, treatment, and disposal methods that minimize pollution and maximize resource recovery. Cities that prioritize sustainable waste management can create a cleaner, healthier, and more resilient environment (Tsai et al., 2020). Safe cities are a critical component of SDG 11. Improper management of hazardous waste can lead to accidents, fires, and other emergencies that pose serious risks to urban populations. By ensuring the safe handling and disposal of hazardous waste, cities can improve their overall safety and resilience to potential disasters.

A CE approach to hazardous waste management emphasizes reuse, recycling, and recovery, supporting SDG 11 by creating opportunities for resource efficiency. This approach means using fewer virgin materials, reducing waste generation, and promoting economic growth within a sustainable framework. Effective hazardous waste management directly addresses SDG 11.6 through environmentally sound practices, particularly in urban areas, contributing to safer and more resilient cities by reducing hazardous materials and preventing accidents (Tsai et al., 2020). Integrating resource recovery and CE principles enhances resource efficiency within cities, while proper management of healthcare waste is crucial for preventing disease spread and maintaining a healthy urban environment. Overall, effective hazardous waste management is essential for creating sustainable cities and communities, aligning with the goals of SDG 11 by reducing environmental impacts, improving public health, ensuring safety, and promoting resource efficiency, thereby supporting a transition to environmentally responsible urban environments (Tsai et al., 2020).

Effective hazardous waste management encourages waste reduction and recycling, aligning with Sustainable Development Goal 12 (SDG 12), which emphasizes the need for sustainable consumption and production patterns. SWM plays a pivotal role in achieving this goal, particularly through Target 12.4, which calls for the environmentally sound management of chemicals and all waste throughout their life cycle. This target is crucial for minimizing adverse impacts on human health and the environment, as highlighted by various sources. Proper handling of hazardous waste is essential, especially in regions like Central Java, where stringent management protocols are necessary due to the toxic and infectious nature of the waste. Regulations in Taiwan mandate the recycling of specific industrial hazardous wastes, promoting resource recovery and reducing reliance on traditional disposal methods like incineration or landfill (Tsai et al., 2020). Innovative approaches, such as repurposing hazardous waste into valuable resources, further support these efforts. The 3Rs strategy—reduce, reuse, and recycle—is central to achieving SDG 12, as it minimizes the need for raw materials and reduces the environmental impact of extraction and manufacturing processes. Life Cycle Assessment (LCA) is a valuable tool for evaluating the environmental impact of waste management systems, aiding in the selection of strategies with minimal negative impact.

The principles of the CE are vital for achieving SDG 12, as they aim to minimize waste and keep materials in use for as long as possible, thereby reducing the need for new resource extraction. Technologies like pyrolysis, gasification, and co-incineration are highlighted for their ability to convert waste into energy, reducing reliance on fossil fuels and minimizing landfill use. WtE technologies provide a range of options for managing waste and recovering energy. The selection of the appropriate technology depends on various factors such as waste composition, availability of infrastructure, economic feasibility, and environmental considerations. The Intergovernmental Panel on Climate Change (IPCC) emphasizes that integrated waste management strategies, such as recycling, composting, and energy recovery, are effective in reducing methane emissions from landfills, a potent greenhouse gas (GHG) contributing to global warming. Effective hazardous waste management is pivotal in achieving the United Nations SDGs, particularly SDG 13, which focuses on climate action. Improper disposal of hazardous waste contributes to (GHG) emissions, thereby exacerbating climate change. Implementing sustainable waste management practices can mitigate these emissions and promote environmental sustainability. Thakur emphasizes that sustainable hazardous waste management is essential for maintaining a safe and eco-friendly environment. They advocate for strategies aligned with circular economic models, focusing on prevention, reduction, recycling, WtE conversion, and advanced treatment technologies. These approaches not only minimize waste generation but also reduce GHG emissions associated with waste disposal processes. In their review, Nizami discuss the role of climate-smart waste management in mitigating global warming. They highlight that integrating waste management strategies with climate change mitigation efforts presents unique opportunities for the waste industry to transform toward achieving sustainable development goals. This integration can lead to significant reductions in GHG emissions from waste-related activities. The IPCC underscores that integrated waste management strategies, including recycling, composting, and energy recovery, are effective in mitigating GHG emissions from the waste sector. These practices not only reduce the volume of waste but also decrease methane emissions from landfills, a potent GHG contributing to global warming.

By strengthening hazardous waste management systems, countries can minimize the adverse impacts of waste on human health and the environment, supporting the broader objectives of the SDGs, including climate action.

In summary, effective hazardous waste management is not just an environmental necessity; it is also a crucial component in achieving broader sustainability goals. By addressing the health, environmental, and economic impacts of waste, proper management practices directly support the achievement of multiple SDGs.

8.5 CHALLENGES AND BARRIERS IN TRANSITIONING TO CIRCULAR ECONOMY AND POLICY RECOMMENDATIONS

The CE emphasizes producing and utilizing products in an environmentally sustainable manner, ensuring that each product is reused repeatedly and nothing is wasted. It balances economic prosperity, environmental sustainability, and social well-being. However, several barriers hinder the practical implementation of CE, including cultural, political, and economic factors such as utilitarian purchasing patterns, ethnocentric attitudes toward waste disposal, and a lack of awareness among consumers and business owners. The deeply entrenched linear economy model, which follows a "take-make-dispose" approach, poses a significant challenge to transitioning to a CE, further complicated by traditional macroeconomic models that align with linear practices. Inefficient waste management practices, low waste collection rates, and inconsistent collection lead to illegal dumping and open burning, exacerbated by low income and education levels, lack of understanding of treatment methods, inadequate disposal facilities, and shared containers. Technological limitations, including the absence of appropriate technology and expertise for sorting algorithms, waste volume measurements, and real-time monitoring systems, also impede the maximization of resource reuse and recycling (Abbasi et al., 2022). The high costs associated with transitioning to a CE model, including investments in new technologies, act as a significant obstacle, particularly for small to medium enterprises, due to low profitability and slow returns on investment. Rapid urbanization leads to the production of construction and demolition debris, causing environmental pollution, and poor urban planning adds to the difficulties in applying CE principles. Ambiguities surrounding the legal framework for CE and SWM, coupled with a lack of social diversity and inclusion in existing policies, necessitate the repeal of existing laws and the enactment of new ones to streamline CE policies.

In business models, a shortage of skilled personnel and a lack of understanding about CE principles among staff members present major obstacles. Issues with existing company policies, financial constraints, technology limitations, scarcity of resources, lack of collaboration, problems with product design, and internal stakeholder challenges further hinder CE adoption. The healthcare sector lags in adopting CE principles due to a lack of comprehensive data about medical items and waste generation, insufficient technology access, product design issues, and challenges with supply chains, economic viability, technical factors, and stakeholder participation. In 2021, the amount of hazardous waste generated in Central Java Province was 2,886,974.024 tons, with the healthcare sector as the largest contributor to hazardous waste generation in the province. Therefore, hazardous waste management in the healthcare sector must receive strong attention from the local government, by making a guideline for hazardous waste management in health facilities available and by providing training for personnel who manages hazardous waste in healthcare facilities.

Policies regarding CE have attracted considerable attention from governments and policymakers, with countries like the Netherlands, Germany, Belgium, and France adopting CE principles to manage solid waste (Aghbashlo et al., 2019). The EU's CE framework serves as a guide for both EU and non-EU countries, promoting environmental sustainability and resource efficiency. The EU is constantly updating regulations and policies to meet emerging trends in SWM. Specific policy areas include EPR, waste management strategies, legislation, stakeholder involvement, macroeconomic policies, and fiscal measures. However, policy implementation challenges persist, particularly in developing nations, which lack awareness and knowledge about CE implementation. Unclear legal frameworks, excessive bureaucracy, poor quality of waste databases, and fragmented governing bodies hinder the smooth implementation of CE policies. Policies should aim to overcome these challenges and promote CE in SWM, with governments providing funding for research, raising awareness, promoting stakeholder participation, and offering tax incentives to encourage CE practices. In Eastern Europe, policies must address specific regional challenges, including the need for clear regulations and reduced bureaucracy, to transition away from traditional waste management practices.

Developed, developing, and underdeveloped nations are at different stages in their transition and adaptation to the CE, facing unique challenges and requiring tailored strategies. Developed nations are leading the way in adopting CE principles, emphasizing innovation, technology, robust policies, and significant investment. Zero-carbon goals drive research into circular technologies, with Western European countries setting ambitious targets that lead to more circular technology research. Robust environmental policies, stringent legal frameworks, and quality standards facilitate the transition to a CE, with developed countries often having specific waste reduction targets and working closely with stakeholders to ensure a shift toward CE. For example, the EU has set the goal for all member states to implement closed-loop recycling in their economies. The European Commission launched the European Green Deal in 2019 to promote a resource-efficient and competitive economy, setting goals for achieving net-zero GHG emissions and decoupling economic growth from resource use. Norway forbade the landfilling of organic waste beginning in 2009, and countries with elevated landfill taxes, like Poland, tend to be more environmentally sustainable due to increased efforts in waste separation. Pay-as-you-throw schemes are used in France, Greece, and Italy to encourage waste reduction and promote recycling, while the Zero Waste Cities Certification focuses on reducing waste generation through community engagement in Slovenia and Belgium.

Developing nations are in the midst of transitioning to CE, facing challenges such as consumer behavior and the need for significant improvements in waste separation and infrastructure. They are slowly adopting CE principles by implementing policies but face barriers that need to be overcome for effective implementation of circularity. Focus areas include waste reduction and enhancing public awareness of the benefits of recycling. East Asia and the Pacific, South Asia, and Europe produce 60% of the global waste, with dry waste (plastic, paper, metals, and glass) making up a major portion of municipal solid waste (MSW) in high-income countries. In low-income countries like Bangladesh, India, and Pakistan, the proportion of food- and agriculture-based refuse in MSW is close to 57–60%, increasing the organic content of MSW.

Underdeveloped nations like Zambia face more foundational challenges, needing to improve basic waste management systems before transitioning to a CE. The focus needs to be on the implementation of basic waste management systems and improving public awareness. The initial focus should be on establishing collection systems, improving access to waste management, and investing in building the knowledge base related to the CE. Waste creation in developing nations averages 0.56 kg per person per day, with organics making up 52% of the total and recyclables making up 26%. Barriers to reuse and recycling in developing nations include a lack of political will and legislative framework, absence of local regulations, inadequate funding, and insufficient training and education. Waste management stakeholders in developing nations sometimes ignore environmental protection and quality regulations, and many policies and operational guidelines fail to take into account the particular characteristics of the environments in developing nations.

Differences in operations and adaptation towards a CE in HSW management between developed and underdeveloped countries are evident. Developed countries, particularly in Western Europe, have sophisticated waste management systems characterized by high rates of waste collection, advanced sorting and recycling technologies, and WtE facilities. These nations actively explore methods to reuse materials without negative externalities and are supported by robust environmental policies, stringent legal frameworks, and quality standards that facilitate the transition to a CE. They implement landfill taxes and other economic measures to discourage landfilling and promote recycling, with EU policies and frameworks serving as guides even for non-EU countries in the region. There is a strong emphasis on implementing closed-loop recycling and reducing primary resource consumption, with many Western European countries setting zero-carbon goals that drive research into circular technologies. Public awareness campaigns and educational initiatives are common, aiming to improve waste separation at the household level. For example, Scandinavian countries like Norway, Denmark, and Sweden are inspirational cases, utilizing advanced smart technologies, efficient logistics, and fostering high levels of public awareness regarding waste management. Countries such as Poland have adopted progressive taxation systems for landfilling to further encourage recycling and waste reduction.

In contrast, underdeveloped countries like Zambia face significant challenges in MSW management due to rising populations and increased economic activity. Waste management systems in Zambia are often collapsed, exacerbated by a lack of financing, equipment, and the presence of uncontrolled, unplanned settlements, which present challenges to local authorities. MSW management is a constitutional mandate for local authorities in Zambia, but the country's focus remains on a linear economy model that only recognizes generation, collection, transportation, and disposal. This model has negative impacts on public health and the environment. The lack of empirical data on generation rates, waste streams, and sources in Zambia makes planning difficult. While there is considerable potential for recycling, especially with plastics and paper comprising a large proportion of waste in cities like Lusaka, metals are the only material currently heavily recycled due to market demands. Zambia aims to become a middle-income nation by 2030, which could be greatly supported by implementing a CE using the 3R systems (reduce, reuse, recycle). The country also seeks to ensure effectiveness and efficiency in the delivery of public services, including waste management. However, Zambia experiences issues such as a lack of experts in waste management, leading to chaotic activities.

In transitioning to CE, developed nations are already well on their way, focusing on innovation and technology, robust policies, and significant investment. They often have specific waste reduction targets and work closely with stakeholders to ensure a shift toward CE. Developing nations are in the process of transitioning, facing challenges such as consumer behavior and the need for significant improvements in waste separation and infrastructure. They are slowly adopting CE principles by implementing policies but need to focus on waste reduction and enhancing public awareness of the benefits of recycling. Underdeveloped nations like Zambia face more foundational challenges, needing to improve basic waste management systems before transitioning to a CE. The focus needs to be on establishing collection systems, improving access to waste management, and investing in building the knowledge base related to the CE.

8.6 SOLID WASTE COMPLIANCES

Compliance requirements in waste management encompass multiple aspects aimed at minimizing environmental and health impacts while promoting resource efficiency. At the foundation of effective waste management lies accurate waste identification and classification. Different regions employ different classification systems to categorize waste based on its source and properties. The United States utilizes the Resource Conservation and Recovery Act (RCRA) for hazardous waste criteria, while China employs the GB 5085.7 system, and India follows the Hazardous and Other Waste Rules, 2016. The globally recognized waste management hierarchy serves as a fundamental framework for prioritizing waste management options, beginning with prevention through improved product design and reduced packaging, followed by preparing items for reuse through restoration. The next priorities are recycling materials into new products and recovering energy through methods like WtE incineration and AD. Disposal, particularly landfilling, is considered the last resort when other methods are not feasible.

Specific regulatory mandates often target particular waste types and management practices. Many jurisdictions have implemented strict requirements for food and organic waste, including mandatory separate collection and landfill bans, as exemplified by Norway's 2009 prohibition on organic waste landfilling. Landfill restrictions are increasingly common, with many regions introducing landfill taxes to discourage this disposal method and promote alternatives. The implementation of waste management regulations varies significantly between developed and developing nations. Developed countries typically maintain robust regulatory frameworks supported by strong legal systems and dedicated environmental protection agencies. These nations benefit from well-established monitoring and reporting systems, leading to higher recycling rates through effective enforcement. In contrast, developing countries often struggle with limited resources for policy implementation, facing challenges in managing household hazardous waste and maintaining sufficient treatment facilities.

Infrastructure and resource availability further highlight the disparity between nations. Developed countries possess advanced waste management infrastructure, including sophisticated recycling and WtE facilities, complemented by comprehensive data collection and monitoring systems. Developing nations, however, face constraints in infrastructure investment, typically dealing with a higher proportion of food and organic waste, and often rely on informal sector involvement in recycling. Cultural and social factors play a crucial role in waste management practices. Research has shown that family attitudes significantly influence waste collection

FIGURE 8.4 Waste management hierarchy from most preferred to least preferred.

practices, while religious and social norms affect disposal behavior. Income and education levels impact participation in waste management programs, with particular consideration needed for base of the pyramid populations in developing nations. The global waste trade raises significant environmental justice concerns. Developed nations often export waste to developing countries, creating environmental and health challenges in recipient nations. Eastern Europe faces particular pressure from EU waste flows, while the management of e-waste presents unique processing safety concerns. China's role in global e-waste management has become particularly significant (Duan et al., 2008). Technological innovation offers promising solutions, particularly for developing nations that have the opportunity for technological leapfrogging. Advanced treatment technologies include WtE systems and various thermal and hydrothermal treatments. The integration of CE principles and improvements in collection and sorting technologies continue to advance the field. Regulatory frameworks incorporate several key components, including permits and authorizations for waste management operations, tracking systems for waste movement monitoring, and standards for storage, treatment, and disposal. EPR has gained prominence, placing end-of-life management responsibility on manufacturers, supported by product stewardship policies. Hazardous waste management requires specific identification methods, such as the US EPA's TCLP, along with stringent treatment requirements and safe disposal regulations. Many regions have also established specific guidelines for recycling and reuse. The field faces ongoing challenges, particularly in implementation. Resource limitations in developing nations, regulatory inconsistencies across regions, and illegal disposal and trafficking remain significant concerns. The management of emerging waste streams presents new challenges that require adaptive regulatory responses. The integration of CE principles focuses on waste minimization and resource efficiency, though implementation varies significantly across different contexts.

Enforcement mechanisms play a crucial role in ensuring compliance with waste management regulations. These include regular inspections, auditing requirements, and penalties for violations. In severe cases, companies or individuals can face legal actions, including criminal charges. The international Basel Convention provides a framework for controlling transboundary movements of hazardous wastes, though its effectiveness depends heavily on member nations' implementation capacity. Several significant gaps exist in current waste management regulatory frameworks. The management of radioactive waste, which requires separate regulation due to its unique risks, often receives insufficient attention. Medical waste regulations, particularly regarding specific treatment methods to prevent infectious disease spread, require further development. The growing challenge of e-waste management, including specific collection, recycling, and disposal requirements, as well as methods for extracting valuable elements, needs more comprehensive regulatory attention.

International cooperation presents both opportunities and challenges in waste management compliance. While collaboration can lead to better compliance through information and technology sharing, particularly in assisting developing nations, achieving a unified approach to combat illegal activity remains challenging (Laurent et al., 2014). The role of technological advancements, such as blockchain, AI, or IoT in tracking and monitoring hazardous waste handling, offers promising solutions but requires careful integration into existing regulatory frameworks. Looking forward, the field of waste management compliance continues to evolve. The emphasis on CE principles, technological innovation, and improved regulatory frameworks suggests a trajectory toward more sustainable practices. However, success will require addressing the persistent challenges of resource limitations, regulatory harmonization, and effective enforcement, particularly in developing nations. The integration of the informal sector into formal waste management systems, strengthening international cooperation, and developing more comprehensive regulatory frameworks for emerging waste streams remain critical areas for future development (González-Sánchez et al., 2023).

8.7 ENERGY AND RESOURCE RECOVERY FROM SOLID WASTE

The global production of HSW has reached approximately 400 million tons annually, equivalent to 13 tons per second, with future projections indicating a significant increase due to industrial growth, urbanization, and rising plastic production, expected to triple by 2060. MSW is projected to grow from 2.1 billion tons in 2020 to 3.8 billion tons by 2050, though specific HSW forecasts remain limited due to varying global definitions and reporting standards. Sweden and Denmark exemplify successful sustainable waste management infrastructure through WTE technologies. Sweden recycles over 99% of their MSW, with 50.3% incinerated for energy recovery and 16% used in AD. Denmark has embraced “hedonistic sustainability,” exemplified by Copenhagen’s innovative incinerator that doubles as a ski slope and WtE education center, sending only 3% of waste to landfills, while 54% generates heat and electricity through incineration, providing 97% of residents with excess heat energy (Reddy et al., 2019).

The goal of WtE is to transform waste from a disposal problem into a resource for energy production. Incineration is a thermal process that involves the controlled

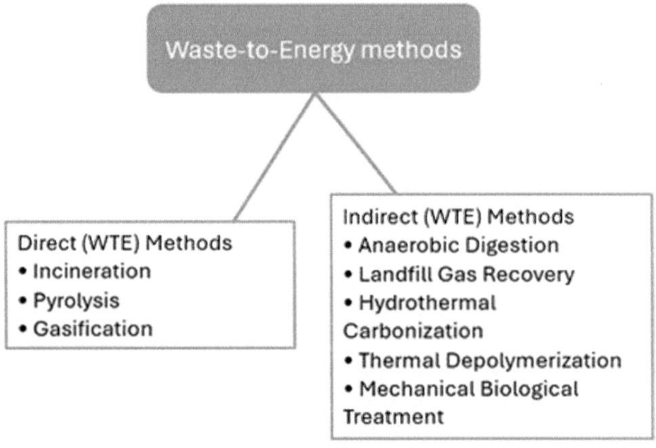

FIGURE 8.5 Classification of waste-to-energy methods.

combustion of waste to produce electricity or heat, significantly reducing waste volume and mass while neutralizing harmful substances with advanced air pollution control equipment. Modern incineration plants, equipped with combustion chambers, recovery boilers, flue gas treatment systems, and steam cycles, generate heat and steam that can be used for electricity production or district heating. This method is considered more controlled and environmentally sound than open burning, as it reduces harmful air pollutants and allows for energy recovery, making it a well-established method for treating mixed household and commercial waste.

Pyrolysis, an advanced thermal treatment, involves heating waste without oxygen to produce liquids or gases that can be used as energy sources, particularly suitable for non-biodegradable materials like plastics, and can yield valuable chemical products. This process aims to reduce waste volume and recover energy, minerals, and chemicals, making it a promising technology for converting waste into valuable products such as syngas, bio-oil, and biochar. Gasification, similar to pyrolysis, converts waste into a burnable gas using a limited amount of air or oxygen, commonly applied to biomass and organic waste to produce syngas, predominantly H_2 and CO. Operating at high temperatures, gasification is a promising technique for valorizing waste material, with syngas used for electricity generation, liquid fuel production, or chemical production.

AD is a biochemical process that breaks down organic matter using microorganisms in the absence of oxygen, converting food and green waste into semi-solid fertilizer (digestate) and biogas, which can be transformed into thermal and/or electrical energy (Hasankhani et al., 2024). Although AD is a potential energy recovery option for urban waste, operating biogas plants with mixed MSW presents challenges in terms of operations, safety, and finances, requiring waste separation. Landfill gas recovery captures methane from landfills, which receive approximately 40% of global waste, and uses it to generate electricity, thereby mitigating the climate impact of landfills (Aghbashlo et al., 2019). This method is crucial for reducing greenhouse gas emissions and is a necessary component of urban waste management systems. Hydrothermal carbonization, though not a major focus, is a method to produce cleaner renewable energy from hazardous waste, using supercritical fluids to treat waste, sewage, and contaminated water, including agricultural and industrial waste under subcritical and supercritical water conditions. This process can achieve high TOC removal efficiency and produce a variety of value-added products, including bio-oil.

Material recovery from various waste streams, including e-waste, tires, and construction waste, supports a CE by reducing the demand for new resources. Sustainable waste management practices are critical for meeting the requirements of SDG 12. Composting is noted as an essential method for treating organic waste, diverting it from landfills and converting it into valuable soil amendments. Adsorption technology is used to treat wastewater and remove pollutants and dyes. Bioremediation is emphasized as a promising technology for treating hazardous waste, contributing to environmental remediation and resource recovery. Global and regional considerations highlight the need for context-specific solutions for SDG 12. Developing countries face significant challenges in managing solid waste due to poverty and a lack of integrated waste management frameworks. In Asia, countries like China and

India contribute significantly to global waste, necessitating effective hazardous waste management (Duan et al., 2008). African nations need to collaborate with developed countries to implement recycling and extended producer responsibility initiatives. Scandinavian countries serve as examples of effective waste management with high recycling and segregation rates.

Effective hazardous waste management is also pivotal for SDG 13, which focuses on climate action. Improper disposal of hazardous waste contributes to GHG emissions, exacerbating climate change. Sustainable waste management practices can mitigate these emissions and promote environmental sustainability. Strategies aligned with circular economic models, including prevention, reduction, recycling, and WtE conversion, minimize waste generation and reduce GHG emissions.

Several limitations impede energy recovery efforts. Collection and infrastructure challenges are particularly acute in developing countries, where less than half of solid waste is collected compared to 98% in high-income nations (Ahirwar and Tripathi, 2021). Technological limitations include the unsuitability of certain technologies for developing nations, such as incineration requiring minimum calorific values of 1900 kcal/kg. Financial constraints encompass high capital costs, operational expenses, and higher Levelized cost of energy compared to other renewables. Socio-environmental concerns include community opposition and pollution risks, while policy and regulatory issues stem from a lack of uniform regulations and insufficient data on waste generation (Dhyani and Bhaskar, 2018).

Resource recovery extends beyond energy to various materials. Cementitious materials can be created from industrial by-products like fly ash and blast furnace slag. Metal recovery occurs from incineration ash, while metal-organic frameworks offer potential in gas storage and catalysis (Duan et al., 2008). Organic waste produces fertilizers and soil conditioners through composting and AD. Other recoverable materials include polymers, catalysts, molecular sieves, and corrosion inhibitors from recycled waste. The effectiveness of resource recovery depends on several factors. MSW composition varies by location and socio-economic conditions, with developing countries typically having higher organic fractions. Preprocessing requirements include sorting, shredding, and drying, while different technologies suit different waste types and desired products. Economic feasibility considers facility costs and recovered material value, while environmental impacts require mitigation through advanced technologies like gasification.

Organizational frameworks and policies play a crucial role. These include legislative frameworks, government incentives, and stricter management practices. Some countries develop specific action plans for areas like landfill gas recovery. Implementation monitoring occurs through agencies like the Central Pollution Control Board (CPCB), with municipalities required to submit annual reports. Financial aid supports project developers in converting landfills to power plants. Future initiatives vary by national development level. Developed nations like the UK, US, and Australia employ advanced WtE technologies and CE strategies, though some policies like the UK's emissions trading scheme for incinerators raise concerns about increased landfill use. The US Department of Defense funds research into plastic-eating microbes and waste-to-fuel conversion. Developing nations like India, China, and Brazil adopt cost-effective solutions through programs like India's Swachh Bharat Mission and

China's incineration-based WtE plants. Underdeveloped nations such as Nigeria, Ethiopia, and Nepal progress through localized initiatives, exemplified by Nigeria's waste-to-biogas projects and Ethiopia's Reppie WtE plant (Jin et al., 2025).

The CE significantly influences HSW management by promoting sustainable practices. Thakur emphasize aligning HSW management with CE models through waste prevention, reduction, recycling, and WtE conversion strategies. González-Sánchez et al. (2023) further explore these connections, highlighting the importance of integrated approaches for environmental protection.

The adoption of CE principles significantly influences HSW management by promoting sustainable practices that minimize environmental and health risks. A critical review by emphasizes the necessity of aligning HSW management strategies with CE models to achieve a safe and eco-friendly environment. The study discusses various approaches, including waste prevention, reduction, recycling, and WtE conversion, highlighting their roles in sustainable HSW management. Furthermore, González-Sánchez et al. (2023) explore the integration of CE principles in waste management, noting that such incorporation facilitates the achievement of SDGs. The research underscores the importance of transitioning from a linear "take-make-dispose" economy to a circular model, which emphasizes resource efficiency and waste minimization (Hakur et al., 2023). However, challenges persist in implementing CE strategies within waste management systems. A bibliometric analysis by González-Sánchez et al. (2023) identifies the need for a comprehensive understanding of existing knowledge and the development of effective policies to overcome these challenges. The study calls for further research to address gaps and enhance the effectiveness of CE practices in HSW management (Houng, 1999).

The rapid growth of industrial sectors, including chemical manufacturing and electronics, has significantly increased hazardous waste generation, contributing to soil, water, and air pollution through heavy metals and persistent organic pollutants. Current projections estimate that global waste generation will rise by 70% by 2050 due to urbanization, industrial expansion, and population growth, exacerbating environmental degradation, resource depletion, and GHG emissions (UNECE, 2022). Inefficient waste management imposes a heavy economic burden, with billions spent on waste treatment and landfilling (González-Sánchez et al., 2023). CE principles, such as industrial symbiosis, closed-loop recycling, and resource recovery, offer solutions by repurposing hazardous by-products into secondary raw materials. CE models emphasize waste minimization, material reuse, and recycling, widely adopted in construction, textiles, and electronics (UNIDO, 2024). However, challenges persist, including limited infrastructure, regulatory enforcement, and the lack of globally standardized metrics to assess progress. Effective implementation of CE in hazardous waste management requires policy enforcement, infrastructure investment, and EPR programs (González-Sánchez et al., 2023). Future efforts must focus on data-driven strategies, international collaboration, and green innovations to achieve long-term sustainability.

Integrating CE principles into HSW management fosters sustainable practices that mitigate environmental impacts and promote resource efficiency. Ongoing research and policy development are essential to address existing challenges and fully realize the benefits of a circular approach to hazardous waste.

8.8 CONCLUSION

This chapter provides an overview of hazardous waste, its diverse forms, and the critical need for effective management strategies that incorporate sustainability and circularity. It emphasizes the dangers of hazardous waste due to its toxic, ignitable, reactive, and corrosive characteristics. The text discusses common types of hazardous waste, such as industrial, medical, electronic (e-waste), and HHW, along with various classification systems.

The chapter details responsible handling, including segregation, labeling, safe storage, transport, personnel training, and emergency preparedness. It also explores sustainable practices like composting, adsorption, and bioremediation and highlights the duties of entities handling hazardous waste as well as the role of the WCO in monitoring international waste movements and the importance of EPR.

The chapter connects effective hazardous waste management to the UN's SDGs, focusing on health, water, sustainable cities, responsible consumption and production, and climate action. The challenges of transitioning to a CE are discussed, along with policy recommendations and differences between developed and developing nations. Compliance requirements, waste identification, and regulatory capacity differences are outlined, as well as opportunities for energy and resource recovery, and their limitations. The chapter stresses the urgent need for a CE model and collaboration among policymakers, industries, and communities to adopt sustainable practices for a healthier planet.

REFERENCES

Duan, H., Huang, Q., Wang, Q., Zhou, B., and Li, J. (2008). Hazardous waste generation and management in China: A review. *Journal of Hazardous Materials, 158*, 221–227. https://doi.org/10.1016/j.jhazmat.2008.01.038

González-Sánchez, R., Smith, J., and Lee, M. (2023). *Economic burdens and policy challenges in hazardous waste management*. Springer. https://doi.org/10.1007/978-3-031-20049-3

Hakur, V., Kumar, A., and Singh, R. (2023). *Hazardous waste management in the circular economy framework*. Springer. https://doi.org/10.1007/978-3-031-31960-7

Li, M., Liu, C., Ding, A., and Xiao, C. (2023). A review on the extraction and recovery of critical metals using molten salt electrolysis. *Journal of Environmental Chemical Engineering, 11*(3), 109746. https://doi.org/10.1016/j.jece.2023.109746

Reddy, S. N., Nanda, S., Kumar, P., Hicks, M. C., Hegde, U. G., and Kozinski, J. A. (2019). Impacts of oxidant characteristics on the ignition of n-propanol–air hydrothermal flames in supercritical water. *Combustion and Flame, 203*, 46–55. https://doi.org/10.1016/j.combustflame.2019.01.028

Tsai, C. H., Shen, Y. H., and Tsai, W. T. (2020). Analysis of current status and regulatory promotion for reusing municipal solid waste incineration bottom ash in Taiwan over the past decade (2010–2019). *Resources, 9*(117), 1–15. https://doi.org/10.3390/resources9100117

United Nations Economic Commission for Europe. (2022). *Global waste generation projections and the need for circular economy*. https://unece.org

United Nations Industrial Development Organization. (2024). *Circular economy approaches in waste management: Trends and challenges*. www.unido.org

Waste Management, 34, 589–606. https://doi.org/10.1016/j.wasman.2013.12.004

9 Circular Strategies for Biomedical Waste Management

Akhil Gokulan, Sahik Naseem Ahmed, Jonathan Jacob, and Sri Charan P

9.1 INTRODUCTION

With the rapid increase in global waste generation, sustainable bio-waste management has become a critical challenge. Circular strategies offer an innovative approach by transforming organic waste into valuable resources, promoting environmental sustainability, and reducing landfill dependency. By integrating practices such as composting, anaerobic digestion (AD), biochar production, and resource recovery, circular system closes the loop on waste, turning it into energy, soil enhancers, and bioproducts. This chapter explores various circular strategies for bio-waste management, highlighting their role in fostering greener economy, reducing carbon footprints, and supporting sustainable agriculture. The concept of circular economy in waste management revolves around reducing, reusing, and recycling organic matter in ways that create economic and ecological benefits. It also fosters innovation in biotechnologies, supports local economies, and promotes resilience in food systems. Through case studies, technological insights, and policy frameworks we aim to present a comprehensive guide to revolutionizing bio-waste handling for a more sustainable future.

9.1.1 Classification of Biomedical Waste

Bio-waste management classification involves categorizing organic waste to ensure proper disposal, recycling environmental impact. It includes food waste, garden waste, paper waste, and animal waste, each requiring specific handling methods. Effective classification promotes sustainability, reduces landfill waste, and supports eco-friendly waste management practices.

9.1.2 Nonhazardous waste (75–85%)

Nonhazardous waste comprises materials that do not pose any risk of infection or toxicity.

- **Leftover Food**: Leftover food wastage, found in vast amounts around the world at several levels such as households, restaurants, and food supply chains,

DOI: 10.1201/9781003637677-9

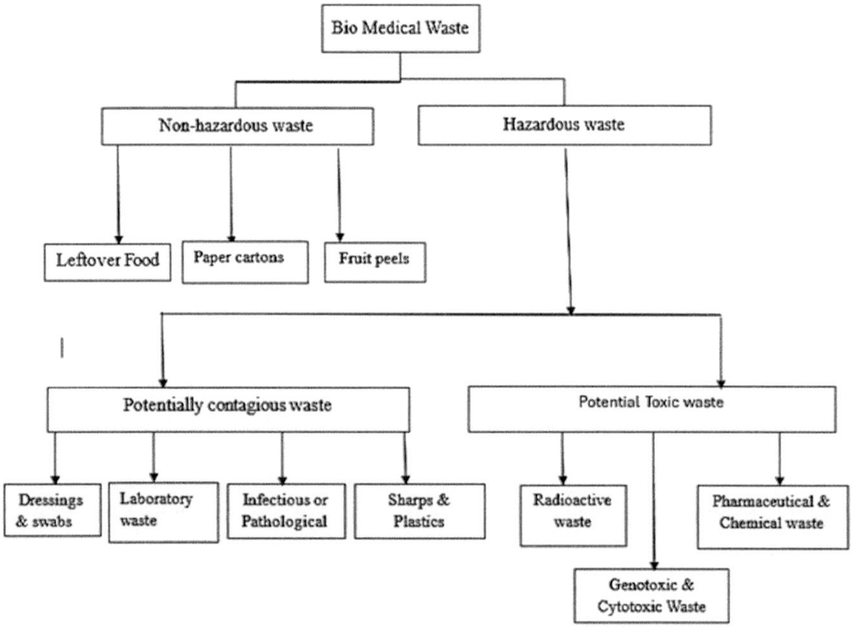

FIGURE 9.1 Block diagram of classification of biomedical waste.

has become one of the most serious global challenges. Quite literally, there are tons of uneaten food littered every day, contributing socially, economically, and environmentally to unfortunate realities. Most homes waste food because lots of food is prepared, but leftovers are not well preserved for consumption later and are eventually thrown away. Much of what restaurants and food service generate is thrown away because of strict adherence to quality indicators, portion size, and health regulations. It is very expensive, wastes taxpayer dollars, and greatly adds to the load of landfills. Simply stated, as deteriorating food produces methane—a very powerful greenhouse gas emanating from landfill sites—it thus contributes to climate change. Furthermore, wasteful food is considered wasteful of resources that are utilized to grow this food, namely, water, energy, and labor. Though specific projects may include food donation and redistribution to combat hunger, they still leave a large percentage of edible food items wasted. Raising awareness, creating better meal planning, and making sustainable practices, such as composting and repurposing leftovers, can further avert food waste and its negative environmental effects.

- **Paper Cartons**: As the market increases their use of cartons to package beverages, dairy, and processed foods, the disposal of cartons is more common than recycling, which poses a significant threat to the ecosystem. The waste of paper cartons has become a threat to the environment. Wastes of wood pulp used in the manufacturing of cartons disrupt the ecosystem. The conveying of nonrenewable resources such as oil and aluminum layer which acts as a coating makes it difficult to dispose of through regular recycling methods. Paper cannot serve the purpose of recycling due to these layers. So rather than using biodegradable methods, burning or throwing it in landfills is the easiest way to do. This transfer causes an increase in pollution and defies resources. In addition, the ecosystem suffers. It is debatable whether a better solution rests in government policy, such as improving recycling infrastructure, or directly targeting the consumer and marketing, suggesting brands switch to more sustainable packaging. For the cans that can be recycled, it is essential for brands to make recycling bins more accessible to help improve the littering problem. It is easy for consumers to decrease the number of containers they use, but the focus should be on switching to compostable materials.

- **Fruit Peels**: Fruit peel wastage is a significant yet often overlooked aspect of food waste, contributing to environmental and resource management challenges. Peels from fruits like bananas, oranges, apples, and mangoes are frequently discarded in large quantities by households, restaurants, and food processing industries. While some fruit peels decompose naturally, improper disposal, such as dumping them in landfills, leads to methane emissions that contribute to climate change. Moreover, wasted fruit peels mean lost nutrients and potential resources, as many peels are rich in fiber, antioxidants, and essential oils that can be repurposed for food, animal feed, cosmetics, or biofuels. Instead of throwing them away, fruit peels can be utilized in various ways, such as making compost, natural cleaning agents, homemade beauty treatments, or even as ingredients in teas and jams. Hazardous chemical waste needs to be gathered in marked containers, labeled correctly, and disposed of by waste management services. Infectious items including bacterial cultures, blood samples, animal tissues, and spent lab equipment like pipettes and petri dishes are all considered biological waste, also referred to as biohazardous waste.

9.1.3 Hazardous Waste (15–25%)

Hazardous biomedical waste consists of materials that could pose a risk of infection, injury, or environmental harm. It is classified into **potentially contagious waste** and **potential toxic waste**.

9.1.3.1 Potentially Contagious Waste (15–18%)

This waste includes materials that may carry infectious agents such as bacteria, viruses, and fungi.

- **Dressings and Swabs**: Used bandages, dressings, and swabs contain blood, pus, or other bodily fluids, making them a potential source of infections. If improperly disposed of, these materials can spread diseases among healthcare workers and waste handlers. Incineration or sterilization methods such as autoclaving are required for safe disposal.

- **Laboratory Waste**: Laboratory waste is defined as a category of waste that is produced in hospital, industrial, research, or academic laboratories. Its potential hazardousness means that this waste can be serious to human health as well as the environment. Laboratory waste refers to hazardous waste material that results from laboratory work or research and can generally be classified into four main categories— chemical, biological, radioactive, and general waste, where each type has its own specific and safe disposal methods.
- Examples are solvents, acids, bases, reagents, or poisons that may accumulate in the course of experiments and other test procedures. Some chemicals are highly corrosive, flammable, reactive or toxic, requiring the use of careful storage and disposal procedures to avoid fires and explosions, or environmental contamination. Before being disposed of, this waste needs to be decontaminated by autoclaving, chemical disinfection, or burning because it presents a danger of disease transmission.
- **Infectious or Pathological Waste**: This category includes human tissues, organs, body fluids, and blood samples collected from surgeries, biopsies, and autopsies. These materials can carry harmful pathogens and must be disposed of through incineration or deep burial in designated sites to prevent disease transmission.
- **Sharps and Plastics**: Sharp objects such as needles, scalpels, broken glass, and plastic syringes pose a significant risk of injuries and infections if not handled correctly. Used sharps must be collected in puncture-proof containers before disposal through incineration or specialized sterilization processes to minimize health hazards.

9.1.3.2 Potential Toxic Waste (5–7%)

Toxic biomedical waste consists of materials that contain harmful chemicals or radioactive substances, which can be hazardous to human health and the environment.

- **Radioactive Waste**: Radioactive waste refers to materials which have become radioactive due to the contamination of radioactive matter and emit ionizing rays that can be harmful to human health and the environment. This specific waste is produced from the use of nuclear power plants, medical procedures like radiation therapy, industries, and even from research labs. Radioactive waste is divided into various categories based on the levels of radioactivity. These include low level waste (LLW), which consists of contaminated clothes and medical devices; intermediate level waste (ILW), which has substantial level of radiation which requires some protection; and high level waste (HLW), which is mainly made up of spent nuclear fuel and also exceedingly radioactive reactor waste. Adequate disposal methods comprise of storage in special repositories, deep geological disposal, or managed decaying until safe radiation levels are achieved. Careless handling of radioactive waste can result in grave outcomes such as radiation, pollution, and permanent damage to the ecosystem.

- **Pharmaceutical and Chemical Waste**: Pharmaceutical waste refers to expired, unused, or contaminated medications, vaccines, and medical chemicals that are hazardous to human health and ecosystems when not disposed of safely. This waste comes from hospitals, pharmacies, research labs, and homes and is divided into hazardous (like chemotherapy drugs and controlled substances), nonhazardous (like over-the-counter medications), and infectious (like antibiotics and biologics). When drugs are disposed of incorrectly—say, by flushing them down the drain or throwing them away in the regular trash—they can pollute the environment, cause antibiotic resistance, and contaminate the water supply. They must be incinerated at extremely high temperatures, chemically neutralized, or dropped off at specialized pharmaceutical take-back programs; all this is to prevent misuse or damage to the environment. Proper management of pharmaceutical waste is essential to protecting public health, maintaining water quality and minimizing toxic waste.
- **Genotoxic and Cytotoxic Waste**: Genotoxic and cytotoxic waste is a class of hazardous substances that can damage genetic material (DNA) or interfere with normal cellular processes, leading to significant health and environmental risks. This type of waste is generated predominantly in hospitals, pharmaceutical industries, investigation laboratories, and cancer treatment centers. Included in this list are drugs used to treat cancers through chemotherapy; agents with activity against viral infections; some hormonal preparations; and laboratory reagents that are produced with mutagenic (DNA altering), carcinogenic (cancer inducing), or teratogenic (causing birth defects) properties. Even small amounts of genotoxic and cytotoxic waste can lead to serious health problems, such as genetic mutations, immune suppression, organ damage, or reproductive toxicity. Improper disposal of these substances can release them into the environment, polluting soil and water supplies and causing lasting environmental impacts.

Biomedical waste management is crucial for maintaining public health and environmental safety. Proper segregation, handling, and disposal methods must be followed to minimize risks associated with infectious, toxic, and hazardous waste materials. Healthcare facilities should implement strict waste management protocols and adhere to regulatory guidelines to ensure safe disposal practices. So as we can see all these wastes can deeply impact the environment and also the amount the waste has been exposed to humans many humanly diseases can also occur so that they can be easily deprived into one another. In hospitals and research labs, bio-waste includes used syringes, surgical gloves, discarded tissues, and cultures of infectious agents, all of which must be carefully segregated and incinerated or treated before disposal to minimize health risks. Improper disposal of biological waste can lead to serious environmental and public health issues, including soil and water contamination, the spread of diseases, and increased methane emissions from landfills. Effective waste management practices, such as composting, bio-digestion, and proper segregation of hazardous materials, can help reduce the negative impact of biological waste. Governments and organizations must enforce strict regulations and promote awareness to ensure safe disposal and sustainable handling of bio-waste, ultimately protecting both the environment and human health.

This chapter discusses the combined role of bio-waste degrading microbes and plant growth-promoting microbes in improving bio-based fertilizer that induces plant growth performance while preserving a bio-circular economy. This review addresses the need for research to discover the potential bacteria that rapidly break down bio-waste and convert it into substrate for fertilizers and valuable products. This chapter also hypothesizes the usage of microbe-processed bio-waste supplemented with Plant Growth-Promoting Microorganisms (PGPM) as a bio-based fertilizer is a revolutionary strategy for reducing massive amounts of bio-waste and improving plant growth and soil health. The current systematic review aimed to address these issues by focusing on (i) microbe-mediated advanced techniques and mechanistic for bio-wastes and sludge waste transformation, (ii) microbe-processed bio-wastes coupled with plant growth-promoting microbes to form bio-based fertilizer, and (iii) the development of sustainable microbe-mediated techniques to mitigate significant volumes of biodegradable waste and sludge waste in the conversion of value-added goods to promote a bio-circular economy. This review mainly focuses on the simultaneous application of bio-based. supplements and plant growth-promoting microorganisms, which advocates for improved efficiency over established biofertilizers. The contribution of this chapter would open up new research vistas in biofertilizer technology, biodegradable waste management, and establishing biofertilizer enterprises. Research is trending toward more sustainable and environmentally friendly solutions to address growing concerns about depletion of natural resources caused by expanding human population, waste production, and waste accumulation. The terms "trash" and "waste" in this sense referred to the evaluation of the process of changing waste into valuable products that foster the growth of plants and enhance soil health. By managing bio-waste, it is possible to build new promising biomaterials to fulfill the global-level fertilizers demand for agricultural crop growth. Additionally, these residues offer a range of beneficial bioactive compounds and nutrients, including proteins, vitamins, cartenoids, and phenolic compounds. Microbes can biodegrade these accessible bio-wastes and convert them into bio-accessible forms of plant nutrients such as phosphorous (P), nitrogen (N), and potassium (K).

9.2 CIRCULAR ECONOMY FRAMEWORK FOR MEDICAL WASTE

The widespread use of personal protective equipment (PPE), such as face masks, gloves, and hand sanitizers, has increased significantly in public spaces to curb the spread of COVID-19. While these measures have been essential in reducing transmission, they have also contributed to a surge in medical waste outside healthcare facilities. Unlike traditional medical waste, which primarily originates from hospitals and clinics, pandemic-related waste is now being generated in everyday locations, including offices, shopping centers, public transportation, and residential areas.

One of the major concerns is the improper disposal of these protective items. Many individuals discard used masks and gloves in open areas, streets, and regular trash bins rather than in designated medical waste disposal units. This careless disposal creates a significant risk of contamination, as these materials can act as carriers of infectious agents. Without proper waste segregation and disposal systems in place, the likelihood of virus transmission increases, particularly for sanitation workers and the general public who may meet contaminated waste. Healthcare facilities, already overwhelmed with increased patient loads during the pandemic, have also struggled to efficiently manage the disposal of medical waste. The sharp rise in waste production has placed additional pressure on existing disposal systems, making it difficult for hospitals to handle medical waste safely and in a timely manner. This issue is even more pronounced in developing nations, where hospitals and local authorities often lack the necessary resources, infrastructure, and technology to manage medical waste effectively.

The growing volume of improperly disposed medical waste poses serious environmental and public health threats. If not managed properly, discarded PPE and other medical supplies can contribute to the spread of infections, contaminating air, water, and soil. Addressing this challenge requires a coordinated effort between governments, healthcare institutions, and the public to implement proper waste management protocols, raise awareness about responsible disposal practices, and invest in sustainable solutions for medical waste treatment. The concept of a circular economy is gaining significant attention from both academics and industry professionals, as reflected in the increasing number of research articles, reports, and working papers published in recent years. Leading consultancy firms, such as Accenture, Deloitte, EY, and McKinsey & Company, have also contributed to the discourse by releasing reports on the topic. This growing interest is driven by the practical implications of a circular economy for business sustainability, making it a key area of research. The primary goal of this research is to create a performance evaluation framework for waste management (WM) that can provide valuable insights for formulating circular economy (CE) strategies and policies, specifically tailored to the medical industry. To achieve this goal, the study adopts an iterative approach to develop the "Dynamic CPI (DCPI)" framework. This innovative framework incorporates several essential features, such as an integrated sustainability assessment, medical industry-specific evaluation, the ability to assess over a defined period, and the use of general waste statistics without requiring additional data collection or research.

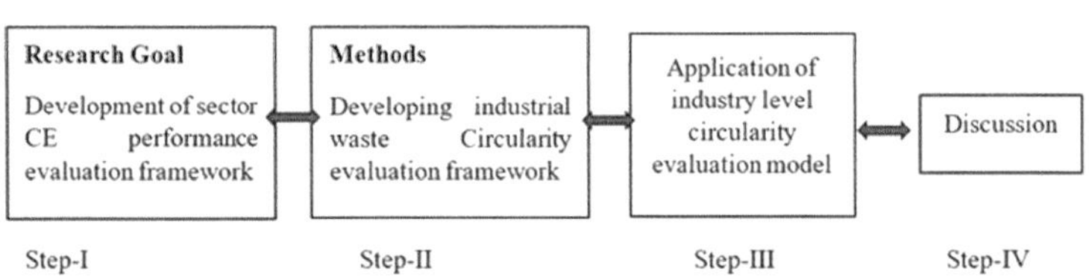

FIGURE 9.2 Circular economy framework for bio-waste.

Step 1: Problem Identification

The first step is to pinpoint the main challenges and opportunities in moving the medical industry toward a CE. This involves evaluating the current waste management practices in the industry and how well they align with circularity principles. A vital part of this step is to identify key elements that define performance specific to the sector, such as regulatory requirements, waste composition, and current recycling or disposal methods. Additionally, reviewing existing industrial waste evaluation systems is essential for understanding current methodologies, their effectiveness, and any gaps that need addressing. Ultimately, this step aims to highlight the significant challenges in achieving circularity at the industry level, including issues like waste segregation limitations, insufficient recycling infrastructure, or policy constraints.

Step 2: Framework Development

After clearly defining the problem, the next step is to create a structured framework for assessing waste management circularity in the medical industry. A crucial element of this framework is the Dynamic Circular Performance Indicator (CPI) equation, which measures circularity by taking into account factors such as waste reduction, recycling rates, and material recovery efficiency. The framework also features a systematic categorization of waste based on the type, hazard level, and potential for reuse or recycling. To ensure accurate assessments over time, a base-year CPI correction is included, allowing for adjustments in line with evolving industry standards and improvements in waste management. The overall design of the framework is intended to be flexible, ensuring its relevance across various contexts within the medical industry.

Step 3: Case Study

To test the framework, a case study is carried out to implement the developed methodology in a real-world environment. This includes selecting a representative medical facility, analyzing its waste management practices, and assessing its performance using the CPI equation.

Step 4: Results

The final step involves analyzing the outcomes of the case study and drawing conclusions about how effective the framework has been. The results offer a clearer understanding of the medical industry's progress toward circularity and pinpoint areas that need further enhancement. Key findings may reveal trends in waste reduction, improvements in recycling rates, and the effects of various waste management strategies. Furthermore, the results assist in crafting recommendations for industry stakeholders, policymakers, and regulatory bodies to boost waste circularity. Ultimately, this step ensures that the framework developed is a valuable resource for the ongoing evaluation and enhancement of waste management practices in the medical sector.

To demonstrate the efficacy of the proposed methodology, local medical waste management data spanning from 2014 to 2021 is utilized as a case study. The analysis of this data forms the foundation of the study, with subsequent findings discussed in the context of existing literature. This enables a comprehensive comparison of the results and validation of the proposed approach. The research framework, including its various components and processes, is visually represented to provide a clear understanding of its structure and functionality.

Municipalities worldwide are increasingly under pressure to adopt sustainable waste management systems. This is especially urgent in the context of managing hospital bio-waste, as it calls for a shift from a traditional linear economy model to a more sustainable CE model. This shift is not only driven by environmental concerns but also the growing need for economic resilience within urban areas. In contrast to the linear model, which primarily emphasizes disposal, CE focuses on resource recovery, reducing landfill dependency, and mitigating environmental impacts—issues that are especially relevant in urban hospital settings where waste generation is both high and complex. The adoption of CE principles in the healthcare sector presents a solution to the mounting challenges of waste management. By promoting recycling, reducing waste, and ensuring that the valuable resources within hospital bio-waste are recovered and reused, CE models align directly with broader urban sustainability goals. These models help municipalities manage the increasing volume of hospital bio-waste, reduce the environmental impact, and address the urgency of stricter environmental regulations. Unlike linear waste disposal, which often leaves bio-waste in landfills, CE prioritizes sustainability by recovering resources and turning waste into valuable products such as biogas or compost, all the while minimizing environmental degradation.

In the face of escalating regulatory pressures and the public's growing demand for greener practices, municipalities are compelled to reconsider their traditional waste management approaches, particularly when it comes to high-risk waste like healthcare bio-waste. Linear waste disposal methods, such as incineration and landfilling, often fail to address the environmental hazards of bio-waste adequately. These methods can also compromise public health and ecological balance. By contrast, CE models provide a more holistic solution by focusing on resource recovery, reducing reliance on landfills, and promoting environmentally friendly disposal methods that minimize the ecological footprint.

Additionally, municipalities are increasingly turning to multi-criteria decision-making (MCDM) tools, such as the Technique for Order Preference by Similarity to Ideal Solution (TOPSIS) method, to evaluate and prioritize waste management strategies based on sustainability criteria. These tools offer a systematic approach to making informed decisions in complex urban settings, considering multiple factors such as cost, environmental impact, and resource recovery. The high costs associated with hazardous waste disposal further incentivize municipalities to adopt CE principles. By embracing sustainable practices, municipalities can not only reduce waste management costs but also enhance their environmental stewardship.

Studies and real-world examples consistently highlight the economic and environmental advantages of transitioning from linear to circular waste management systems, particularly in sectors like healthcare where waste generation is substantial and complicated. With the adoption of CE models, municipalities can incorporate advanced waste segregation and treatment technologies. These technologies not only ensure the safe disposal of bio-waste but also facilitate the recovery of valuable resources, such as energy from biogas or organic fertilizers from composting. TOPSIS has been used to assess the trade-offs between conflicting criteria like cost-efficiency, environmental impact, and resource recovery, providing municipalities with actionable insights into how best to implement CE principles.

This transition to CE models also supports global sustainability targets, enabling municipalities to manage the growing waste volumes and meet stringent environmental regulations. This should define the range of areas where they can be effectively implemented and utilized without harming the environment in the name of sustainability. Additionally, various options and structural frameworks should be analyzed and supported through proper government policies, while also addressing the scope of multivariate analysis.

9.2.1 Importance of Policy Frameworks

The effective implementation of a CE in managing healthcare waste depends significantly on robust policy frameworks that establish clear guidelines, regulations, and incentives for sustainable waste management practices. These policies are essential for helping hospitals, municipalities, and other healthcare institutions shift from conventional waste disposal methods to more sustainable, circular approaches. Numerous case studies have shown that supportive policies greatly influence the success of CE adoption by ensuring the availability of proper infrastructure, financial backing, and stakeholder involvement. Moreover, when policy frameworks are paired with MCDM tools like the TOPSIS, they provide a structured, data-driven method for waste management. These tools assist decision-makers in assessing various sustainability factors, including environmental impact, cost-effectiveness, resource recovery, and regulatory compliance, to identify the most effective waste management strategies. By systematically evaluating different waste management options, MCDM tools help ensure that policies are in line with sustainability objectives and support informed decision-making. Furthermore, incorporating MCDM tools into policy frameworks optimizes waste management systems by balancing economic, environmental, and social factors. This strategy not only enhances efficiency but also ensures adherence to national and international regulatory standards, thus minimizing environmental risks linked to improper healthcare waste disposal. Ultimately, the collaboration between well-designed policies and advanced decision-making tools bolsters the shift toward a CE in healthcare waste management, fostering sustainability and long-term environmental advantages.

9.2.2 Integration of Innovation and Technology

The success of CE models in bio-waste management from hospitals significantly relies on advanced technologies incorporated to enhance waste treatment in terms of efficiency and sustainability. These technologies—for instance, autoclaving and AD—are very important in minimizing the impacts on the environment by means of hazardous waste reduction and recovery of some resources. In AD, microorganisms are used to convert organic hospital waste into biogas, which can act like fuel and make Rivera development and help reduce dependency on fossil fuels.

In Athens, a clear example of successful integration takes place, where hospital bio-waste has been included in the broader municipal waste systems. This approach has improved sustainability by guaranteeing proper treatment and utilization of the waste while creating massive savings for hospitals and local authorities. Moreover, integrating hospital waste disposal practices with broader municipal waste management improves regulatory compliance while at the same time enhancing operational efficiencies. Further, by steeping over to adopt such advanced technological solutions that integrate into municipal waste systems, also better position the cities for tackling sustainable and economic management challenges in the management of healthcare bio-waste.

9.2.3 Global Adoption of CE Principles in Hospital Waste Management

Several cities globally have successfully adopted CE principles for managing hospital bio-waste. In Germany, the "Closed Substance Cycle Waste Management Act" enforces waste segregation at the source and supports recycling, leading to increased recycling rates in hospitals and contributing to sustainability. Similarly, Spain's "España Circular 2030" strategy encourages public–private partnerships to advance CE in hospital waste management. These collaborations help overcome logistical challenges, improve waste system efficiency, and foster a culture of sustainability.

9.2.4 Technological Advancements in Waste Management

Technological advancements in waste management have significantly enhanced sustainability efforts, particularly through CE principles. One such innovation is AD, which is widely used in Europe to convert organic waste into biogas. This process not only reduces the volume of waste sent to landfills but also generates renewable energy, making it a practical example of the circular economy (CE) in action. Figure 9.3 illustrates the character development of the matter. Additionally, Finland has developed advanced waste management systems that efficiently sort bio-waste to produce both energy and fertilizers. By repurposing waste into valuable resources, these systems minimize landfill use while simultaneously generating renewable energy and recycled materials, contributing to a more sustainable and resource-efficient economy.

9.2.5 Role of Digital Tools

Digital tools play a crucial role in transforming waste management systems, making them more efficient and sustainable. Smart waste tracking technologies help optimize the entire process by providing real-time insights into waste flows. This allows municipalities to adjust collection schedules, ensuring waste is processed at the right facilities without unnecessary delays. Internet of Things (IoT)-based intelligent systems further enhance efficiency by monitoring waste bin levels, preventing overflow, and optimizing collection routes. Additionally, data analytics help cities analyze waste

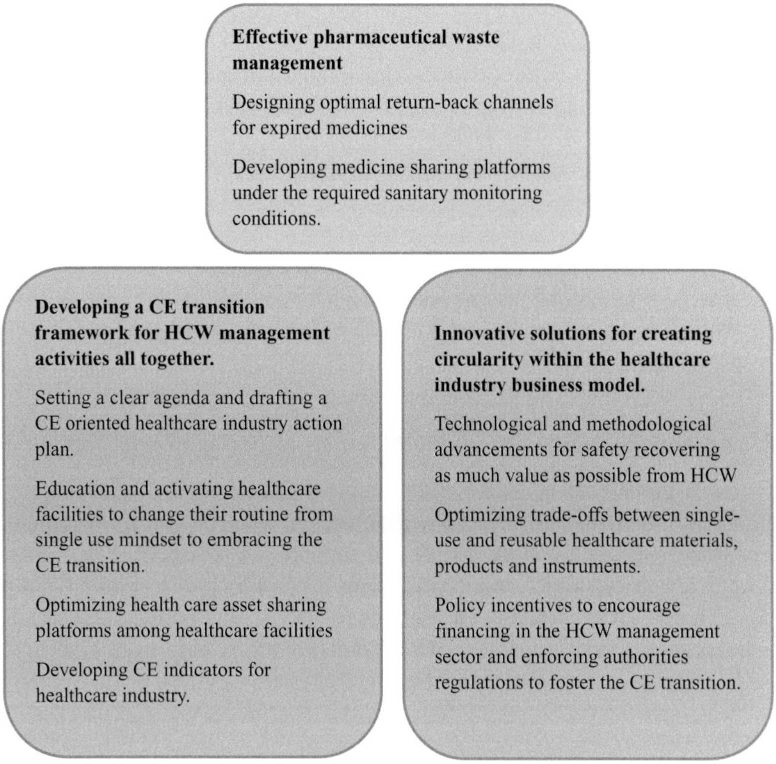

FIGURE 9.3 Block diagram for CE flow analysis.

trends, allowing for better decision-making, cost reduction, and overall improved waste management strategies. By leveraging these digital innovations, municipalities can create smarter, more sustainable waste systems.

Case Study in the Netherlands: The University Medical Center Utrecht has embraced innovative waste management by using the Steril Wave system, which decontaminates medical waste directly on-site. This process transforms hazardous waste into inert material, making it safe for recycling or disposal, significantly reducing environmental impact. Similarly, several hospitals in the Netherlands have implemented the Pharma Filter system, which processes both hospital waste and wastewater on-site. This system efficiently converts organic waste into biogas, providing a sustainable energy source while aligning with CE principles. By adopting these advanced technologies, hospitals are not only improving waste management but also contributing to a greener, more sustainable healthcare system.

9.3 OBSOLESCENCE AND RECOVERY OF MEDICAL PRODUCTS

Unlike a linear economy, where materials and energy embedded in products are lost upon disposal, a CE is intentionally designed to be restorative and regenerative.

In this model, manufactured products move through continuous loops within the system, undergoing multiple technological processes such as reuse, reprocessing, repair, repurposing, and recycling. These cycles help maximize the value of materials while significantly reducing waste generation. A CE effectively minimizes resource consumption, waste production, and energy leakage by both slowing and closing material and energy loops. Slowing loops involves extending the lifespan of products by designing them for durability and creating systems that encourage reuse, ensuring that items remain in service for as long as possible. Closing loops, on the other hand, focuses on deriving value from waste by finding innovative applications for used materials. This concept is exemplified by industrial symbiosis, where the waste output from one industry is repurposed as raw material for another. The CE framework integrates principles from various disciplines, including industrial ecology, biomimicry, natural capitalism, and the performance economy, reflecting a multidisciplinary convergence of ideas aimed at sustainable resource management.

The adoption of CE principles typically occurs at either a macro-level or a micro-level. Macro-level strategies involve nationwide, regional, or transnational policies that promote circular economic practices, such as China's Circular Economy Promotion Law and the European Union's Circular Economy Action Plan. These policies create large-scale frameworks that encourage sustainable production and waste management. At the micro-level, CE implementation is more focused, targeting specific industries, products, or materials. Several studies have explored the application of circular principles in medical device design and innovative business models, demonstrating how regulatory measures and incentives can drive micro-level adoption. Regulations and financial incentives play a crucial role in encouraging businesses to develop and implement circular economy strategies within their operations, leading to sustainable and cost-effective solutions.

In the healthcare sector, medical devices vary in their suitability for circular economy strategies based on their complexity and reusability. Some low-complexity medical devices, such as intravenous catheters, tubing, syringes, and needles, are designed for single use due to their difficulty in cleaning and sterilization. For these devices, recycling to recover base materials presents the most viable CE strategy. On the other end of the spectrum, high-complexity medical devices, such as imaging equipment, are engineered for longevity and are maintained through cycles of repair, refurbishment, and regular maintenance, ensuring their prolonged use. Between these two extremes are medium-complexity medical devices, which can benefit from circular design principles when coupled with effective reprocessing protocols. For such devices, maintaining their integrity and usability for as long as possible is key to maximizing both economic and environmental benefits. Notably, although high-complexity devices offer significant cost savings and environmental advantages through refurbishment, the widespread use of low-complexity devices in healthcare cannot be overlooked. Given their high consumption rates, these disposable items contribute significantly to healthcare expenditures and environmental pollution, making them an important area for targeted recovery and CE integration.

9.3.1 Barriers to the Adoption of a Circular Economy

The widespread reliance on single-use disposable medical devices in modern healthcare systems is driven by multiple factors, with one of the primary drivers being the perception that disposables offer greater safety than reusable alternatives. However, despite their extensive use, there is no definitive evidence demonstrating that single-use devices reduce healthcare-acquired infections. The risk of such infections is influenced by numerous factors, making it challenging to attribute infection rates solely to the type of medical devices used. Additionally, proving a significant reduction in infection rates due to a single intervention would require studies with enormous sample sizes, which are impractical to conduct. For instance, demonstrating a 10 percent relative reduction (0.2 percent absolute) in surgical site infections would necessitate nearly 500,000 patient enrollments. Historical trends suggest that the decline in surgical site infections from 4–6 percent in the late 1980s to approximately 2 percent in 2009 can be primarily attributed to standardized evidence-based protocols rather than the use of single-use devices. These protocols, which enhance patient immune defense mechanisms through measures like glycemic control and maintaining normothermia, have been far more effective in reducing infection rates. Furthermore, Table 9.1 shows the impact that eliminating healthcare-acquired infections is a commendable goal, and it is ultimately unrealistic, as some infections remain unavoidable despite best practices. Overemphasis on achieving zero infections may also lead to unintended negative consequences, such as unnecessary interventions that introduce other risks.

The entrenched use of single-use disposables is not solely driven by concerns over infection control; multiple stakeholders, including device consumers, manufacturers, regulators, accreditors, and professional standards organizations, play a role in maintaining the throwaway culture in healthcare. Hospitals and other medical institutions, as device consumers, often prefer a linear supply chain because it simplifies operations, reduces liability, and minimizes the risks associated with human error in reprocessing reusable devices. The adoption of single-use alternatives is often seen as a convenient way to eliminate the need for complex sterilization protocols. For example, single-use blood pressure cuffs were introduced to reduce the necessity of cleaning, despite limited evidence suggesting that properly reprocessed reusable cuffs pose any significant infection risk. Similarly, highly complex medical instruments like endoscopes, bronchoscopes, and ureteroscopes have traditionally been reprocessed and reused, with occasional contamination issues being addressed through improved cleaning protocols and validation procedures. However, in recent years, even these high-cost, intricate devices have seen the introduction of single-use alternatives, despite the substantial economic and environmental costs associated with such a shift.

Many single-use medical devices, ranging from basic consumables to more sophisticated equipment, can be safely reprocessed and reused. However, hospitals have largely avoided in-house reprocessing due to concerns over liability, regulatory complexity, and the financial burden of establishing and maintaining sterilization infrastructure. As a result, third-party commercial vendors have emerged to provide reprocessing services, allowing hospitals to outsource both the legal responsibility and technical challenges associated with reusing medical devices. This industry has

TABLE 9.1
Analysis Showing the Adoption of CE

Risk Level	Category (Infection Risk)	Contact Tissue	Reprocessing Requirements	Examples	category (Safety)	Regulatory Requirements
Low	Noncritical	Intact skin	Low-level disinfection	Stethoscopes, blood pressure, cuffs, pulse oximetry probes	Class 1	General control
Intermediate	Semi critical	Mucous membranes	Intermediate or high-level disinfection	Endoscopes, laryngoscope, tongue blades vaginal specula	Class 2	General controls, performance standards, post market
High	Critical	Blood and normally sterile tissues	Sterilization with steam, ethylene oxide	Surgical instruments, implants, scalpels	Class 3	General controls, premarket notification approval

extended the lifespan of many single-use devices, offering cost savings to hospitals and reducing environmental impacts by lowering waste generation and resource consumption. In 2018, reprocessing companies in North America and Europe prevented approximately 7,100 tons of medical waste from being generated and saved healthcare institutions more than $470 million. Despite the clear financial and environmental benefits of reprocessing, only a small fraction of the single-use devices that the FDA has approved for reuse are reprocessed, leaving significant potential for improvement in this area.

Original equipment manufacturers (OEMs) have played a central role in reinforcing the linear economic model by designing medical devices with built-in obsolescence and engaging in practices that hinder reusability. Their business models prioritize high-volume sales, making single-use disposables more financially lucrative than reusable alternatives. As a result, manufacturers have deliberately promoted the use of disposables by arbitrarily labeling certain devices as single-use, even when they could be safely reprocessed. Other tactics include unnecessarily shortening shelf-life expiration dates, using fragile materials to limit durability, and engineering products with features that obstruct reprocessing. For example, manufacturers may apply glue to critical components to prevent disassembly, introduce unnecessary holes or creases that make thorough cleaning difficult, or integrate electronic chips and proprietary software updates that render capital equipment incompatible with reprocessed devices. Additionally, several major OEMs have acquired third-party reprocessing companies, only to subsequently reduce the number of new FDA applications for additional reprocessable devices, effectively restricting competition and maintaining their dominance in the disposable medical device market.

The regulatory and oversight landscape for medical devices is fragmented and complex, with various organizations holding overlapping or ill-defined responsibilities. This lack of clarity has contributed to the proliferation of single-use disposables, as many regulators and accrediting bodies have avoided taking decisive action. Hospitals and healthcare providers often resort to single-use products simply to minimize the risk of regulatory violations, citations, or liability claims. The FDA is responsible for approving the reprocessing of medical devices, with regulatory requirements varying depending on the level of risk a given device poses to patients. Meanwhile, the Centers for Disease Control and Prevention (CDC) establishes sterilization and disinfection guidelines for healthcare facilities, largely based on the Spaulding classification system, which categorizes medical devices according to their invasiveness and required level of decontamination. Despite these responsibilities, both the FDA and CDC have effectively delegated much of the regulatory burden to manufacturers, creating a situation that incentivizes single-use labeling. To market a device as reusable, manufacturers must demonstrate its ability to be safely reprocessed, which involves costly and time-consuming validation procedures. However, there is no corresponding requirement for labeling devices as single-use, making it far easier and more financially attractive for manufacturers to promote disposables rather than reusable alternatives. As a result, regulatory gaps and conflicting interests have contributed to the dominance of single-use disposables in modern healthcare, reinforcing an unsustainable and wasteful model that undermines efforts to transition toward a CE.

9.4 MAIN FACTORS DETERMINING OPPORTUNITIES FOR RECOVERY

The recovery rate of biomedical waste stands crucial for healthcare system sustainability and allows the implementation of a closed-loop supply chain (CLSC). The healthcare industry needs efficient waste recovery systems because medical waste keeps growing at the same time that environmental regulations and concerns rise. Recovery opportunities are influenced by multiple factors, including waste management processes, technological progress, policies, cost-effectiveness, environmental responsibility, infrastructure capacity, stakeholder participation, and public awareness. Developing an efficient sustainable waste recovery framework requires complete knowledge of these essential variables because they determine both environmental outcome and resource utilization effectiveness.

9.4.1 Composition and Segregation of Biomedical Waste

Biomedical waste contains multiple materials, including plastics, glass, metals, textiles, and electronic components. Waste recovery success depends mainly on how waste materials are separated and sorted for disposal at their origins and on their absence of hazardous contaminants. Sufficient material separation enables the preservation of items that can return to use or recycling processes without harmful substance or infectious material intermingling.

Syringes and IV bags together with face masks and surgical instruments alongside gloves represent a major part of medical waste which consists primarily of plastics. The recycling process transforms some plastic varieties into raw substance materials suitable for producing new medical products after sterilization protocols. When glass wastes consisting of vials and ampoules undergo proper sorting they can be transformed through processing for new uses. Medical staff treating stainless steel surgical tools and titanium implants can return them to use after appropriate sterilization treatments and manufacturing procedures. Medical device electronic waste including monitors and batteries can be successfully refurbished before their materials like gold and silver and rare earth elements are extracted through recycling.

Physical recovery operations from hospitals depend on three key elements: systematic segregation protocols with color-coded recycling containers as well as RFID tagging systems with staff educational programs. The improper segregation of waste generates higher levels of contamination that make material recovery less feasible. Any biomedical waste recovery plan depends on a well-designed segregation system.

9.5 TECHNOLOGICAL ADVANCEMENTS IN WASTE PROCESSING AND RECOVERY

Advanced technologies have transformed biomedical waste management by enhancing safety and recovery efficiency. Autoclaving, gamma irradiation, and plasma gasification sterilize waste effectively, making materials recyclable. Chemical processes like pyrolysis and depolymerization break down complex polymers for reuse, while mechanical recycling of uncontaminated plastics and metals reduces reliance on virgin resources.

Smart systems—RFID, IoT sensors, and AI-driven sorting robots—enable precise tracking and sorting, boosting recovery rates and reducing manual labor. Blockchain is being explored for secure, auditable sterilization and reuse records. These innovations promote both environmental and economic sustainability but require significant investment, regulation, and skilled personnel.

9.5.1 Regulatory and Policy Frameworks

Biomedical waste recovery is governed by strict safety, hygiene, and ethical regulations. The World Health Organization (WHO) guidelines set global standards, while the EU's Circular Economy Action Plan and U.S. FDA/EPA regulations guide safe recovery and reuse. EPR policies encourage manufacturers to design recyclable medical products, and governments support initiatives with funding and tax incentives.

However, inconsistent global regulations hinder progress. Harmonizing international standards is essential to scale recovery operations and enable cross-border collaboration.

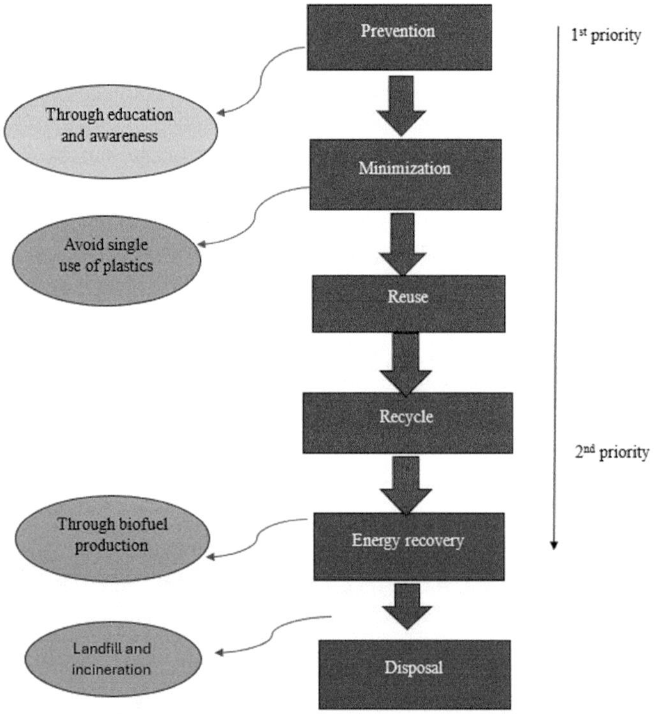

FIGURE 9.4 Block diagram of bio-waste intensity.

9.5.2 Economic Viability and Cost-Benefit Analysis

Financial feasibility is key to adopting biomedical waste recovery. While traditional disposal methods like incineration and landfilling are costly, recycling medical plastics and refurbishing tools are more cost-effective. Recovering valuable materials offsets waste management costs and helps healthcare systems lower operational expenses.

The extraction of energy from biomedical waste has now become a profitable business opportunity. The conversion of medical waste through waste-to-energy technologies, such as gasification and pyrolysis, generates electricity and biofuels, which create financial motives for hospitals and waste management operators to advance recovery methods.

The financial benefits of biomedical waste recovery do exist but hospitals face significant initial capital expenses together with strict regulations that require adherence as challenges. Governing authorities together with private investors need to create sustainable funding schemes promoting ongoing implementation of biomedical waste recovery techniques despite existing obstacles.

9.6 ENVIRONMENTAL AND SUSTAINABILITY CONSIDERATIONS

Waste management is an important aspect of environmental sustainability. Improper disposal and excessive waste generation are some of the significant contributors to pollution, resource depletion, and climate change. Waste accumulation in landfills, oceans, or urban areas has negative environmental and health impacts. Methane, for example, a potent greenhouse gas, is emitted from landfills, thereby exacerbating global warming. Plastics wastes, perhaps one of the most tenacious kinds of pollutants, are difficult to degrade within a human's lifespan as well as to any ecosystem—the terrestrial or the marine—and will likely stay here until long after this writing has passed. Other related concerns arise when it becomes common for there to be such large usage of one-time products or other items which cannot decay. Sustainability in waste management is centered on the 3Rs (reduce, reuse, and recycle), aimed at minimizing the generation of waste and maximizing resource efficiency. Reducing waste at the source can be achieved by using sustainable consumption patterns, including using eco-friendly packaging, reducing food waste, and opting for reusable alternatives. Reusing materials prolongs the product lifecycle, which means fewer virgin resources will be needed, hence reducing the environmental footprint. Recycling, when done efficiently, converts waste into valuable raw materials that can be used to manufacture new products, thus conserving natural resources and reducing energy consumption. However, ineffective recycling systems and lack of proper waste segregation remain challenges in many parts of the world. Some innovative waste management solutions include composting organic waste, promoting CE models, and investing in waste-to-energy technologies as sustainable alternatives to traditional disposal methods. Composting transforms organic waste into nutrient-enriched soil amendments, lightening the landfill burden and encouraging sustainable agriculture. A CE approach focuses on designing products with durability, repairability, and recyclability so that materials remain in use as long as possible. Waste-to-energy technologies further provide an opportunity to generate renewable energy from non-recyclable waste, hence reducing dependence on fossil fuels and managing waste sustainably. Governments, businesses, and individuals all have significant roles to play in solving the problems of waste. Businesses can adopt green manufacturing processes, use sustainable packaging, and promote responsible waste disposal. Individuals can contribute by practice consumption, proper segregation of waste, and supporting recycling efforts. Education and awareness campaigns further help in creating a culture of environmental responsibility.

9.6.1 Design Strategies for Recovery

Healthcare facilities encounter an essential operational challenge from biomedical waste management because it generates influential environmental risks and severe public health dangers. The traditional BMW management practices of incineration along with landfill dispose methods result in multiple environmental consequences which include detrimental pollution effects and greenhouse gas releases and valuable resource exhaustion. Working family members transferred their earnings to their home countries. The strategies target BMW content extraction from BMW sources which leads to opportunities for materials recycling and environmentally friendly disposal practices.

An efficient medical waste recovery system requires sustainable material selection, easy-to-use medical product design, knowledgeable collection management, state-of-the-art processing technology, and sorting capabilities to function effectively. A well-developed reverse logistics system lets businesses incorporate recovered materials successfully into their supply chain operations. Healthcare facilities employing such design methods simultaneously reduce their waste production and operational costs and promote a sustainable healthcare system. This chapter presents multiple design strategies that lead to optimal material recycling efficiency before circular supply chain integration. Sustainable material selection and product design for recovery are fundamental to BMW's recovery-focused design, as they involve choosing environmentally sustainable materials that reduce end-of-life complications. The combination of medical products from various hazardous materials and non-recyclable composites makes their reuse and recycling post-use recovery more difficult. The selection of specific materials leads to enhanced recovery process performance when they either decompose or easily enter recycling streams or reuse cycles

9.6.2 Selection of Sustainable and Recyclable Materials

Simplification of product recovery and reprocessing operations stands as a primary design consideration for medical products according to current strategy. Through their eco-friendly properties both Polylactic Acid (PLA) and Polyhydroxyalkanoates (PHA)function as alternative plastic materials which are sustainable polymers. These materials do not require landfills because their decomposition happens by natural methods or through bio-compost processing. Manufacturers utilizing Polyethylene terephthalate (PET), polypropylene (PP), and polycarbonate (PC) for their devices can reinforce the recycling of these premium plastic materials to enable their reuse in production workflows.

Medical implant materials such as titanium along with stainless steel represent crucial items for retrieval purposes. Developing proper sterilization methods and refurbishing steps enables organizations to safely reuse items or recapture them to create new industrial materials. The purposeful design of medical tools and equipment by CRMS materials enables manufacturers to improve the recovery capacity of biomedical waste.

9.6.2.1 Modular and Disassemblable Product Design

Recovery-oriented design achieves success more easily when its components and modules easily detach from one another. Medical devices use different materials which remain difficult to separate during their recycling process. The success rates of recovery efforts become higher when production uses modular design methods to create standardized assembly components that maintain easy separability.

The assembly of surgical tools happens through snap-fit and screw-based methods instead of using adhesive binding to enable facile separation of plastic from metal parts. The standardization of medical device parts across products allows efficient interchanging of replacement parts which leads to reduced total replacement demands. Medical devices employ digital identification strategies including RFID tags or QR codes to improve both material tracking and waste disposal route retrieval.

9.6.2.2 Smart Collection and Segregation Systems

The effectiveness of waste management depends on proper waste segregation methods that happen when waste collection teams first approach the materials. When misclassification occurs to BMW the recyclable materials become contaminated and end up as hazardous waste. Through proper collection and segregation procedures waste recovery materials sustain their value so that they can be reprocessed successfully.

9.6.2.3 Color-Coded and AI-Assisted Waste Segregation

WHO together with other regulatory bodies mandates healthcare facilities to implement color-coded waste segregation bins that doctors must use to separate BMW types. The effective waste segregation method fails to produce correct waste sorting results because of human mistakes. The precise classification of waste streams by robots powered with AI technology happens through material identification before machines finalize accurate waste stream set-ups.

Waste management teams gain real-time data from IoT sensors installed on smart waste bins about waste volume and their composition with contamination rates. The system enables waste collection planning improvements that lead to optimum processing operation efficiency.

9.6.2.4 Digital Tracking and Monitoring of Waste

The management of BMW throughout its life duration requires blockchain and RFID technology as essential tools for complete tracking. These technological solutions enable hospitals and waste management companies to obtain better visibility in their waste management procedures. Waste tracking systems supported by computers identify product variations and check waste sorting quality to offer real-time data about operational compliance standards.

9.6.2.5 Recovery and Reprocessing Technologies

After sorting procedures are performed the BMW materials meet the requirements necessary for recovery methods. Medical waste recycling functions through the partnership of medical sterilization methods and mechanical recycling units together with chemical practices and metallurgical recovery systems for waste material reintroduction into production streams.

9.6.2.6 Sterilization for Reuse

Medical products and protective equipment become reusable after the implementation of safe sterilization procedures without needing fresh materials. The sterilization process of autoclaving applies high-pressure steam generation to destroy microorganisms that exist in surgical equipment and glass items. Medical waste made of plastic can become reusable through combined microwave system and plasma sterilization treatments.

Supercritical CO_2 technology works as an effective decontaminating tool by maintaining cool conditions which protects temperature-sensitive medical equipment during sterilization processes. The utilized sterilization methods enable medical facilities to conduct safe operations with large amounts of suitable reusable BMW.

9.6.3 Advanced Recycling Methods

The difficulty in treating biomedical waste derives from contamination challenges, yet modern technological approaches deliver high-quality recycling outcomes. Medical plastics require initial shredding before mechanical recycling can produce new products for non-tainted medical applications. Two essential methods of recycling complex polymer composites exist through depolymerization and pyrolysis for optimal outcomes. Manufacturers isolate fundamental monomers from plastic waste through chemical extraction which enables them for brand-new manufacturing processes. Efficient techniques to recover metals from medical devices lead to implant retrieval possibilities. Medical waste X-ray films have silver contents which get separated through processes that parallel titanium separation from orthopedic implants. The processing generates clean metal materials that serve to build new products, thus minimizing the necessity for raw material beginning sources.

Bangalore Service Partners manage reverse logistics systems alongside CLSC operations, which are crucial for BMW's success in optimizing biomedical waste recovery. Efficiently designed reverse logistics enable planned transportation of medical waste through stages that facilitate recycling and reuse, with centralized collection improving operational efficiency and lowering transport costs. Medical device manufacturers support return programs where used products are repaired and redistributed, benefiting hospitals through rebate incentives for surplus equipment contributions. This comprehensive supply chain management minimizes waste and enhances resource control. Economically, recovery-oriented design helps healthcare institutions reduce disposal and material costs, while generating revenue through the sale of recovered metals and high-grade plastics. Environmentally, it cuts down on incineration and landfill use, reducing emissions like dioxins and furans, and conserving energy in manufacturing. Facilities designed with sustainability in mind not only support international environmental directives but also align with regulatory frameworks, promoting both ecological and financial gains in the healthcare sector.

9.7 CONCLUSION

Proper segregation of biomedical waste minimizes health and environmental risks and promotes sustainability through CE principles. This emphasizes the necessity for sustainable waste management strategies such as composting and repurposing food waste, paper cartons, and fruit peels, which are no longer deposited in landfills. It also underlines the proper handling of hazardous waste such as infectious and toxic materials by incineration, sterilization, and chemical disinfection. The CE framework is proposed as a solution to enhance sustainability through recycling, biogas production, and efficient disposal methods. Strong policy frameworks, financial incentives, and advanced technologies, such as autoclaving, AI-powered sorting, and plasma gasification, are emphasized in enhancing waste management systems. Case studies from Germany, Spain, and Finland illustrate how sustainable waste disposal practices can reduce landfill waste and enhance energy recovery. Several barriers are identified, however, in the adoption of a CE such as dependence on single-use disposables, complexities in regulatory policies, and lack of finances. Hence, this calls for more policy interventions and further research. Opportunities in the recovery of biomedical waste include segregation, technology advancements, and the use of digital tools in recovery to improve waste recovery rates and reduce costs while minimizing damage to the environment. Strategies for design will be proposed that include sustainable product development, smart segregation systems, and modular medical devices to improve waste recovery and circular supply chain integration. Case studies from real-world applications of circular waste management models in Tokyo, Kerala, and Stockholm show the success of robotic sorting, decentralized autoclaving, and waste-to-energy conversion. This calls for urgent application of CE principles in the management of biomedical waste, seeking collaboration between the government, industry, and citizens to implement a sustainable waste solution. Ultimately, the document calls for a shift from traditional waste management practices to a CE model that integrates advanced technologies, regulatory support, and public participation, to transform biomedical waste into valuable resources through innovations in biomedical waste management within a CE framework. It effectively demonstrates that transitioning from linear to circular waste management models is essential for achieving sustainability, reducing environmental harm, and enhancing economic resilience. The adoption of advanced waste segregation techniques, resource recovery technologies, and supportive policy frameworks can revolutionize biomedical waste handling, ultimately leading to a more sustainable future for healthcare industries and global waste management systems. By integrating scientific advancements, regulatory frameworks, and active public participation, societies can transform bio-waste from an environmental burden into a valuable resource, paving the way for a cleaner and healthier future.

10 Circular Practices to Regulate Tannery Waste

Burada Durga Sai and Kamal Aryan

10.1 INTRODUCTION

Due to growing public concerns over the pollutants discharged by tanneries and their adverse effects on water and air quality, the leather industry—a significant economic contributor globally and especially in Tamil Nadu, India—has been under pressure to mitigate its environmental impacts. Traditional leather processing methods, rooted in a linear take-make-dispose model, are marked by high water consumption, extensive chemical usage, and substantial waste generation, making it one of the most polluting industries (Kanagaraj et al., 2006; Thanikaivelan et al., 2005). In response to the increasing demand for sustainable tannery waste management, the circular economy (CE) offers a promising framework to test waste recovery technologies, optimize resource use, and enhance environmental sustainability (Muthu, 2017). This chapter explores how CE principles can be applied in the leather industry through strategies like reducing, reusing, and recycling to foster a closed-loop system. It will focus on key interventions such as wastewater treatment and water reuse to reduce water demand and minimize effluent pollution (Chandran et al., 2018), recovery of valuable tanning agents like chromium to lower waste and raw material costs (Lofrano et al., 2013), and transformation of solid tannery waste into valuable by-products like biogas and composites (Chandran et al., 2018). Additionally, the chapter will consider the use of renewable energy sources—such as solar and biogas—to reduce reliance on fossil fuels and decrease carbon emissions. The comprehensive implementation of CE in Tamil Nadu's leather sector will require policy interventions, stakeholder collaboration, technological innovation, and public awareness to drive systemic transformation. By integrating CE strategies, the leather industry can significantly reduce its carbon footprint, improve resource efficiency, and transition from being a major polluter to a model of environmental responsibility (Figure 10.1).

10.1 Sources of Wastes in the Tannery Industry

The leather industry is recognized as one of the most waste-intensive sectors, producing significant quantities of solid, liquid, and gaseous waste, each posing unique environmental challenges. Studies have shown that these wastes are generated throughout various stages of leather processing—ranging from soaking and liming to tanning

DOI: 10.1201/9781003637677-10

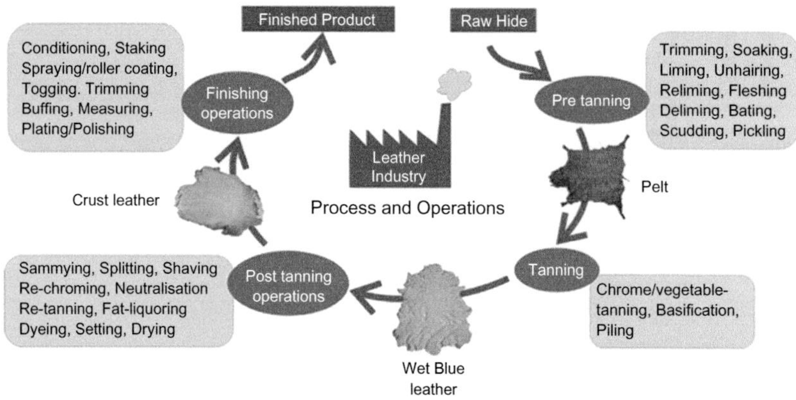

FIGURE 10.1 Advanced technological options for leather processing.

and finishing—each contributing to different forms of pollution (Thanikaivelan et al., 2005; Lofrano et al., 2013). Solid waste includes fleshing, trimmings, and hair, while liquid waste contains high concentrations of organic matter, sulfides, and chromium, and gaseous emissions include volatile organic compounds (VOCs) and odors. Effective waste management and environmental protection strategies must begin with a thorough identification of waste sources across the leather production chain to implement targeted control measures and minimize ecological impact (Kanagaraj et al., 2006).

10.1.1 Solid Wastes

The tanning industry generates substantial amounts of solid waste, with approximately 730 kg of waste produced for every 1,000 kg of raw hide processed, while only around 270 kg is converted into finished leather (Kanagaraj et al., 2006). This solid waste arises from various stages of leather processing. Skin trimmings and fleshing, produced during the early preparation phases, are rich in moisture (50–80%) and contain high levels of protein and fat, making them prone to decomposition. Shavings and buffing dusts, generated during finishing, often contain residual chromium from chrome tanning processes, presenting significant environmental challenges during disposal (Thanikaivelan et al., 2005). Additionally, keratin waste from depilation processes adds to the organic load. Tanned leather waste, such as chrome shavings and trimmings, typically contains 30–60% water and 3.5–4.5% chromium (as Cr_2O_3), further complicating safe disposal due to its hazardous nature (Lofrano et al., 2013). If these wastes are not effectively managed, they contribute significantly to landfill burden and long-term environmental degradation.

10.1.2 Liquid Wastes

Eather manufacturing is a water-intensive process, requiring approximately 60 m^3 of water per ton of hide processed, with significant volumes of liquid waste generated primarily during pre-tanning and tanning stages (Kanagaraj et al., 2006). These wastewaters vary in composition depending on the production stage, with pre-tanning effluents being particularly toxic, showing fluctuating values of pH, chemical oxygen demand (COD), total dissolved solids (TDS), chlorides, and sulfates (Lofrano et al., 2013). Tanning process effluents typically contain a complex mix of leached proteins, residual chemicals, and degradation products from hides and skins. One of the most polluting stages is dehairing, which involves high levels of sodium sulfide, contributing to 84% of BOD, 75% of COD, and 92% of total suspended solids (TSS) in the effluent (Thanikaivelan et al., 2005). Post-tanning processes further increase contamination through elevated levels of TDS, COD, and heavy metals, particularly chromium. The sludge and sediments resulting from wastewater treatment processes often contain 45–65% moisture, 30% organic matter, and up to 2.5% trivalent chromium, posing serious environmental risks. If not treated adequately, tannery effluents become major pollutants of aquatic ecosystems, endangering biodiversity and human health (Lofrano et al., 2013).

10.1.3 Air Pollution and Gaseous Pollutants

The tanning process, which involves the decomposition of hides and the use of various chemical treatments, releases several gaseous pollutants into the atmosphere. Key pollutants include ammonia, which is emitted during the deliming process and is a severe respiratory irritant, and hydrogen sulfide, a toxic gas with a strong odor. Additionally, VOCs are released, consisting of a complex mixture of hydrocarbons, amines, and aldehydes, all of which contribute significantly to air pollution and odorous emissions (Ali et al., 2017; Garcia et al., 2020). These emissions pose serious health risks to tannery workers and surrounding communities by degrading air quality and causing respiratory and other health issues (Zhou et al., 2020).

10.1.4 Chemical Effluents and Toxicity

The tanning industry relies heavily on a variety of chemicals, many of which pose significant health and environmental risks. Among the most commonly used are chromium, phthalates (including BBP, DEHP, and DBP), and heavy metals such as lead chromate and cadmium sulfate. Chromium, primarily used in chrome tanning, is not fully absorbed, with only 50–70% of it being utilized, leading to significant waste and environmental contamination (Nasser et al., 2020). Phthalates, known for their reproductive toxicity, have faced increasing restrictions due to their harmful effects, while heavy metal pigments, despite their color stability, remain a serious environmental hazard (Rai et al., 2019). Additionally, phenol, which is used in leather finishing, is restricted to a limit of 0.1% in finished products due to its persistence and estrogenic-like activity (Smith et al., 2021). Biocides, such as Methylisothiazolinone (MIT), Methylchloroisothiazolinone (CIT), and o-Phenylphenol (OPP), are used to protect against microbes in water-based finishing chemicals but are highly restricted due to their carcinogenic properties (Yang et al., 2018). Other chemicals, like N-Methyl Pyrrolidone, a known reproductive toxin, and formaldehyde, used as a cross-linker, have also faced increasing regulatory restrictions because of their carcinogenic and toxic nature (Singh et al., 2020). Furthermore, chromium (III) has the potential to induce apoptosis in proteins under certain conditions, emphasizing the need for precise chemical management in tanning processes (Ali et al., 2017).

10.1.5 Environmental and Regulatory Issues

The tanning industry presents severe environmental challenges due to the generation and improper management of various types of waste, including solid, liquid, and gaseous forms. Untreated tannery effluents often contain hazardous substances such as heavy metals, particularly chromium, and organic pollutants that are discharged into water bodies, causing significant water pollution and ecological harm (Thanikaivelan et al., 2005). The release of toxic gases like ammonia and hydrogen sulfide during different stages of leather processing further contributes to air pollution, deteriorating ambient air quality and posing health risks to tannery workers and nearby communities (Chandra et al., 2021). Moreover, the industry lacks environmentally safe solid waste disposal systems, leading to the accumulation of hazardous waste materials that exacerbate land and water contamination (Ramasami and Rajamani, 2004). The coexistence of organic waste with toxic chemicals like formaldehyde and phthalates necessitates stringent waste treatment protocols, as their unchecked release can have devastating environmental and public health consequences (Sreeram and Ramasami, 2003). While regulatory mechanisms have been introduced to control pollution levels, there is a pressing need for continuous innovation in waste treatment technologies and the adoption of CE practices to enhance sustainability in the leather sector. Effective waste management strategies are crucial to mitigate environmental degradation and to support the industry's transition toward more sustainable and ecologically responsible operations (UNIDO, 2012).

10.2 EFFECTS OF THE TANNING INDUSTRY ON THE ENVIRONMENT

The tanning industry plays a vital role in the global economy by supplying essential raw materials to leather manufacturers; however, it is also recognized as one of the most environmentally hazardous industries due to its heavy reliance on natural resources, high levels of solid waste generation, and pollutant emissions. Its operations significantly affect the environment through water and air pollution, soil degradation, and threats to plant, animal, and human health (Thanikaivelan et al., 2005). The major pollutants associated with tanning include chlorides, sulfides, chromium, tannins, sulfates, dyes, pesticides, and various finishing agents, many of which are toxic, persistent, and bioaccumulative (Sreeram and Ramasami, 2003). The disposal of untreated or poorly treated tannery effluents into agricultural land has led to the accumulation of heavy metals in crops, thereby endangering food safety and human health (Lofrano et al., 2013). In addition to water pollution, the release of airborne pollutants such as VOCs and hydrogen sulfide from tanning facilities contributes to deteriorating air quality and respiratory problems among nearby communities (Chandra et al., 2021). To effectively address these environmental and public health

TABLE 10.1
CPCB Standards for Effluents vs. Actual Effluents

Output Effluents	Actual Effluents in Industry	CPCB Standards
pH	4.4–8.8	6.0–9.0
TSS	80–150 Kg	100 Kg
BOD	50–100 Kg	30 Kg
COD	155–240 Kg	250 Kg
Total Cr	3–7 Kg	2.0 Kg

concerns, the implementation of comprehensive, multistep wastewater treatment systems and advanced air purification technologies is critical. These interventions are necessary not only for emission mitigation but also to ensure compliance with increasingly stringent environmental regulations.

10.2.1 Water Pollution and Resource Utilization

The global water crisis is intensifying and has emerged as one of the most pressing challenges of the 21st century, with water demand increasing annually by approximately 1% (UNESCO, 2020). Among the contributors to this crisis is the tanning industry, which consumes and pollutes vast quantities of water throughout the leather production process. China, the world's leading leather producer, utilizes around 140 million tons of water annually, yet only 2% of this is directly used in tanning operations, with the rest contributing to wastewater (UNIDO, 2015). Similarly, in Brazil—especially in the southern and São Paulo metropolitan regions—industrial activities, including tanneries, are major sources of water pollution (Oliveira et al., 2019). Italy's Veneto, Toscana, Lombardy, and Campania regions, known for dense tannery clusters, discharge substantial amounts of tannery effluents into local water systems, such as the heavily polluted Sarno River Basin (Di Mauro et al., 2014).

Tannery wastewater is characterized by high levels of pollutants, especially chromium (III), sulfides, ammonium chloride, dyes, and organic matter. These contribute significantly to the COD, biological oxygen demand (BOD_5), and total dissolved solids (TDS) in effluents. According to Chakraborty et al. (2021), only about 20% of the chemicals applied during tanning are retained by the hides, while the remaining 80% are discharged as pollutants. Pre-tanning and soaking stages are particularly detrimental, with soaking alone accounting for roughly 50–55% of the overall pollution load (Oran et al., 2020). As a result, improper treatment and disposal of these effluents pose severe environmental and health risks, necessitating advanced treatment technologies and strict regulatory enforcement.

10.2.2 Air Pollution and Atmospheric Contaminants

Tannery operations emit a range of hazardous air pollutants, including hydrogen sulfide (H_2S), VOCs such as hydrocarbons, amines, and aldehydes, all of which pose significant risks to both air quality and human health (Mohan et al., 2019). If not

TABLE 10.2
Water Consumption and Pollution Production in the Main Steps of the Tannery

	Wastewater and Pollutant Production (kg/ton hide)				
	Pre-tanning	**Tanning**	**Post-tanning**	**Finishing**	**TOTAL**
Wastewater (m3)	8–30	2–5	4–10	1–5	15–50
Suspended Solids	70–126	5–10	5–11	0–3	80–150
COD	120–70	10–20	25–40	0–10	155–240
BOD5	40–70	2–7	8-20	0–3	50–100
Chromium(III)	-	2–5	1–-2	-	3–7
Sulfides	4–10	-	-	-	4–10
N-NH4+	3.2–4.5	05–1.0	0.3–0.5	-	4–6
Total Kjeldalh Nitrogen	9.5–15.0	0.5–1.0	1–2	-	11–18
Chlorides	90–150	20–60	5–10	-	115–220
Sulfates	5–30	30–55	10–40	-	45–125

properly controlled, these emissions can contribute to environmental degradation and respiratory ailments among workers and nearby populations. Moreover, the oxidation of trivalent chromium (Cr III) to its more toxic form, hexavalent chromium (Cr VI), in the presence of oxygen further exacerbates air pollution, heightening the risk of severe health impacts such as lung irritation and long-term carcinogenic effects (Sreeram and Ramasami, 2003).

10.2.3 Soil Contamination and Toxic Metal Accumulation

Tannery sludge is a major environmental concern due to its high content of toxic heavy metals such as chromium (Cr), cadmium (Cd), lead (Pb), zinc (Zn), and nickel (Ni), which contribute significantly to soil and groundwater contamination (Rao et al., 2020). With nearly one million tons of sludge generated annually by the leather industry, improper disposal practices lead to heightened soil toxicity and groundwater leaching, often resulting in the accumulation of these metals in agricultural fields through irrigation with industrial effluents. In regions like Kanpur, India, continuous irrigation using untreated tannery wastewater has caused substantial heavy metal buildup in soils, leading to a 25–40% decline in crop productivity (Sharma et al., 2016). Chromium, especially in its hexavalent form [Cr(VI)], is highly mobile and toxic, impairing soil fertility, crop germination, and overall plant development. Furthermore, Cr(VI) disrupts key microbial processes such as carbon mineralization and nutrient cycling, thereby destabilizing the soil ecosystem. As plants do not naturally absorb chromium, the presence of Cr(VI) induces severe phytotoxic effects due to its ability to penetrate cell membranes and disrupt cellular functions (Kotas and Stasicka, 2000).

10.2.4 Adverse Effects on Plant Growth and Photosynthesis

The toxic effects of chromium on plant health are evident through inhibited seed germination, reduced root elongation, and impaired seedling development. Exposure to chromium, particularly in its hexavalent form, often leads to chlorosis due to a significant reduction in chlorophyll content, which negatively affects photosynthesis and overall plant productivity. Additionally, chromium-contaminated soils experience a decline in microbial populations and metabolic activity, including decreased microbial respiration. This disruption of soil microbiota further compromises soil fertility, ultimately stunting plant growth and reducing agricultural output (Sinha and Pandey, 2003; Zayed and Terry, 2003).

10.2.5 Health Hazards and Social Impacts

Tannery wastewater and sludge, which contain a mix of stones, heavy metals, and toxic chemicals, pose significant health risks to both tannery workers and nearby communities. Prolonged exposure to hazardous substances such as chromium, cadmium, and lead has been linked to chronic respiratory issues, dermatological conditions, and heavy metal poisoning. Case studies from tannery hubs such as Kanpur (India), São

Paulo (Brazil), and industrial regions in China and Italy have documented elevated levels of toxic exposure among local populations, resulting in serious health concerns (Dutta et al., 2018; Khatri et al., 2020). Furthermore, contamination of local water bodies exacerbates public health challenges, contributing to gastrointestinal illnesses, renal dysfunction, and heightened cancer risks due to the leaching of hexavalent chromium (Cr(VI)) and other pollutants into drinking water sources (Gupta and Sinha, 2006; Bai and Sutano, 2002).

10.2.6 Mitigation Strategies and Waste Treatment Technologies

To address the significant environmental challenges posed by the tanning industry, a wide array of physicochemical, biological, and advanced treatment technologies have been developed and implemented. Among biological treatments, anaerobic digestion stands out for its dual benefits: breaking down organic matter and generating biogas, a renewable source of energy (Durai and Rajasimman, 2011). Sequencing Batch Reactors (SBRs) have shown high efficiency in removing both organic and inorganic pollutants, making them suitable for treating complex tannery effluents (Arulazhagan et al., 2010). Constructed wetlands provide an energy-efficient and eco-friendly alternative, using natural filtration and microbial processes to purify wastewater (Brix, 1997).

Advanced oxidation processes (AOPs)—such as UV/H_2O_2, electro-Fenton (EF), and ozonation—are particularly effective in degrading recalcitrant and toxic compounds that resist conventional treatments (Oturan and Aaron, 2014). Membrane filtration methods like reverse osmosis and nanofiltration offer precise removal of dissolved solids, rendering the treated effluents suitable for reuse (Madaeni, 1999). Adsorption techniques, especially those using activated carbon or low-cost biosorbents, have proven effective in removing heavy metals such as Cr(III) and Cr(VI) (Naiya et al., 2009). Electrocoagulation (EC) and catalytic treatments further enhance pollutant removal by reducing COD and extracting heavy metals (Mollah et al., 2004).

The choice and integration of these treatment methods depend on the specific characteristics of the tannery effluent and desired treatment outcomes. A multi-stage treatment strategy is often necessary to achieve regulatory compliance and environmental sustainability. When properly implemented, these technologies significantly mitigate the ecological footprint of the leather industry and foster circular and sustainable waste management practices.

10.3 BY-PRODUCT TREATMENT METHODS OF THE TANNING INDUSTRY

The tanning industry generates millions of tons of solid, liquid, and gaseous waste annually, posing significant environmental risks if not properly managed. However, recent advancements in waste management have enabled the recovery, reuse, and conversion of tannery by-products into valuable resources, contributing to more sustainable industrial practices (Lofrano et al., 2013). These by-products typically include solid waste—such as fleshings, trimmings, shavings, hair, and sludge—liquid waste in the form of chemically laden wastewater rich in organic matter and heavy metals, and gaseous emissions like VOCs, ammonia, hydrogen sulfide, and methane. The integration of sustainable approaches, such as low-water tanning technologies and effluent stream segregation, has significantly improved waste minimization by allowing cleaner streams to be treated separately, thus easing the burden on central treatment systems (Chandran et al., 2017). Additionally, the adoption of polyethylene glycol-based preservation techniques has been shown to effectively reduce salt usage and prevent chloride accumulation in nearby water bodies. Enzymatic depilation and hair-saving techniques have also demonstrated a considerable reduction in acid sulfide levels and COD, further enhancing the environmental sustainability of leather production processes.

10.3.1 Recovery and Reuse of Chromium

Chromium is one of the most critical pollutants in the tanning industry, primarily due to its widespread use in chrome tanning, which accounts for approximately 80–90% of global leather production. The most common tanning agent, basic chromium sulfate, is only partially absorbed—between 50% and 70%—by hides during the tanning process. The unabsorbed portion is discharged into the environment, posing significant ecological risks (Buljan and Kral, 2015). Chromium exists in two primary oxidation states: trivalent chromium (Cr(III)), which is used in tanning, and hexavalent chromium (Cr(VI)), which is highly toxic. Under improper waste disposal conditions, Cr(III) can oxidize to Cr(VI), intensifying its environmental and health hazards (Song et al., 2019).

Several remediation and recovery technologies have been developed to address the chromium pollution in tannery wastewater. Optimization of the chrome fixation process can improve uptake efficiency and reduce residual chromium levels in effluents. One of the most commonly employed recovery methods is **chemical precipitation**, in which chromium-containing effluents are treated with alkaline agents such as lime or sodium hydroxide. This reaction forms insoluble chromium hydroxide, which can be filtered and reused, though it generates large volumes of sludge requiring appropriate disposal (Lofrano et al., 2013).

Adsorption techniques have gained traction for their effectiveness and low cost. Adsorbents such as cement kiln dust and various bio-based materials have demonstrated high chromium removal capacities. For instance, cement kiln dust has been shown to reduce chromium concentrations from 22 mg/L to as low as 0.6 mg/L, utilizing an industrial by-product in a CE approach (Venu et al., 2020).

Electrochemical treatments, including electrocoagulation and electrooxidation, represent advanced methods for chromium removal. These processes use electrical currents to destabilize and remove chromium ions from wastewater while producing less sludge compared to traditional methods. They offer a more sustainable alternative for chromium recovery and are gaining popularity for their environmental benefits (Mohora et al., 2012).

In addition, **microbial-assisted bioremediation** is emerging as a promising technology, utilizing chromium-resistant bacterial strains to detoxify tannery sludge and recover valuable chromium salts. This method offers a biologically sustainable option for treating contaminated waste and aligning with CE principles.

10.3.2 Biological Treatment of Tannery Wastewater

Recent advancements in green technologies for treating tannery effluents include bioremediation and phytoremediation. These biological methods rely on the ability of microorganisms or plants to degrade or eliminate contaminants.

10.3.2.1 Bioremediation

Bioremediation involves the breakdown of organic pollutants in tannery wastewater by bacteria, fungi, and algae. Notable fungal species identified for this purpose include *Penicillium commune* and *Cladosporium perangustum*, which can reduce COD and total chromium levels by more than 90%. Additionally, microalgae such as *Chlorella vulgaris* have shown significant effectiveness in absorbing heavy metals and organic pollutants.

10.3.2.2 Phytoremediation

Phytoremediation refers to the use of plants to absorb or alter toxic substances found in tannery waste. These plants have shown promise in removing chromium, cadmium, and other harmful metals from wastewater. Plants like *Sesuvium portulacastrum* absorb chromium and detoxify effluents. These plants accumulate chromium and other contaminants in their biomass, effectively cleaning the water without requiring chemical treatments.

10.3.3 Utilization of Solid Tannery Waste

Tanneries generate a range of solid wastes, including fleshings, trimmings, shavings, buffing dust, and hair. Rather than being discarded in landfills, these by-products can be repurposed into valuable materials. One of the major challenges in tannery waste management is stabilizing toxic by-products to prevent environmental contamination. To address this, the use of calcium/zinc stabilizers, often utilized in polymer applications to inhibit thermal degradation and hydrogen chloride release, has been explored. Recent research has shown that blending these stabilizers with bio-based carboxylates enhances heat resistance, improving the recyclability of mixed waste leather products (Vinci et al., 2020). This innovation offers a sustainable approach for reusing tannery waste in industrial applications that align with CE principles.

10.3.3.1 Leather Waste in the Construction Sector

Leather by-products, such as shavings and buffing dust, have been studied as alternative raw materials for the construction industry. Research has demonstrated the potential to produce sound insulation panels and composite boards from these materials. Experimental studies indicate that the sound absorption properties of leather-based panels are comparable to those of cork panels and carpets (Kumar and Soni, 2018), making them viable options for sustainable construction materials.

10.3.3.2 Biofuel Production from Tannery Waste

Organic waste generated during the leather-making process, particularly fleshings and hair, has been evaluated as a potential biofuel source. Studies have shown that the calorific values of these tannery wastes are comparable to traditional biomass fuels, such as wood pellets, suggesting they have excellent potential for thermal energy generation (Ekechukwu et al., 2019).

10.3.3.3 Hydrolyzed Proteins as Animal Feeds and Fertilizers

Proteins derived from tannery waste, such as fleshings and trimmings, can be hydrolyzed using enzymes or chemicals to break them down into amino acids and peptides. These hydrolyzed proteins can then be used as animal feed or fertilizers, contributing to agricultural sustainability and providing an alternative to synthetic inputs (Chandra et al., 2017).

10.3.4 Collagen Extraction through Enzymatic and Chemical Hydrolysis

Collagen, a protein found in waste materials from tanneries, such as fleshings and trimmings, can be hydrolyzed enzymatically or chemically for use in various industries, including cosmetics, pharmaceuticals, and food processing.

10.3.4.1 Enzymatic Hydrolysis

Collagen fibers can be broken down into bioactive peptides by proteolytic enzymes like alcalase and pepsin. This method is environmentally friendly and operates under mild conditions, preserving the bioactivity of collagen peptides. These bioactive peptides are highly valued in applications such as anti-aging skin creams and medical-grade wound dressings (Hernandez et al., 2020).

10.3.4.2 Stabilization Effect

Calcium/zinc thermal stabilizers derived from tannery waste have been evaluated for their effectiveness in preventing thermal degradation in polyvinyl chloride (PVC) resins. These stabilizers enhance heat resistance, which helps maintain the integrity of PVC materials (Singh and Sharma, 2021).

10.3.4.3 Chemical Hydrolysis

Chemical hydrolysis involves the treatment of collagen with acid or alkaline agents to dissolve it into gelatin. While this method is effective, it requires special safety precautions due to the use of hazardous chemicals, such as hydrochloric acid (HCl) and sodium hydroxide (NaOH). The gelatin extracted from collagen is widely used in the food industry as a stabilizer in dairy products and in pharmaceuticals and biomaterials, particularly in tissue engineering and regenerative medicine, for applications like capsule coatings (Patel et al., 2019).

10.3.4.4 Circular Economy in the Tanning Industry

Traditionally, the tanning industry has been associated with high levels of waste and environmental pollution. However, with the increasing adoption of sustainable practices, the demand for CE principles has grown significantly. These principles focus on reducing waste, enhancing resource efficiency, and promoting the reuse and recycling of by-products. Recent developments in green tanning agents, alternative chemicals, and waste valorization techniques have paved the way for more sustainable practices in the leather industry (Williams and Green, 2021).

The implementation of vegetable-based tanning agents, which provide chromium-free tanning with comparable durability and quality, has been particularly promising. Moreover, the introduction of waste-to-value technologies, such as biochar production from tannery sludge and gelatin recovery from fleshing, has the potential to significantly reduce the environmental impact of leather production. Innovations in biodegradable tanning auxiliaries are also driving sustainability efforts in leather processing (Johnson and Zhang, 2022).

10.4.1 Extraction and Processing of Tannery Waste for Thermal Stabilizers

Innovations in the CE focus on converting tanning waste into value-added products, such as thermally stabilizing agents. Fleshing waste, a by-product of tanning, is processed into calcium/zinc thermal stabilizers through fat extraction, chemical synthesis, and performance evaluation. These processes help recover useful compounds while reducing environmental risks associated with raw tannery waste disposal (Singh et al., 2021).

10.4.1.1 Fat Extraction from Fleshing Waste

Tannery waste primarily consists of fatty acids, proteins, and residual tanning agents, which can be valorized at different stages. The first step is fat extraction, typically achieved through mechanical separation, solvent extraction, filtration, and distillation. This results in purified fatty acids, which can then be converted into carboxylates for the production of calcium/zinc stabilizers. This process minimizes waste disposal issues and creates value-added products (Patel et al., 2020).

10.4.1.2 Synthesis of Calcium/Zinc Carboxylates

Calcium/zinc carboxylates are produced through a controlled chemical reaction between purified fatty acids and calcium and zinc salts. Parameters such as pH, temperature, and reaction time are carefully controlled to optimize the efficiency and stability of the final product. The resulting carboxylates are used as heat stabilizers for

PVC applications, enhancing the durability and service life of PVC products (Kumar and Singh, 2021).

10.4.2 Reactive vs. Proactive Circular Strategies in SMEs

The classification of CE adoption in the tanning sector, based on certain criteria, reveals varying strategies among small and medium enterprises (SMEs). Some SMEs adopt reactive strategies, primarily driven by external pressures, such as compliance with environmental regulations, leading to investments in wastewater treatment, emission controls, and restrictions on chemical use. These SMEs also focus on cost-saving measures, such as recycling waste materials like chrome and leather offcuts and producing eco-certified leather to meet the growing global demand for sustainable products. On the other hand, proactive strategies involve SMEs moving beyond compliance and cost savings to adopt advanced CE approaches, including forming partnerships for material exchange programs, utilizing organic by-products in industries like cosmetics and biofertilizers, and investing in green innovations like biodegradable tanning agents and natural dyes. Digitalization also plays a crucial role, with IoT, blockchain, and AI-based systems helping SMEs enhance resource efficiency, track waste, reduce operational costs, and improve energy monitoring. Additionally, anaerobic digestion, a widely used method for treating tannery wastewater, has gained prominence for its ability to generate bioenergy and produce organic fertilizers while reducing solid waste disposal. The process involves four stages—hydrolysis, acidogenesis, acetogenesis, and methanogenesis—and can be optimized through co-digestion with agricultural residues to improve methane yield and reactor stability. Advanced systems like Expanded Granular Sludge Bed (EGSB) have also shown great potential for removing sulfides and organic pollutants from tannery waste, leading to better energy recovery and resource efficiency in the industry.

10.4.4 Types of Anaerobic Reactors for Tannery Wastewater

Several anaerobic reactor configurations are used for treating tannery effluents, each offering distinct advantages.

Among these, UASB reactors are commonly employed in tannery wastewater treatment due to their low energy consumption, high methane production, and ability to adapt to varying wastewater compositions. The integration of membrane bioreactors (AnMBR) further improves effluent quality, facilitating water recycling and reuse (Smith et al., 2023).

10.4.5 Thermal Processing of Leather Waste

Thermal processing techniques such as pyrolysis and gasification offer effective solutions for converting discarded leather from tanneries into valuable energy sources, thereby preventing landfilling. Pyrolysis involves the thermal degradation of leather waste in an inert environment, producing bio-oil, biochar, and syngas. This method

TABLE 10.3
Pros and Cons of Different Anaerobic Reactor Types

Reactor Type	Advantages	Limitations
UASB (Up-flow Anaerobic Sludge Blanket)	High treatment efficiency, compact design	Sensitivity to sulphide inhibition
EGSB (Expanded Granular Sludge Bed)	Improved mixing, higher loading rates	Requires careful sludge retention
CSTR (Continuous Stirred Tank Reactor)	Simple design, effective for high COD	High sludge production
AnMBR (Anaerobic Membrane Bioreactor)	Superior effluent quality, water reuse potential	High operational costs

not only provides fuel and useful by-products but also helps immobilize heavy metals, reducing the leaching of chromium and mitigating environmental hazards (Jones et al., 2022). Fast pyrolysis, typically carried out at temperatures around 500°C, is especially suitable for producing bio-oil, making it ideal for liquid fuel production. Gasification, on the other hand, involves partial oxidation of leather waste at high temperatures to produce syngas, a hydrogen-rich gas mixture, which can be used for electric power generation. Advanced gasification technologies further enhance resource recovery by enabling selective purification of syngas and effective use of all by-products, maximizing waste minimization (Lee and Park, 2021).

10.4.6 Sustainable Effluent Reduction Strategies in the Leather Tanning Industry

Improved process optimizations and advanced treatment methods can significantly reduce and control pollution in leather tanning units. Techniques such as better wastewater management, chemical dosing, enzymatic treatments, and waste recovery methods help minimize effluent loads and maintain production efficiency, thereby reducing the ecological footprint. Water-intensive leather tanning processes benefit from low-water technologies that control washing effects through mechanical actions, such as drum speed and float ratio, which lower water consumption. Separating wastewater streams based on composition, like sending liming effluents back to the same process, further reduces pollution. Additionally, modern preservation methods using biodegradable biocides and lower salt concentrations help mitigate chloride pollution in tannery effluents. Enzymatic depilation and controlled sulfide dosages minimize sulfide pollution, while ammonium-based deliming processes can be replaced with organic acid-based or carbon dioxide deliming methods to reduce nitrogen pollution. Strategies to reduce chrome pollution include better chrome fixation and recycling techniques, while solid waste, such as fleshings and shavings, can be repurposed into biofuels or composite materials, contributing to waste valorization and reducing landfill use. By adopting these practices, tanneries can improve resource efficiency, reduce their carbon footprint, and align with sustainable development goals.

10.4 TANNERY PROCESS AND ITS ENVIRONMENTAL IMPACTS: A CASE STUDY—VELLORE, TAMIL NADU, INDIA

The leather industry serves as a vital economic driver in many developing nations, offering employment opportunities and contributing significantly to export revenues. However, its environmental footprint has been a persistent concern. This case study focuses on the tannery processes in Vellore District, Tamil Nadu, India—home to a significant portion of the country's leather production. It examines the intricate stages of leather processing, the associated environmental challenges, and the technological advancements adopted to mitigate these impacts.

Animal hides are tanned into leather through a series of processes such as pre-tanning, tanning, and post-tanning. As by-products of the processes, the chemicals used become pollutants in the form of proteins, hair, salt, lime sludge, and sulfides. In terms of water consumption, for every ton of raw hides processed, between 15,000 and 40,000 liters are utilized. Hence, the wastewater produced is highly polluted and serves to contaminate both surface and ground waters. Furthermore, solid wastes such as raw trimmings, fleshings, and shavings are collected at various points of production.

To counteract these environmental problems, tanneries adopted cleaner technologies since the 1990s. Quality chemicals, optimal dosing, water use efficiency, and improved waste management systems like reverse osmosis (RO) treatment are some examples of such advances. The RO technology enables recycling of treated water in tannery operations while evaporating brine to recover salt. Such a strategy largely lowers pollution levels and enables many tanneries to achieve zero liquid discharge (ZLD).

Around 630 tanneries are in the Vellore District, contributing to 40% of India's total leather exports. However, the industry has been historically polluting groundwater through the discharge of untreated or poorly treated effluents.

The improvement of leather industry in India for the past five decades is significant. The data on improvement of the industry in the past three decades is shown in Table 10.4.

Real progress has been made in the industrial front for pollution control during the past three decades. This included the installation of an effluent treatment plant (ETP) in the early 1990s for handling the problems of wastewater. The advent of modern technologies like RO has also enhanced environmental benefits through water recycling and solid waste minimization.

TABLE 10.4
Leather Industry Improvement in India (Value in Million Rupees)

Item	1972	1991	1998	2001	2011
Value of Production	3,000	59,570	110,560	160,000	2,80,000
Export	1,830	32,170	69,560	92,120	2,33,323
Domestic Consumption	1,170	27,400	41,000	67,880	92,320

To summarize, the leather industry is important not only for Tamil Nadu but also for the whole country. However, it is really worrying that this activity is an important source of environmental degradation. Efficient waste treatment methods with cleaner technologies have also relieved some of these pressures, showing movement along the path toward cleaner practices in the industry.

10.5 TANNING INDUSTRIES IN INDIA

The Indian leather industry ranks second in leather goods exports, third in leather clothing exports, and fourth in leather and saddlery goods exports, including harnesses. Between 2020 and 2022, India's leather exports totaled $3.68 billion, highlighting the sector's significant role in the national economy. The country manufactures approximately 2 billion sq.ft of leather annually, with more than 2,000 tanneries, making it one of the leading producers of finished leather worldwide. Despite its economic importance, the leather industry is associated with serious environmental issues, primarily the generation of waste contaminated with toxic substances such as chromium and sulfide, which lead to significant water and soil pollution. This pollution is particularly pronounced in areas with high concentrations of tanneries, affecting local communities' health. To address these challenges, regulatory measures and countermeasures have been implemented, including ETPs to better manage wastewater. However, more actions are needed to tackle hazardous waste disposal and ensure sustainable practices in the sector. Striking a balance between economic growth and environmental sustainability remains one of the primary challenges in the leather industry.

10.6 CONCLUSION

- The tanning industry is resource-intensive and highly polluting, generating vast amounts of solid, liquid, and gaseous wastes. Adopting CE principles offers a sustainable approach to minimize environmental impacts and improve resource efficiency.
- Waste is generated at every stage of leather processing, including fleshings, trimmings, and chromium-laden shavings. Wastewater contains heavy metals, sulfates, and high COD levels, while air emissions include VOCs, ammonia, and hydrogen sulfide.
- Tanneries contribute to severe water and soil contamination due to the discharge of toxic effluents, heavy metals, and chemical residues. Air pollution from solvent emissions and greenhouse gases further worsens environmental and human health risks.
- Technologies like chromium recovery, bioremediation, and anaerobic digestion help reduce pollution while enabling resource reuse. Solid waste is repurposed into biofuels, fertilizers, and construction materials to support sustainability.
- Tannery waste valorizations transform by-products into valuable applications, such as collagen extraction for pharmaceuticals and thermal stabilizers for polymers. Digital innovations, regulatory support, and industry collaboration are key to achieving long-term sustainability.

REFERENCES

Ali, H. H., Al-Zyoud, F. H., and Hashim, R. (2017). Environmental impacts of the leather industry: A critical review. *Environmental Monitoring and Assessment*, *189*(4), 186. https://doi.org/10.1007/s10661-017-5914-6

Arulazhagan, P., Vasudevan, N., and Yeom, I. T. (2010). Biodegradation of polycyclic aromatic hydrocarbons by a halotolerant bacterial strain Staphylococcus aureus. *Chemosphere*, *80*(8), 929–936. https://doi.org/10.1016/j.chemosphere.2010.05.033

Bai, R. S., and Sutano, T. (2002). Health hazards due to tannery effluent pollution in the water sources of Vellore, Tamil Nadu. *Indian Journal of Environmental Health*, *44*(3), 209–215

Brix, H. (1997). Do macrophytes play a role in constructed treatment wetlands? *Water Science and Technology*, *35*(5), 11–17. https://doi.org/10.1016/S0273-1223(97)00047-4

Buljan, J., and Kral, I. (2015). *Introduction to treatment of tannery effluents*. UNIDO.

Chakraborty, S., Roy, M., and Das, P. (2021). Treatment of tannery wastewater: A review. *Environmental Science and Pollution Research*, *28*(13), 16030–16045. https://doi.org/10.1007/s11356-020-11612-9

Chandra, B., Singh, R., and Verma, V. (2021). Air quality degradation from leather tanneries: Health impacts and mitigation strategies. *Environmental Monitoring and Assessment*, *193*(5), 281. https://doi.org/10.1007/s10661-021-09002-3

Chandran, M., Aravind, J., and Thanikaivelan, P. (2018). A zero discharge system for integrated leather tannery: Economic and environmental benefits. *Environmental Science and Pollution Research*, *25*(4), 3237–3248. https://doi.org/10.1007/s11356-017-0712-0

Di Mauro, C., Maglionico, M., and Petrucci, G. (2014). Impact of tanneries on water quality in Italy's Sarno River. *Environmental Technology*, *35*(23), 2970–2978. https://doi.org/10.1080/09593330.2014.923997

Durai, G., and Rajasimman, M. (2011). Biological treatment of tannery wastewater—a review. *Journal of Environmental Science and Technology*, *4*(1), 1–17. https://doi.org/10.3923/jest.2011.1.17

Dutta, S., Bandyopadhyay, M., and Ghosh, T. (2018). Occupational health risks in the leather industry: A case study from Kanpur, India. *International Journal of Occupational Safety and Health*, *8*(2), 92–98. https://doi.org/10.3126/ijosh.v8i2.20884

Ekechukwu, O. V., Udemba, E. I., and Asiedu, A. B. (2019). Potential of tannery wastes as biofuels for energy production. *Renewable and Sustainable Energy Reviews*, *111*, 1–9. https://doi.org/10.1016/j.rser.2019.04.020

Garcia, A. P., Marques, M., Costa, L. M., and Lima, D. (2020). Emissions from leather industry processes: A review on air pollution and mitigation technologies. *Science of the Total Environment*, *698*, 134278. https://doi.org/10.1016/j.scitotenv.2019.134278

Gupta, S., and Sinha, R. (2006). Health impacts of heavy metal pollution in water: A case study of chromium in tanning industry. *Journal of Environmental Biology*, *27*(2), 405–408.

Hernandez, M. E., Smith, R., and Patel, A. (2020). Enzymatic hydrolysis of collagen for use in cosmetics and pharmaceuticals. *Journal of Applied Biotechnology*, *15*(2), 123–135.

Johnson, P., and Zhang, L. (2022). Waste-to-value technologies in leather production: Biochar and gelatin recovery. *Sustainable Manufacturing Review*, *18*(1), 45–58.

Kanagaraj, J., Velappan, K. C., Babu, N. K. C., and Sadulla, S. (2006). Solid wastes generation in the leather industry and its utilization for cleaner environment—A review. *Journal of Scientific & Industrial Research*, *65*, 541–548.

Khatri, M., Tyagi, I., and Soni, A. (2020). Occupational exposure to toxic metals in tanneries: A global overview. *Environmental Science and Pollution Research*, *27*(25), 31764–31775. https://doi.org/10.1007/s11356-020-09176-z

Kotas, J., and Stasicka, Z. (2000). Chromium occurrence in the environment and methods of its speciation. *Environmental Pollution*, *107*(3), 263–283. https://doi.org/10.1016/S0269-7491(99)00168-2

Kumar, S., and Soni, S. (2018). Utilization of leather waste in the construction industry: A review. *Materials Today: Proceedings*, *5*(1), 2354–2360. https://doi.org/10.1016/j.matpr.2017.08.022

Kumar, V., and Singh, R. (2021). Synthesis and applications of calcium/zinc carboxylates as stabilizers in PVC. *Polymer Materials Science & Engineering*, *19*(5), 1234–1242.

Lofrano, G., Meriç, S., Zengin, G. E., and Orhon, D. (2013). Chemical and biological treatment technologies for leather tannery chemicals and wastewaters: A review. *Science of the Total Environment*, *461–462*, 265–281. https://doi.org/10.1016/j.scitotenv.2013.05.004

Madaeni, S. S. (1999). The application of membrane technology for water disinfection. *Water Research*, *33*(2), 301–308. https://doi.org/10.1016/S0043-1354(98)00212-7

Mohan, S. V., Modestra, J. A., Amulya, K., Butti, S. K., and Velvizhi, G. (2019). A circular bioeconomy approach for integrated tannery wastewater treatment and resource recovery. *Bioresource Technology*, *289*, 121603. https://doi.org/10.1016/j.biortech.2019.121603

Mohora, E., Gavrilescu, M., Păduraru, C., and Vlad, M. (2012). Electrochemical methods for wastewater treatment. *Environmental Engineering and Management Journal*, *11*(5), 1035–1045.

Mollah, M. Y. A., Schennach, R., Parga, J. R., and Cocke, D. L. (2004). Electrocoagulation (EC)—Science and applications. *Journal of Hazardous Materials*, *B84*, 29–41. https://doi.org/10.1016/S0304-3894(01)00176-5

Muthu, S. S. (Ed.). (2017). *Sustainability in the Textile Industry: A Collection of Papers from the 10th Asia-Pacific Conference*. Springer. https://doi.org/10.1007/978-981-10-2639-3

Naiya, T. K., Bhattacharya, A. K., and Das, S. K. (2009). Adsorption of Cd(II) and Pb(II) from aqueous solutions on activated alumina. *Journal of Colloid and Interface Science*, *333*(1), 14–26. https://doi.org/10.1016/j.jcis.2009.01.006

Nasser, A., Abed, S. S., and Ghosh, S. (2020). Environmental impact of chromium and its management in leather production. *Journal of Environmental Management*, *254*, 109849. https://doi.org/10.1016/j.jenvman.2019.109849

Oliveira, R. R., Almeida, C. A., and Xavier, A. (2019). Industrial wastewater pollution and its impact on water quality in São Paulo, Brazil. *Water Environment Research*, *91*(10), 886–895. https://doi.org/10.1002/wer.1153

Oran, C., Cinar, O., and Yetis, U. (2020). Pollution load estimation of leather tanning processes. *Journal of Cleaner Production*, *247*, 119114. https://doi.org/10.1016/j.jclepro.2019.119114

Oturan, M. A., and Aaron, J. J. (2014). Advanced oxidation processes in water/wastewater treatment: Principles and applications. A review. *Critical Reviews in Environmental Science and Technology*, *44*(23), 2577–2641. https://doi.org/10.1080/10643389.2013.829765

Patel, R., Sood, M., and Kumar, P. (2019). Chemical hydrolysis of collagen for the extraction of gelatin: Applications in food and pharmaceuticals. *Journal of Food Science and Technology*, *55*(9), 3541–3549.

Patel, R., Sood, M., and Kumar, P. (2020). Valorization of tannery waste: Fat extraction and synthesis of value-added products. *Journal of Environmental Chemistry*, *14*(3), 215–222.

Rai, P. K., Zhang, M., and Sharma, R. (2019). Hazardous chemicals in leather production: Environmental impacts and regulatory measures. *Science of the Total Environment*, *687*, 928–937. https://doi.org/10.1016/j.scitotenv.2019.06.104

Ramasami, T., and Rajamani, R. (2004). Cleaner leather production technologies. *Journal of the Society of Leather Technologists and Chemists*, *88*(1), 3–8.

Rao, P., Sarma, P. N., and Subrahmanyam, Y. V. (2020). Sustainable management of tannery sludge: Challenges and opportunities. *Journal of Environmental Management*, *270*, 110905. https://doi.org/10.1016/j.jenvman.2020.110905

Singh, J., and Verma, V. (2020). Formaldehyde and its carcinogenic risks in the leather industry. *Environmental Toxicology and Pharmacology*, *71*, 103240. https://doi.org/10.1016/j.etap.2020.103240

Singh, R., Gupta, M., and Sharma, L. (2021). Circular economy in the leather industry: Waste-to-value innovations. *Sustainable Manufacturing Journal*, *23*(1), 45–56.

Singh, V., and Sharma, K. (2021). Stabilization of PVC resins with tannery-derived thermal stabilizers. *Materials Science and Engineering Journal*, *32*(3), 204–210.

Sinha, S., and Pandey, K. (2003). Impact of hexavalent chromium on germination and early seedling growth of *Vigna radiata* L. Wilczek. *Water, Air, and Soil Pollution*, *145*, 67–75. https://doi.org/10.1023/A:1023640520543

Smith, R., Chandra, S., and Gupta, D. (2021). Phenol toxicity in the leather industry: Regulatory standards and environmental implications. *Toxicology Reports*, *8*, 312–319. https://doi.org/10.1016/j.toxrep.2021.04.015

Song, Y., Tian, X., Guo, Q., and Zhao, Y. (2019). Reduction and detoxification of hexavalent chromium by microorganisms: Mechanisms and application. *Frontiers in Environmental Science & Engineering*, *13*(3), 1–12. https://doi.org/10.1007/s10653-025-02618-2

Sreeram, K. J., and Ramasami, T. (2003). Sustaining tanning process through conservation, recovery and better utilization of chromium. *Resources, Conservation and Recycling*, *38*(3), 185–212. https://doi.org/10.1016/S0921-3449(02)00151-9

Thanikaivelan, P., Rao, J. R., Nair, B. U., and Ramasami, T. (2005). Recent trends in leather making: Processes, problems, and pathways. *Critical Reviews in Environmental Science and Technology*, *35*(1), 37–79. https://doi.org/10.1080/10643380590521436

UNESCO. (2020). *United Nations World Water Development Report 2020: Water and Climate Change*. United Nations Educational, Scientific and Cultural Organization.

UNIDO. (2015). *Water Management in the Tanning Industry: Technical Cooperation Programme Report*. United Nations Industrial Development Organization.

United Nations Industrial Development Organization (UNIDO). (2012). *Future Trends in the World Leather and Leather Products Industry and Trade*. UNIDO. www.unido.org/resources/publications

Venu, H., Abarna, A., and Subramanian, R. (2020). Application of industrial waste as adsorbent in the removal of heavy metals from tannery wastewater. *Journal of Environmental Chemical Engineering*, *8*(1), 103546. https://doi.org/10.1016/j.jece.2019.103546

Vinci, G., Manfredini, T., and Martinelli, M. (2020). Development of a bio-based stabilizer blend for waste leather recycling: Heat resistance and recyclability improvements. *Journal of Cleaner Production*, *267*, 122098. https://doi.org/10.1016/j.jclepro.2020.122098

Williams, T., and Green, L. (2021). Circular economy approaches in the leather industry. *Environmental Sustainability Journal*, *10*(4), 289–301.

Yang, J., Zhang, Z., and Lin, Y. (2018). Carcinogenic risks associated with biocide usage in the leather industry. *Environmental Health Perspectives*, *126*(12), 127002. https://doi.org/10.1289/EHP345

Zayed, A. M., and Terry, N. (2003). Chromium in the environment: Factors affecting biological remediation. *Plant and Soil*, *249*(1), 139–156. https://doi.org/10.1023/A:1022504826342

Zhou, S., Wang, Y., and Zhang, Z. (2020). Health risk assessment of air pollutants in leather manufacturing industries. *Journal of Environmental Management*, *264*, 110440. https://doi.org/10.1016/j.jenvman.2020.110440

11 Circular Strategies for Handling Greywater

Rajyalakshmi G, Mahajan Gargi Ravindra, Jayakrishna Kandasamy, and Deepa Aakepati

11.1 INTRODUCTION

Water is essential for the survival of all living beings, and having enough safe water is vital for healthy growth. According to the United Nations, approximately 4,400 children under the age of five lose their lives each day because of diseases associated with polluted water and poor sanitation. Presently, one-third of the global population experiences varying levels of water shortages, and one in six individuals does not have clean water available within a one-kilometer radius (Edwin et al., 2014). Moreover, half of the population in developing countries lacks proper sanitation facilities. Recent figures show that about 800 million people experience water stress, defined as having less than 1,700 cubic meters of renewable water resources per person each year. This number is expected to rise to three billion in Asia and Africa by 2025. To combat water stress, especially in developing nations, the reuse of wastewater is becoming more popular. This trend is driven by water shortages caused by low rainfall and high evaporation rates (as seen in Australia) or the rising demand for freshwater from growing populations (like in Japan). Additionally, environmental and economic factors encourage wastewater reuse, as it can lower costs related to wastewater treatment by reducing the amount of water that needs processing (Vuppaladadiyam et al., 2019).

Greywater refers to the wastewater generated from domestic activities such as bathing, showering, washing dishes, and laundry. It accounts for approximately 75% of the total sewage produced in households. This type of water is often considered ideal for reuse due to its relatively lower levels of contamination compared to municipal wastewater, as it typically lacks significant amounts of fecal matter and urine.

When evaluating the potential for greywater reuse, several factors must be considered. First, it is essential to analyze the specific characteristics of the greywater, including its chemical and biological composition (Khanam and Patidar, 2022). Understanding these properties helps determine whether pretreatment is necessary before the water can be safely reused. Additionally, it is crucial to assess the health risks posed by microorganisms that may be present in greywater. These pathogens can lead to various health issues if not adequately managed. Moreover, consideration must also be given to the environmental implications of greywater reuse, particularly regarding the accumulation of harmful organic compounds and metals in the soil and

DOI: 10.1201/9781003637677-11

groundwater. Over time, these substances can adversely affect soil quality and water sources, highlighting the importance of careful management and treatment processes when reusing greywater (Rakesh et al., 2020).

Greywater can be categorized into two primary types: untreated greywater, which is collected at a low cost and immediately utilized for purposes like toilet flushing or outdoor irrigation, where it undergoes natural filtration through plant roots and soil microorganisms; and treated greywater, which is gathered through automated systems that redirect, purify, and distribute the water, making it safe for storage and suitable for tasks such as dishwashing and laundry. While some critics of greywater systems raise concerns about the potential for disease transmission, supporters argue that a significant portion of household water usage does not require potable water, and that, when properly managed, greywater systems serve as a valuable supplementary source of water for homes.

11.1.1 Essential Features of Greywater

Understanding these essential features of greywater is crucial for developing effective management strategies that promote its safe and beneficial reuse.

Greywater is produced from everyday household activities, including bathing, showering, dishwashing, and laundry. In contrast to blackwater, it does not typically contain significant levels of fecal matter or urine (He et al., 2022).

Composition—This type of wastewater generally consists of a combination of organic materials, soaps, detergents, and various household chemicals. The specific makeup can differ depending on the activities that generate it.

Pollution Level—Greywater is usually less contaminated than blackwater, making it a more viable option for reuse. However, it may still harbor pathogens and other contaminants that necessitate treatment prior to any reuse.

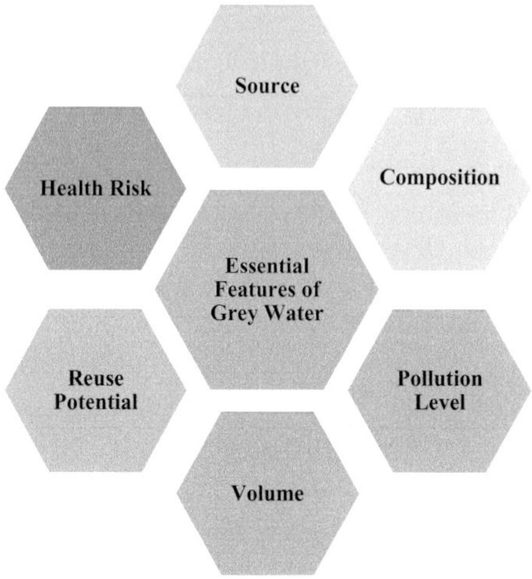

FIGURE 11.1 Essential features of greywater.

Volume—A considerable share of total household wastewater is greywater, often accounting for approximately 50–75% of the sewage produced in an average household.

Reuse Potential—Owing to its relatively lower levels of contamination, greywater can be repurposed for several applications, including irrigation, toilet flushing, and other non-potable uses, thereby helping to conserve freshwater resources.

Health Risks—Despite being less hazardous than blackwater, greywater can still present health risks due to the presence of microorganisms. Effective treatment and management are crucial to minimize these risks.

Treatment Requirements—The extent of treatment needed for greywater depends on its intended application. Basic filtration may be adequate for irrigation, whereas more sophisticated treatment methods might be required for purposes like toilet flushing.

Environmental Impact—Improper handling of greywater can result in environmental concerns, such as soil degradation and groundwater contamination. Therefore, recognizing its characteristics and potential hazards is essential for sustainable reuse.

Regulatory Standards—Many jurisdictions have established specific guidelines and regulations for greywater reuse, outlining acceptable treatment levels and usage. Adhering to these standards is crucial for ensuring safety and environmental integrity.

Socioeconomic Benefits—Recycling greywater can help decrease water consumption and reduce utility expenses, offering economic advantages for households and communities, particularly in areas facing water scarcity.

11.1.2 Source of Greywater

Greywater primarily comes from showers, bathtubs, and sinks, excluding kitchen sinks where food scraps may be present (Oteng-Peprah et al., 2018). The detailed sources of grey water generation are shown in Figure 11.2.

11.1.3 Greywater Sources and Their Constituents

The average greywater generation ranges from 90 to 120 liters per person per day, whereas in low-income countries facing water scarcity, it drops to approximately 20 to 30 liters per person per day. The makeup of greywater can differ significantly based on various factors, such as the size of the household, the ages of its members, lifestyle choices, water consumption habits, socioeconomic status, cultural practices, the types and quantities of household cleaning products used, and the duration of greywater storage before reuse. Bathroom greywater typically comprises a mixture of soaps, shampoos, toothpaste, personal hygiene products, shaving remnants, skin cells, hair, body oils, lint, and occasional small amounts of urine and feces. In contrast, laundry greywater is characterized by elevated levels of detergent-related substances, including sodium, phosphorus, surfactants, nitrogen, bleach, oils, paints, solvents, and nonbiodegradable fibers from fabrics (Chaillou et al., 2011). Greywater

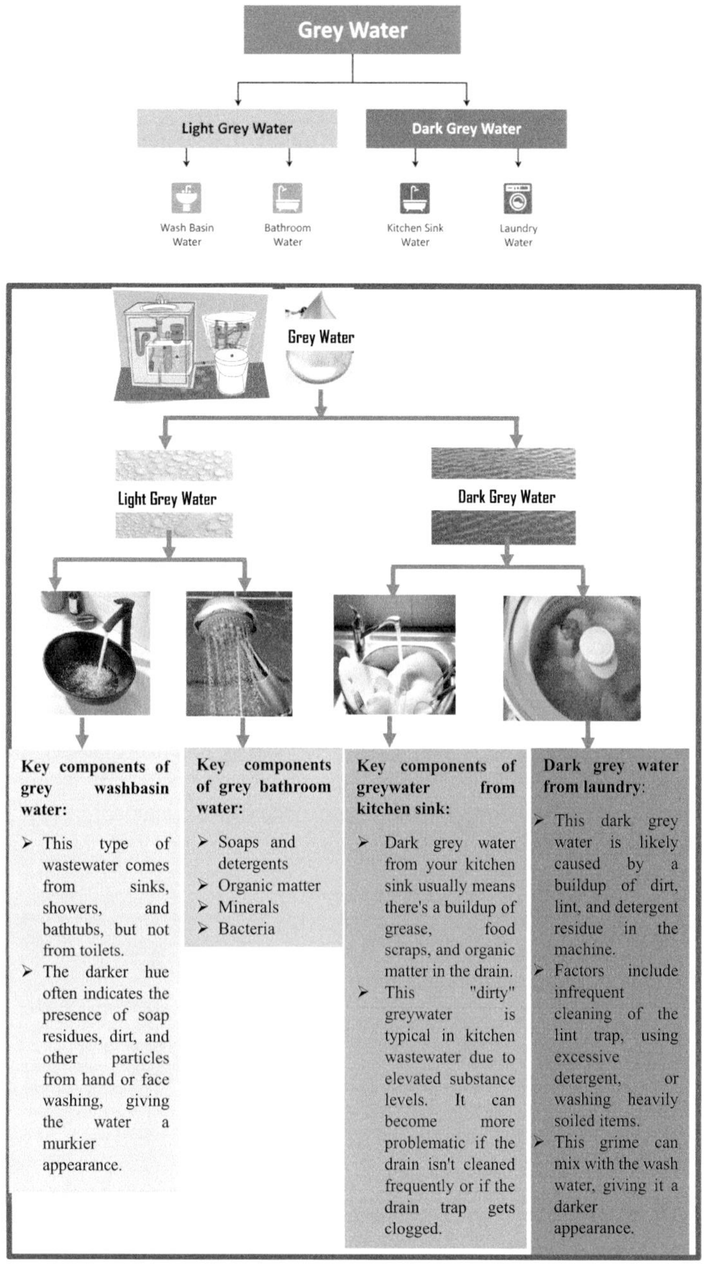

FIGURE 11.2 Sources and characteristics of greywater.

from washing machines includes bleach, foam, elevated pH levels, hot water, nitrates, oil, grease, phosphates, salinity, soaps, suspended solids, and turbidity. Kitchen sink greywater is rich in food scraps, significant amounts of oil and fat, and dishwashing detergents. Dishwasher greywater contains bacteria, foam, food remnants, high pH, hot water, odors, oils, organic matter, salinity, soaps, suspended solids, and turbidity. Dark greywater from wash basins is typically classified as greywater, which comes from sinks, showers, and bathtubs, excluding toilet water. The darker color often signals a buildup of soap scum, dirt particles, and other residues from washing hands or faces, resulting in a murkier appearance than usual (Arden and Ma, 2018).

11.1.4 Possible Applications of Treated Greywater

Greywater can be effectively reused for various purposes, significantly contributing to water conservation efforts. One common application is irrigation, where it is used for watering plants in gardens, landscapes, and certain crops, thereby reducing the demand for freshwater resources. Additionally, treated greywater can be repurposed for toilet flushing in specific systems, offering an alternative to potable water for this non-potable use. In industrial settings, some processes may utilize treated greywater for various functions, further demonstrating its versatility (Boano et al., 2020). However, successful greywater reuse requires careful consideration of several important factors. First, proper treatment is essential to eliminate contaminants, employing methods such as filtration and disinfection tailored to the intended application. Moreover, adherence to local regulations is crucial, as these laws often dictate reuse practices, system design, and quality standards to ensure safety and effectiveness. Lastly, public perception plays a significant role; concerns about potential health risks associated with greywater reuse necessitate clear communication and educational initiatives to inform and reassure the community about its safety and benefits (Al-Ismaili et al., 2017).

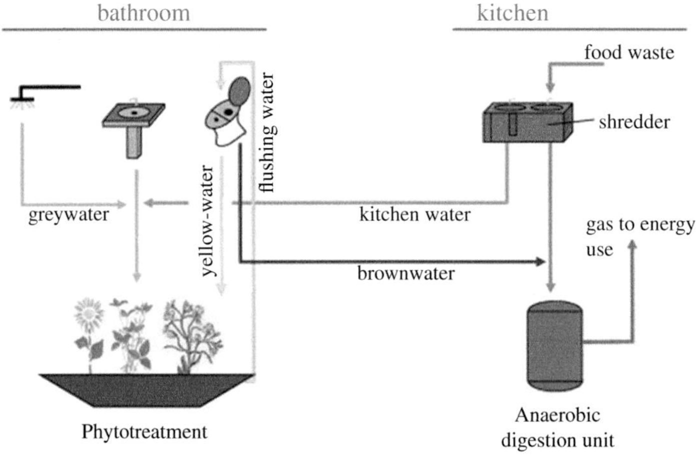

FIGURE 11.3 Applications of greywater.

11.1.5 Benefits of Greywater

Implementing greywater systems can lead to significant cost savings on both freshwater and sewage expenses, alleviating pressure on municipal water supplies and sewage systems, particularly during droughts or periods of water rationing. These systems enable the creation of lush landscapes in areas where irrigation water is scarce and allow for the recycling of nutrients from kitchen wastewater, which can enhance soil fertility. Moreover, greywater is generally simpler to recycle and process compared to blackwater, as it breaks down more quickly in natural settings. These systems can be set up in newly constructed homes or added to older ones, and they can include technologies to recover heat from greywater, such as from shower runoff. Additionally, certain local governments offer tax breaks to encourage the installation of greywater systems, and numerous public golf courses and parks use treated greywater for watering purposes, leading to cost reductions for the community (Shaikh and Ahammed, 2020).

11.1.6 Drawbacks of Greywater

Improper management of greywater can pose significant health risks, and some municipalities impose costly and complicated permitting and inspection processes for legal operation of these systems. In certain areas, the indoor use of greywater may even be prohibited, further complicating implementation. Additionally, the installation of advanced filtration and treatment systems can be prohibitively expensive, and claims made by retailers regarding the effectiveness of these systems may often be exaggerated or misleading. Furthermore, many professionals have a limited understanding of greywater systems in comparison to conventional plumbing practices. Regular maintenance is also necessary, which can lead to the need for costly replacements of specific components over time (Shaikh and Ahammed, 2020).

11.1.7 Facts and Figures of Greywater

- Oasis Design estimates that between 660,000 and 1.77 million households in California use greywater, with the number expected to rise to 8 million across the United States. Oregon enacted legislation in 2009 to promote greywater usage.
- While thorough research focused on health risks associated with greywater is lacking, there are no recorded cases of illnesses associated with exposure to greywater.
- In many developing countries, the use of greywater is prevalent and often goes unregulated.

One practical application of greywater reuse is for flushing toilets and urinals, as many regions currently utilize potable water for this purpose. Studies suggest that reusing greywater for toilet flushing could lead to a 30% reduction in overall household water consumption. Successful examples of greywater reuse can be found in

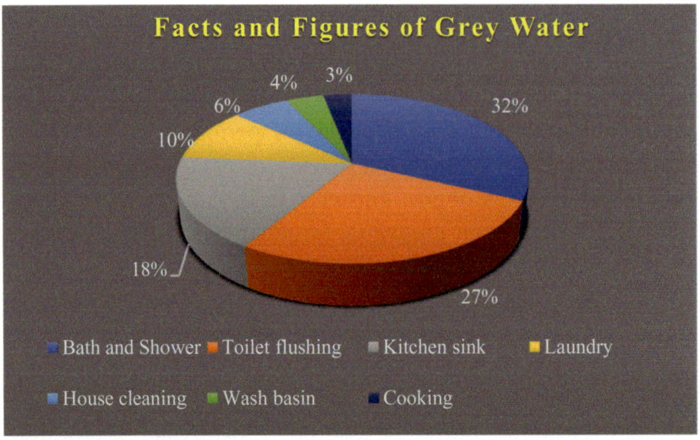

FIGURE 11.4 Facts about greywater.

Germany, where systems have been developed to meet health standards while maximizing water recovery. Research has shown that not only greywater from bathrooms but also wastewater from washing machines and rainwater can contribute to sufficient supplies for non-potable uses in large buildings (Edwin et al., 2014).

Outdoor applications for greywater include irrigation for lawns, sports fields, parks, and gardens, as well as uses such as vehicle washing, fire protection, and feedwater for boilers. Additionally, greywater can play a role in the creation and maintenance of wetlands. Another approach to managing greywater is through infiltration into the ground, which helps to integrate it into the urban hydrological cycle (Marjoram et al., 2014).

However, several challenges accompany the reuse of untreated greywater. One significant concern is the potential for disease transmission due to microorganisms present in the water, especially when used for toilet flushing or irrigation. The aerosolization of pathogens during toilet flushing poses a health risk, as it can lead to inhalation or hand-to-mouth contact. Research has highlighted these health risks and proposed frameworks for establishing new health guidelines for greywater reuse.

Various countries and states are developing guidelines for the safe reuse of treated wastewater for non-potable purposes. In California, for example, the maximum allowable level of total coliforms in reclaimed water for toilet flushing is set at 2.2 per 100 ml. Florida has established stricter limits, requiring no detectable fecal coliforms in reclaimed water for similar uses. The World Health Organization recommends a limit of fewer than 1,000 fecal coliforms per 100 ml for agricultural irrigation. In Germany, regulations include limits for total coliforms, fecal coliforms, and *Pseudomonas aeruginosa* (Byrne et al., 2020).

Concerns about soil and water pollution due to various contaminants from greywater—such as elevated detergent levels—must also be addressed. These pollutants, often originating from household cleaning and personal care products, can negatively impact plant health and soil quality.

TABLE 11.1
Characteristics of Greywater from Various Sources (Byrne et al., 2020)

Source	This study Lp^{-1} d^{-1} mean (sd)	Literature[a] Lp^{-1} d^{-1} mean	Literature[b] Lp^{-1} d^{-1} mean	Literature[c] Lp^{-1} d^{-1} mean
Total water consumption	114 (27.34)	50	91.56	127.5
Drinking and cooking use	3.5 (0.84)	10	6.59	1.8
Toilet flushing	30 (7.19)	7.5	18.31	37.1
Gardening irrigation/ others	2.5 (0.6)	NIL	2.20	
Washing and cleaning of house	7 (1.68)	3.5	6.68	
Total GW production	71 (17.03)	29	57.77	88.6
Shower and bath	35 (8.39)	15	25.82	52.3
Hand basin	5 (1.2)	NIL	NIL	5.3
Laundry	19 (4.56)	10	17.03	17.2
Kitchen/ dishwashing	12 (2.88)	4	14.92	13.8

Greywater is reusable water that comes from daily household activities such as laundry, dishwashing, and showers, but it does not include sewage or toilet waste (blackwater); it is considered less contaminated than regular wastewater, containing lower levels of organic matter and pathogens, making it potentially reusable for irrigation purposes after proper treatment due to its relatively cleaner nature (Edwin et al., 2014).

11.2 ENVIRONMENTAL IMPACTS AND HEALTH RISKS OF GREYWATER

Improperly managed greywater irrigation over time can cause issues like runoff and water pooling, leading to potential environmental harm. Additionally, greywater tends to be alkaline, which can negatively affect acid-loving plants like camellias and azaleas.

11.2.1 Factors Related to Public and Environmental Health

The composition of greywater can differ greatly due to several factors, including the number of people living in a household, their ages and lifestyles, existing health issues, the source of the water, and the cleaning products used, such as soaps and detergents. This wastewater may harbor harmful substances, including pathogens (like bacteria, viruses, and protozoa) from contaminated laundry, along with chemicals found in various household items, such as detergents, bleaches, and cleaners. Additionally, it can contain debris like dirt, lint, food scraps, hair, skin cells, and small amounts of urine, feces, and blood. Consequently, it is essential for users and homeowners to be aware of the potential health risks and environmental consequences tied to inadequately designed, installed, or maintained greywater systems (Gross et al., 2005).

In terms of public health, greywater can harbor significant levels of pathogens, especially when household members are experiencing gastrointestinal illnesses (Benami et al., 2016). These organisms can be transmitted through direct contact with greywater, inhalation of spray from irrigation, and contact with broken skin. Indirect transmission may occur through contaminated items, interaction with pests, or contact with family pets. To mitigate these risks, several guidelines should be followed: untreated greywater should only be used for subsurface irrigation to limit human and animal exposure; specific setback distances from buildings, wells, and water sources must be established to avoid contamination, and avoid using greywater for irrigating edible fruits and vegetables.

From an environmental perspective, greywater can introduce various chemical and physical contaminants that could harm plant health and soil quality. Prolonged use of greywater without proper management can lead to negative consequences such as pooling and runoff. To minimize environmental impacts, it is advisable to use environmentally friendly cleaning products, opt for those with low levels of boron, phosphorus, and salt, and utilize rinse water from washing machines, which contains fewer detergents. Avoiding harsh chemicals, bleaches, and dyes is also important, as these can damage beneficial soil organisms and plants. Additionally, greywater's slightly alkaline nature can be detrimental to acid-loving plants like azaleas and camellias (Al-Hamaiedeh et al., 2010). If left untreated, greywater can cause serious environmental issues, including soil contamination, groundwater pollution, and eutrophication of water bodies due to nutrient overload, which can harm aquatic life. This is particularly concerning when greywater is discharged directly into natural environments without adequate treatment, as it contains soaps, detergents, oils, and bacteria from domestic wastewater.

11.2.2 Key Impacts of Untreated Greywater

Untreated greywater poses considerable environmental and public health risks due to the presence of pathogens, nutrients, and various chemicals. When discharged directly into the environment without adequate treatment, it can contribute to a range of serious issues.

Greywater can harbor harmful pathogens, presenting health risks when used without appropriate treatment, with the level of risk varying based on its source and application. Health hazards include the transfer of disease-causing organisms through dirty hands, breathing in irrigation mist, and contact with open wounds and the involvement of pest vectors like rodents and insects, as well as family pets.

Specifically, the risk associated with different greywater sources varies; laundry rinse water and sub-surface irrigation pipes are considered low risk, while kitchen greywater, hosing the garden, and vegetable gardens pose high risks. In contrast, using greywater for ornamental plants is categorized as low risk. First, the pathogens found in greywater can lead to waterborne diseases, which can have detrimental effects on human health. These diseases may arise from the consumption of contaminated water sources or through exposure to infected surfaces. Second, the nutrient load, particularly nitrogen and phosphorus, can promote algal blooms in aquatic systems. These blooms can deplete oxygen levels in the water, leading to hypoxic conditions that threaten fish and other aquatic organisms. Furthermore, the toxins produced by certain algal species can be harmful to both wildlife and humans (Khajvand et al., 2022).

Additionally, the introduction of untreated greywater into soil can result in contamination, affecting soil health and potentially entering the food chain through crops. This contamination can disrupt plant growth and reduce agricultural productivity. Finally, the overall impact on aquatic ecosystems can be profound. Disruption of these ecosystems can lead to a decline in biodiversity, affecting not only fish but also other organisms that rely on healthy water bodies for survival (Vuppaladadiyam et al., 2019).

11.3 IMPORTANCE OF DECENTRALIZED GREYWATER TREATMENT

Greywater can be a vector for pathogenic microorganisms, creating significant health risks when utilized without adequate treatment, with the severity of these risks contingent on its origin and method of application. Potential pathways for pathogen transmission include contact with contaminated hands, which can facilitate the spread of harmful organisms, and inhalation of aerosolized particles from irrigation spray, which can also introduce pathogens into the respiratory system. Moreover, exposure through broken skin increases susceptibility to infections, while pest vectors—such as rodents, insects, and even domestic pets—can further disseminate these pathogens. The risk assessment of greywater varies considerably based on its source; for instance, greywater from laundry is typically deemed low risk due to lower pathogen concentrations, whereas kitchen greywater, which often contains food residues and higher levels of contaminants, represents a high risk. Additionally, using greywater for hosing gardens is associated with significant health hazards, particularly when it comes to edible crops, while sub-surface irrigation systems and ornamental plants are generally considered low risk, highlighting the importance of careful management and treatment strategies to mitigate health concerns effectively (Boyjoo et al., 2013).

Vertical flow constructed wetlands (VFCWs) are engineered systems that utilize vegetation, soil, and microbial processes to effectively remove micropollutants, including personal care products and pharmaceuticals, from greywater. In these systems, water flows vertically through a substrate, allowing for enhanced contact with plant roots and microbial communities, which facilitate the degradation and assimilation of contaminants (Wichmann and Otterpohl, 2009).

Membrane bioreactors (MBRs) represent an advanced treatment technology that integrates biological treatment with membrane filtration. This system allows for the

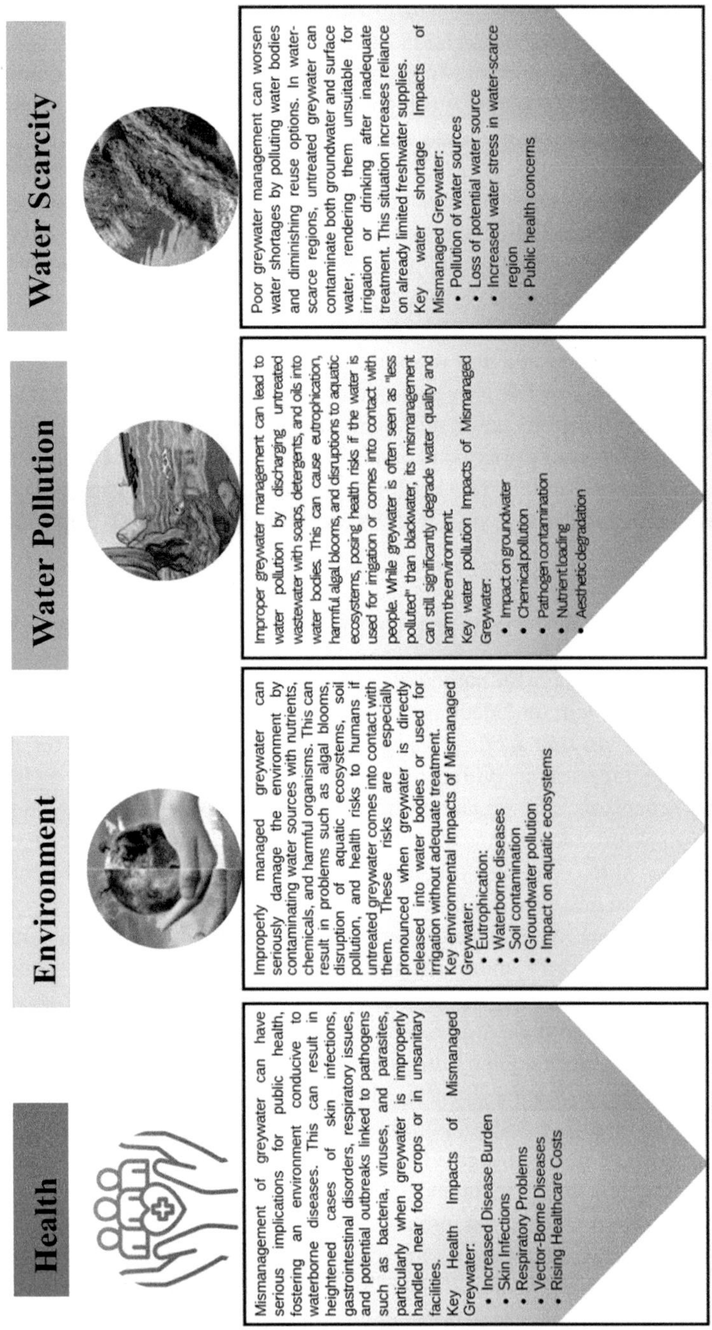

FIGURE 11.5 Key impacts of greywater.

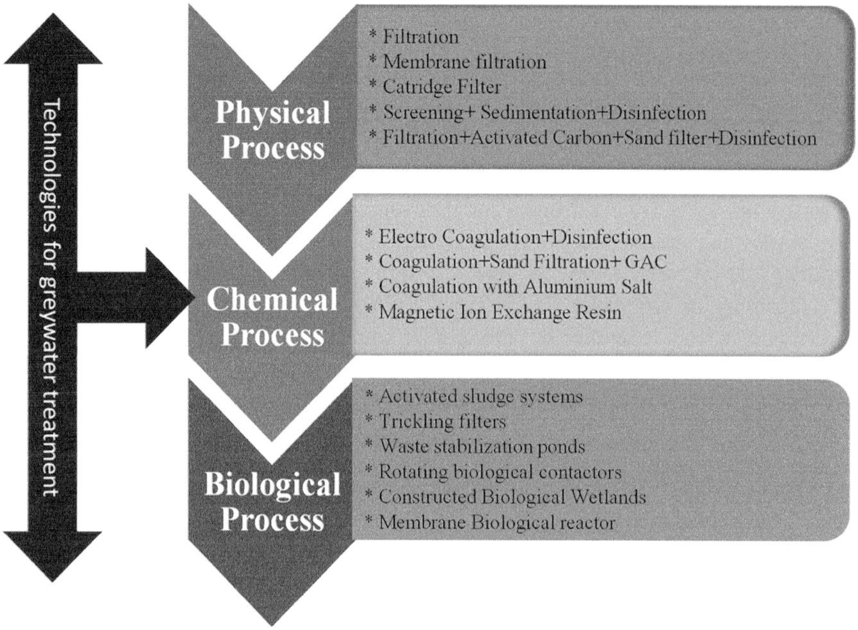

FIGURE 11.6 Key technologies for treatment of greywater.

treatment of greywater without the need for extensive cleaning and flushing of the input water, significantly reducing the overall water consumption and operational energy requirements. MBRs are particularly effective in producing high-quality effluent suitable for various reuse applications, as they effectively remove suspended solids, pathogens, and a wide range of organic compounds.

Constructed wetlands serve as a viable option for decentralized greywater treatment by mimicking natural wetland processes. These systems can be designed to treat varying volumes of greywater, utilizing a combination of physical, chemical, and biological processes to filter and decompose contaminants. They are particularly advantageous in rural or remote areas where conventional wastewater treatment infrastructure may be impractical (Abdel-Shafy et al., 2014).

Waste stabilization ponds are another method for decentralized greywater treatment, where natural processes are harnessed to treat wastewater. These shallow ponds utilize sunlight and microbial activity to promote the breakdown of organic matter, with the added benefit of sedimentation to remove solids. Waste stabilization ponds are low-maintenance and cost-effective, making them suitable for small communities or isolated developments (Boano et al., 2020).

Electrochemical technologies have emerged as innovative solutions for decentralized greywater treatment (Khajvand et al., 2022). These methods employ electrical currents to drive chemical reactions that can effectively degrade pollutants and pathogens in greywater. Electrochemical processes can enhance the removal of a wide array of contaminants, offering a compact and energy-efficient treatment option that can be integrated into various water management systems (Ansari et al., 2022).

11.4 RESOURCE RECOVERY FROM GREYWATER

Resource recovery from greywater involves the extraction of valuable resources such as water, nutrients (like nitrogen and phosphorus), and occasionally heat energy from wastewater produced by household activities such as bathing, dishwashing, and laundry. This process enables the reuse of greywater for applications like irrigation, toilet flushing, or preheating fresh water, thereby decreasing the reliance on fresh water sources and lessening environmental impact. Greywater recycling involves treating wastewater generated from appliances like showers, bathtubs, and sinks so that it can be reused within a property for non-potable applications, such as toilet flushing (Khajvand et al., 2022).

The primary focus of greywater recovery is the extraction of water, which can be reused for various non-potable applications. However, depending on the specific treatment methods employed, nutrients such as nitrogen and phosphorus can also be separated and repurposed as fertilizers for gardening. Additionally, in more sophisticated systems, it is possible to capture heat energy from the greywater. This captured heat can then be utilized to preheat incoming cold water, enhancing energy efficiency and reducing overall energy consumption in water heating processes.

11.4.1 Key Aspects of Greywater Resource Recovery

Water Reuse:

One of the key advantages of greywater management is the recovery of treated greywater for non-potable applications. This includes uses such as irrigation for landscaping, toilet flushing, and supporting industrial cooling systems. By implementing such practices, significant reductions in the demand for freshwater resources can be achieved, contributing to sustainable water management (Besson et al., 2021).

Nutrient Recovery:

Greywater is rich in essential nutrients, particularly nitrogen and phosphorus. These nutrients can be effectively extracted and repurposed as organic fertilizers for agricultural purposes. This process not only promotes plant growth but also diminishes reliance on synthetic fertilizers, fostering more environmentally friendly agricultural practices.

Heat Energy Recovery:

Advanced greywater treatment systems can incorporate heat exchangers to capture thermal energy from warm greywater. This recovered heat can be utilized to preheat incoming cold water for various domestic applications, such as washing and bathing. By enhancing the energy efficiency of water heating processes, this technique contributes to overall energy savings and reduces greenhouse gas emissions associated with heating water.

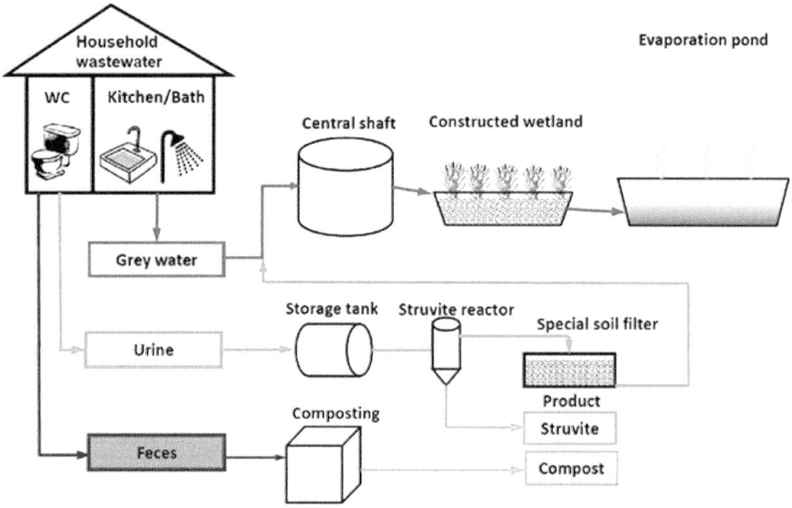

FIGURE 11.7 Development of self-sustaining wastewater treatment.

Water conservation through the reuse of treated greywater for non-potable purposes greatly decreases the consumption of freshwater. This practice not only alleviates pressure on natural water sources but also minimizes the volume of wastewater that needs to be processed at treatment facilities. Additionally, it can lead to cost savings by reducing dependence on municipal water supplies, ultimately lowering water bills for consumers (Diaz-Elsayed et al., 2019).

Ensuring strict quality standards is vital for the safe reuse of greywater, with regulatory bodies establishing comprehensive guidelines that specify acceptable contaminant levels essential for effective implementation. In Australia, frameworks such as the Australian Guidelines for Water Recycling, the "Code of Practice for the Reuse of Greywater" in Western Australia, and the "Urban Greywater Design and Installation Handbook" provide crucial guidance for these initiatives. Successful examples of greywater reuse can be seen in various households and commercial buildings that have effectively integrated greywater systems for applications like irrigation and toilet flushing, showcasing significant potential for water conservation and alleviating pressure on urban drainage systems. However, the adoption of greywater reuse faces challenges, including public misconceptions, technical treatment difficulties, and health concerns that must be addressed. While initial installation costs can be substantial, the long-term savings and environmental benefits often justify the investment, supported by government programs that offer tax incentives and grants to encourage adoption. Furthermore, advancements in monitoring and control technologies are essential for optimizing treatment processes, while community engagement and education play a crucial role in fostering public trust and acceptance of greywater reuse practices.

11.5 STRATEGIES FOR THE CIRCULAR ECONOMY OF WATER

Our contemporary lifestyles have increasingly become reliant on water, placing significant stress on local water resources that are often limited. As populations in urban areas continue to grow and climate change exacerbates water scarcity, the demand for water is projected to rise further. Meanwhile, existing linear water management systems consume substantial amounts of clean water and energy, and current wastewater treatment methods do not effectively recover valuable nutrients (de Marsily, 2007).

To address these challenges, a transition to a more circular water system is essential. This approach involves closing the loops in water usage to reduce both the extraction and pollution of local water bodies. Additionally, repurposing treated wastewater can provide a new source of water, easing the burden on local resources and lowering overall costs. Implementing a circular water system is particularly advantageous for urban environments, where the proximity of water reuse to its source can maximize efficiency and benefit.

The European Water Platform advocates for the development of a "water smart" society through a water management system based on three core principles: minimizing the societal impact on water resources; recognizing and communicating the intrinsic value of water for communities, economies, and ecosystems; and ensuring the long-term resilience, stability, sustainability, and security of water resources (Khajvand et al., 2022).

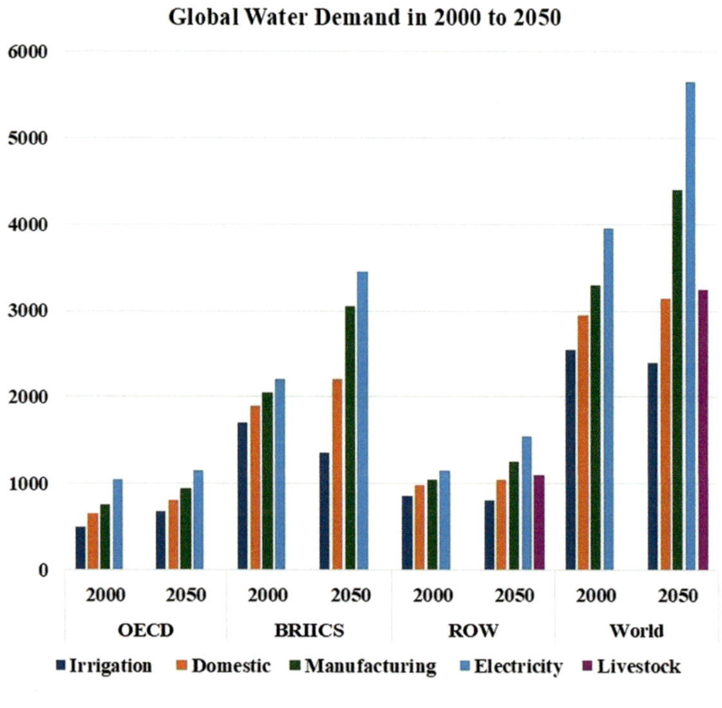

FIGURE 11.8 Global water demand in 2000 and 2050.

To promote a circular economy of water, several strategies can be employed. One key approach is reducing water consumption through efficiency measures, such as installing low-flow fixtures in residential and commercial settings, which can significantly lower water usage while maintaining essential functionality. Another critical strategy involves reusing treated wastewater for various applications, including irrigation for agriculture and landscape maintenance; for instance, California utilizes treated wastewater to irrigate golf courses and parks, thereby decreasing dependence on freshwater sources (Ribarova et al., 2024).

Cascading water use across industries is another effective method, where the output from one process is utilized as the input for another. This can be observed in industrial parks where water used in manufacturing is treated and repurposed for cooling systems in adjacent facilities. Additionally, recovering nutrients from wastewater is essential in a circular water economy; techniques like anaerobic digestion can extract valuable nutrients such as nitrogen and phosphorus, which can then be applied as fertilizers in agriculture, reducing reliance on synthetic alternatives. Rainwater harvesting is also a valuable strategy, enabling the collection and storage of rainwater for various uses, including irrigation and toilet flushing. Many urban areas implement rainwater harvesting systems to mitigate stormwater runoff and provide an alternative water supply. Furthermore, water-sensitive urban design encourages landscaping and infrastructure that manage stormwater in a natural manner, improving water quality and lessening the demand for treated water. For example, the incorporation of green roofs and permeable pavements in urban settings helps to absorb rainwater, reducing runoff and promoting groundwater recharge (Guldenaar, 2020).

To incorporate circular economy principles into the water sector, various strategies can be employed across all stages of water management. For example, circular design models can include the implementation of green infrastructure, such as green roofs, which reduce runoff into urban sewer systems and help alleviate the urban heat island effect. Establishing local water loops in settings like apartment complexes or industrial zones enables the recycling and reuse of water within those areas. When designing wastewater treatment facilities, it is crucial to incorporate water recycling capabilities from the outset, along with provisions for future advanced treatment technologies, ideally in conjunction with nature-based ecosystems.

Use and life extension models focus on minimizing drinking water consumption through effective demand management strategies. By adopting a fit-for-purpose approach, non-potable water can be utilized for applications such as cleaning, toilet flushing, and cooling systems, ensuring that the correct type of water is used for its intended purpose. Improving water efficiency is also essential, which includes minimizing leaks. Innovative technologies in water and crop management can further reduce agricultural water use. Water can be reused with minimal or no treatment for similar or different applications; for example, repurposing shower water for toilet flushing establishes a circular system within households. Additionally, rainwater harvesting can be employed for irrigation, and individuals may opt for service-based access to hot water rather than owning a boiler, with necessary upgrades included in the service (Bakker et al., 2016).

In terms of value recovery models, once water has been fully utilized, it can become a source of secondary materials. Recovering nitrogen and phosphorus from sewage

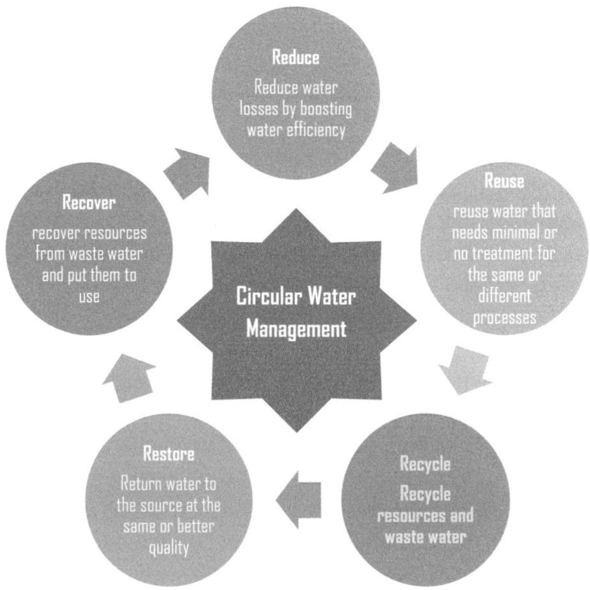

FIGURE 11.9 The 5rs for circular water management.

sludge can lead to the creation of high-quality fertilizers and ammonia. Moreover, biogas produced from the anaerobic digestion of sewage sludge can simultaneously generate renewable electricity and heat, enabling wastewater treatment facilities to achieve energy self-sufficiency while potentially supplying surplus energy to the grid (Delgado et al., 2024). Circular support models highlight the importance of information and communications technology (ICT) in providing valuable data on water usage, quality, and reuse. This information can be shared through platforms that connect various solutions and initiatives, enhancing transparency regarding water consumption among users within the same water basin. Access to this data facilitates predictive maintenance analysis and improves the detection of leaks and spills (Shangdiar, 2024).

11.6 CONCLUSION

- Greywater is relatively clean wastewater produced from household activities and has distinct characteristics that make it different from other types of wastewater. Understanding its makeup is vital for effective treatment and reuse, promoting resource recovery and sustainable water management.
- While greywater can be beneficial, improper management can lead to environmental and health risks. Contaminants may pollute soil and water, and pathogens can pose health threats. Therefore, strict treatment methods are necessary to ensure safe use.

- Decentralized systems for treating greywater provide a sustainable way to manage water resources. By processing greywater near its source, communities can lessen the burden on centralized systems, encourage local water reuse, and improve urban water management resilience.
- Recovering resources like nutrients and water from greywater is key to a circular economy. Techniques that retrieve these resources enhance sustainable water management and decrease the need for synthetic fertilizers and freshwater.
- Implementing circular economy strategies for greywater involves effective treatment, reuse, and resource recovery. By using innovative technologies and practices, communities can maximize greywater's benefits, reduce waste, and foster a sustainable water cycle that supports both the environment and public health.

REFERENCES

Abdel-Shafy, H. I., and Al-Sulaiman, A. M. (2014). Assessment of physico-chemical processes for treatment and reuse of greywater. *Egyptian Journal of Chemistry*, *57*(3), 215–231.

Al-Hamaiedeh, H. D. (2010). The impact of greywater reuse in irrigation on public health and safety. *Electronic Journal of Geotechnical Engineering*, *15*, 1131–1138.

Al-Ismaili, A. M., Ahmed, M., Al-Busaidi, A., Al-Adawi, S., Tandlich, R., and Al-Amri, M. (2017). Extended use of grey water for irrigating home gardens in an arid environment. *Environmental Science and Pollution Research*, *24*, 13650–13658.

Ansari, K., Shrikhande, A., Malik, M. A., Alahmadi, A. A., Alwetaishi, M., Alzaed, A. N., and Elbeltagi, A. (2022). Optimization and operational analysis of domestic greywater treatment by electrocoagulation filtration using response surface methodology. *Sustainability*, *14*(22), 15230.

Arden, S., and Ma, X. (2018). Constructed wetlands for greywater recycle and reuse: a review. *Science of the Total Environment*, *630*, 587–599.

Bakker, P., Nahon, C., Auguste, L., and Oosterveer, P. (2016). *Business guide to circular water management: spotlight on reduce, reuse and recycle*.

Benami, M., Gillor, O., and Gross, A. (2016). Potential health and environmental risks associated with onsite greywater reuse: a review. *Built Environment*, *42*(2), 212–229.

Besson, M., Berger, S., Tiruta-Barna, L., Paul, E., and Spérandio, M. (2021). Environmental assessment of urine, black and grey water separation for resource recovery in a new district compared to centralized wastewater resources recovery plant. *Journal of Cleaner Production*, *301*, 126868.

Boano, F., Caruso, A., Costamagna, E., Ridolfi, L., Fiore, S., Demichelis, F. et al. (2020). A review of nature-based solutions for greywater treatment: applications, hydraulic design, and environmental benefits. *Science of the Total Environment*, *711*, 134731.

Boyjoo, Y., Pareek, V. K., and Ang, M. (2013). A review of greywater characteristics and treatment processes. *Water Science and Technology*, *67*(7), 1403–1424.

Byrne, J., Dallas, S., Anda, M., and Ho, G. (2020). Quantifying the benefits of residential greywater reuse. *Water*, *12*(8), 2310.

Chaillou, K., Gérente, C., Andrès, Y., and Wolbert, D. (2011). Bathroom greywater characterization and potential treatments for reuse. *Water, Air, and Soil Pollution*, *215*, 31–42.

de Marsily, G. (2007). An overview of the world's water resources problems in 2050. *Ecohydrology & Hydrobiology*, *7*(2), 147–155.

Delgado, A., Rodriguez, D. J., Amadei, C. A., and Makino, M. (2024). Water in circular economy and resilience (WICER) framework. *Utilities Policy*, *87*, 101604.

Diaz-Elsayed, N., Rezaei, N., Guo, T., Mohebbi, S., and Zhang, Q. (2019). Wastewater-based resource recovery technologies across scale: a review. *Resources, Conservation and Recycling*, *145*, 94–112.

Edwin, G. A., Gopalsamy, P., and Muthu, N. (2014). Characterization of domestic gray water from point source to determine the potential for urban residential reuse: a short review. *Applied Water Science*, *4*, 39–49.

Gross, A., Azulai, N., Oron, G., Ronen, Z., Arnold, M., and Nejidat, A. (2005). Environmental impact and health risks associated with greywater irrigation: a case study. *Water Science and Technology*, *52*(8), 161–169.

Guldenaar, J. (2020). *Towards a city-level perspective on the stimulating and hindering factors for a successful circular economy transition: Two cases of Dutch frontrunner cities* (Master's thesis).

He, Z., Li, Y., and Qi, B. (2022). Recent insights into greywater treatment: a comprehensive review on characteristics, treatment technologies, and pollutant removal mechanisms. *Environmental Science and Pollution Research*, *29*(36), 54025–54044.

Khajvand, M., Mostafazadeh, A. K., Drogui, P., and Tyagi, R. D. (2022). Management of greywater: environmental impact, treatment, resource recovery, water recycling, and decentralization. *Water Science & Technology*, *86*(5), 909–937.

Khajvand, M., Mostafazadeh, A. K., Drogui, P., Tyagi, R. D., and Brien, E. (2022). Greywater characteristics, impacts, treatment, and reclamation using adsorption processes towards the circular economy. *Environmental Science and Pollution Research*, 29, 1–38.

Khanam, K., and Patidar, S. K. (2022). Greywater characteristics in developed and developing countries. *Materials Today: Proceedings*, *57*, 1494–1499.

Li, F., Wichmann, K., and Otterpohl, R. (2009). Review of the technological approaches for grey water treatment and reuses. *Science of the Total Environment*, *407*(11), 3439–3449.

Marjoram, C. (2014). *Graywater research findings at the residential level* (Master's thesis, Colorado State University).

Oteng-Peprah, M., Acheampong, M. A., and DeVries, N. K. (2018). Greywater characteristics, treatment systems, reuse strategies and user perception—a review. *Water, Air, and Soil Pollution*, *229*(8), 255.

Rakesh, S., Ramesh, P. T., Murugaragavan, R., Avudainayagam, S., and Karthikeyan, S. (2020). Characterization and treatment of grey water: a review. *IJCS*, *8*(1), 34–40.

Ribarova, I., Vasilaki, V., and Katsou, E. (2024). Review of linear and circular approaches to on-site domestic wastewater treatment: analysis of research achievements, trends and distance to target. *Journal of Environmental Management*, *367*, 121951.

Shaikh, I. N., and Ahammed, M. M. (2020). Quantity and quality characteristics of greywater: a review. *Journal of Environmental Management*, *261*, 110266.

Shangdiar, S. (2024). 12 Circular resource economy recovery on from industrial wastewater. *Circular Economy: Applications for Water Remediation*, *209* , 153–184.

Vuppaladadiyam, A. K., Merayo, N., Prinsen, P., Luque, R., Blanco, A., and Zhao, M. (2019). A review on greywater reuse: quality, risks, barriers and global scenarios. *Reviews in Environmental Science and Bio/Technology*, *18*, 77–99.

12 Circular Strategies for Effective Plastic Recycling

Shiven Shanmugananthan, Pratyush Pandey, Neelesh Mishra, and Nishkarsh Singhal

12.1 CIRCULAR ECONOMY AND PLASTICS

12.1.1 Introduction to Circular Economy

From what we have learnt so far, the circular economy is all about designing and redesigning products at the end of their life cycle to keep materials in use for as long as possible. It stands in complete contrast to the traditional linear economy model. In the linear model, raw materials are extracted, transformed into products, used by consumers, and then discarded as waste. This linear flow of resources is unsustainable as it depletes finite resources, generates massive amounts of waste, and contributes to environmental degradation, including pollution and climate change (Kirchherr et al., 2017).

Looking at plastic waste metrics from the past few decades (Figure 16.1), it is clear that unmonitored production and inefficient use of plastics in general have led to significant accumulation of plastic waste in landfills and the environment, causing severe ecological and health issues. This unsustainable practice has resulted in marine pollution, wildlife harm, and challenges in waste management (Figure 16.1), stressing the urgent need for improved regulations, innovation in recycling, and a shift towards a circular economy.

12.1.2 Contrasting with the Linear Economy

The basic difference between linear and circular economies is their approach towards resource allocation and waste management. The linear economy treats resources as a linear flow of materials with little regard for the environmental consequences or even future sustainability. It is built on the assumption of abundant resources and unlimited waste management, which very clearly could not be further from the truth. Circular economy, on the other hand, recognises the finite nature of resources and the need to minimise waste production. It views its resources as a valuable asset to be kept in circulation for as long as possible, creating a closed-loop system that minimises

DOI: 10.1201/9781003637677-12

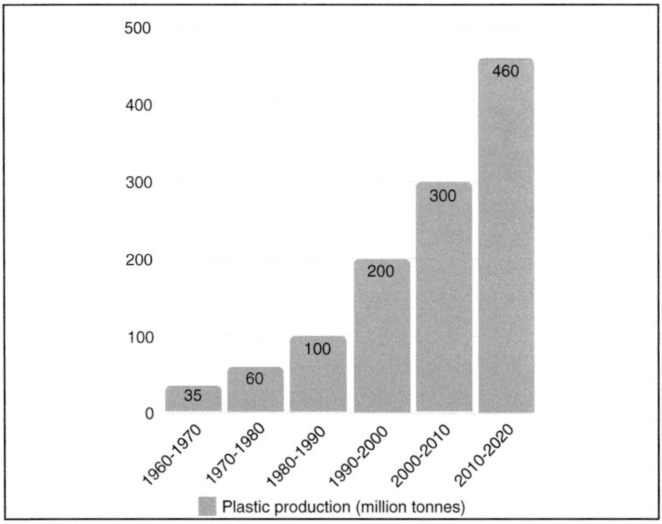

FIGURE 12.1 Global plastics productions.

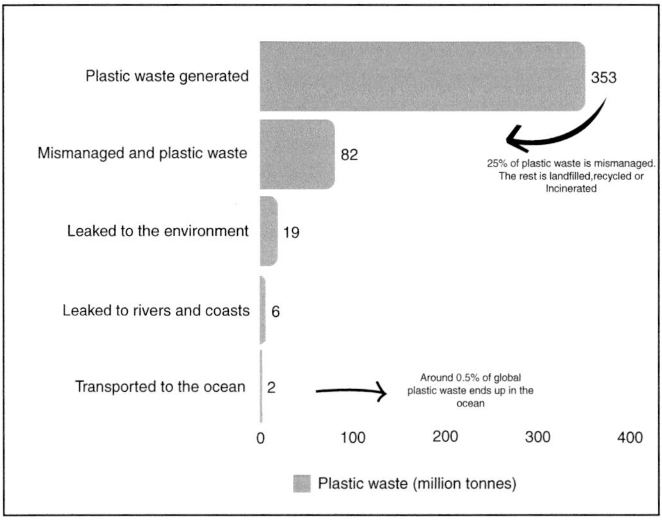

FIGURE 12.2 Pathway to global plastic waste to ocean (Ritchie, 2023).

environmental impact and maximises resource efficiency to ensure a bright present and an even brighter future (this is not a global warming joke). It requires collaboration across the entire supply chain, starting from raw material production, product designers and manufacturers to consumers and policymakers, and finally ending at scrap dealers and recyclers to create a system that is both economically viable and environmentally responsible (Korhonen et al., 2018).

12.1.3 Role of Plastics in Circular Economy

Plastics occupy a complex yet large position within the circular economy; their versatility and low cost have warranted their widespread use, but their durability and persistence in the environment have contributed to pollution and also warranted a fundamental shift in how we manage them (Ragaert et al., 2017).

The core principle is to move away from the linear model and towards a closed-loop system. To tackle plastic waste, a comprehensive strategy involves reducing and eliminating waste at the source, improving recycling systems, promoting reuse, and shifting to sustainable feedstocks. This begins with rethinking product design to prioritise durability, reuse, and minimal plastic content, while promoting alternatives such as reusable containers and bags to replace single-use plastics. Improving recycling involves standardizing plastic materials to simplify processing, investing in infrastructure to handle diverse plastic types, and advancing technologies like chemical recycling to recover raw materials. Extending the lifespan of plastic products includes designing for disassembly, encouraging business models based on refillable or reusable packaging, and supporting upcycling and creative reuse. Lastly, transitioning to sustainable feedstocks means developing bioplastics from renewable sources such as plants or algae and using captured carbon dioxide in plastic production, thus reducing reliance on fossil fuels.

12.1.4 IKEA's Circular Economy: Focusing on Plastics

IKEA's circular economy strategy prioritises minimising the environmental impact of plastics. Their multifaceted approach includes reducing virgin plastic use, designing for recyclability, exploring bioplastics, phasing out problematic plastics, collaborating across the value chain, and engaging consumers.

Reducing Virgin Plastic: IKEA is making an effort to reduce the amount of plastic that comes from virgin fossil fuels. Recycled PET (rPET) is being used more often in items like kitchen fronts (like the KUNGSBACKA series), and additional recycled plastic streams are being investigated for a range of uses. Their objective is to completely switch to recycled and renewable plastic sources (Al Saad et al., 2023).

Designing for Recyclability: IKEA is implementing design principles that make their plastic components easier to recycle. Material simplification involves reducing the number of different plastic types used in a single product to streamline sorting and recycling, avoiding complex composite materials that are difficult to separate. Complementing this, design for disassembly focuses on creating products that can be easily taken apart, allowing plastic components to be readily recovered for recycling.

Exploring Bioplastics: Research is underway where the potential of bioplastics derived from renewable resources, like sugarcane or corn, is being tested as alternatives to traditional plastics. While keeping in mind the challenges around bioplastics (e.g., land use and end-of-life management), they explore their use in specific applications where appropriate. They are also careful to ensure that any bioplastics they use are sustainably sourced and can be properly processed in existing recycling streams or composted where applicable.

Phasing Out Problematic Plastics: IKEA is actively phasing out certain types of plastics that are known to be problematic, such as PVC, which contains harmful additives and can be difficult to recycle. Efforts are also being put in to reduce or eliminate single-use plastic packaging where possible, opting instead for more sustainable alternatives like recycled cardboard.

Collaboration and Advocacy: IKEA understands that creating a truly circular economy for plastics requires collaboration. They actively participate in industry initiatives, work with recycling companies, and engage in policy advocacy to improve recycling infrastructure and promote sustainable plastic management practices. They recognise the need for standardised recycling systems and are working with others to advance these systems.

Consumer Engagement: IKEA recognises that consumer behaviour is crucial. They aim to provide clear information to customers about how to recycle or dispose of plastic components in their products. They are also exploring initiatives to encourage customers to return used IKEA products, including those with plastic parts, for recycling or reuse. They also encourage and promote start-ups which come up with innovations in packaging technology to work towards recyclability. It is important to acknowledge that IKEA's journey towards circularity with plastics is in progress. They acknowledge the complexities and challenges involved but are still committed and innovating to achieve improvement in this critical area [3].

Swedish Furniture Giant IKEA is making serious strides to work towards a circular economy by rethinking how they use plastics and maximising their efficiency. They have so far been successful in cutting down their use of virgin plastics and producing products that are sustainable and recyclable while also being user-friendly. Most importantly, they are bringing customers into the process, making it easier for them to understand and evaluate the need and ease of recycling by themselves. While IKEA has come a long way, their journey is far from over, and there are still challenges ahead.

12.2 SUSTAINABILITY ISSUES RELATED TO PLASTIC

Plastic's versatility and low cost have led to its unmonitored and inefficient use, but its persistence in the environment poses significant sustainability challenges. This subtopic examines key issues including resource depletion, pollution (land, water, and air), greenhouse gas emissions, impacts on ecosystems and biodiversity, and human health concerns related to microplastics and chemical additives. It highlights the urgent need for sustainable plastic management practices, including reducing consumption, improving recycling, and developing alternative materials (Zhang, 2024).

12.2.1 Environmental Impact

The current linear model that dominates the plastics industry presents several critical environmental challenges, primarily due to the non-degradability of plastics,

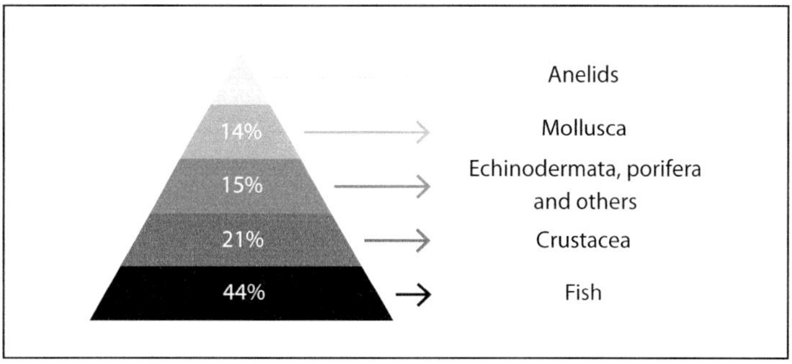

FIGURE 12.3 Microplastic distribution in water (Zhang et al., 2024).

which leads to their persistent accumulation in ecosystems across the globe. Plastic waste is now found even in remote areas, far from its origin, with approximately 70% ending up in landfills or the natural environment, particularly in oceans. This widespread pollution poses serious threats to wildlife and ecosystems, as plastics do not break down easily and persist for long periods. Marine pollution is especially severe—animals often become entangled in plastic debris or ingest it, mistaking it for food. This ingestion leads to the buildup of plastics and toxins within marine species, triggering a domino effect of microplastic accumulation through the food chain. Consequently, this pollution extends its impact to human health, as many fish consumed by humans have been found to contain alarming levels of plastic in their stomachs.

12.2.2 Resource Depletion and Emissions

The production of plastics is heavily reliant on fossil fuels, with approximately 4–8% of global oil production consumed both as raw materials and during the conversion process, contributing significantly to the depletion of natural resources and the emission of greenhouse gases in an already resource-constrained and carbon-intensive global economy. Moreover, plastic manufacturing is highly energy-intensive; studies indicate that it can require between 109.2 and 115.2 MJ per kilogram—an energy input nearly five times the calorific value of the resulting polymer—underscoring the unsustainable energy demands of the current plastic production system.

12.2.3 Waste Management Challenges

Due to the immense scale of plastic production and the corresponding generation of plastic waste, a significant portion is either landfilled or incinerated—both of which are inefficient and pose additional environmental risks. In regions like Europe, a large percentage of plastic waste is directed towards landfilling or incineration for energy recovery, while only a smaller fraction is actually recycled. One of the major challenges to efficient recycling is the complex heterogeneity of mixed plastic waste, which necessitates meticulous sorting to achieve acceptable recycling outcomes. For example, the separation of laminated flexible structures, such as those used in bottles

TABLE 12.1
Percentage of Plastic Waste Based on Management Process (Ritchie et al., 2023)

Country/Area	Recycled (Percentage) 2019	Incinerated (Percentage) 2019	Mismanaged (Percentage) 2019	Landfilled (Percentage) 2019
United States	4%	19%	4%	73%
India	13%	4%	46%	36%
China	13%	24%	27%	36%
Sub-Saharan Africa	6%	1%	64%	30%
Oceania	7%	12%	7%	75%
Middle East and North Africa	5%	1%	64%	30%
Europe	12%	38%	6%	44%
Asia (excl. China and India)	8%	19%	34%	39%
Americas (excl. USA)	10%	1%	29%	60%
Other	-	-	-	-
World	9%	19%	22%	49%

for cosmetic products like moisturisers and lotions, presents considerable difficulties due to their multi-material composition.

12.3 THE NEED FOR A CIRCULAR ECONOMY APPROACH

The current linear model is evidently unsustainable, as seen by the points stated above. The transition from a linear to a circular economy—where materials are kept in use for as long as possible—is essential for addressing the significant environmental and economic issues associated with plastic production and disposal, including pollution, harm to wildlife, human health risks, resource depletion, and waste management challenges. A circular economy promotes waste reduction, reuse, and recycling, offering a more sustainable approach to managing plastics. Central to this transition is the recognition and acceptance of waste plastic as a viable secondary raw material in industrial processes, enabling its integration into the plastic value chain. This shift is increasingly driven by growing awareness of the environmental impact of plastic waste and the economic pressures resulting from diminishing natural resources.

12.3.1 Health Concerns

12.3.1.1 Chemical Leaching

While it may seem so, plastics are not simple inert polymers. They usually contain complex mixtures of chemical additives, including plasticisers, stabilisers, pigments, and flame retardants. These chemicals can leach out of plastic products and into the environment, particularly under conditions of heat, stress, or prolonged use, which are very common in today's environmental conditions. This leaching can occur from food packaging, water bottles, toys, and even medical devices, exposing humans to a cocktail of potentially harmful substances (Chen et al., 2019).

- **Endocrine Disruption:** Many of these chemicals, such as BPA and phthalates, are known endocrine disruptors. They can mimic or interfere with the body's natural hormones, potentially leading to a range of health problems, including developmental issues, reproductive abnormalities, metabolic disorders, and an increased risk of certain cancers. The developing foetus and young children are particularly vulnerable to the effects of endocrine disruptors [6].
- **Other Health Impacts:** Research suggests that exposure to these chemicals may also be linked to other health issues, such as respiratory problems, neurological effects, and immune system dysfunction. The long-term health consequences of exposure to this complex mix of chemicals are **still being investigated**, but the potential for harm is a significant concern.

12.3.1.2 Microplastics: The Invisible Invasion Within

Microplastics are created when plastic is weathered into ever-tinier pieces as it makes its way through the processes of nature. These microscopic plastic particles, which are almost always unseen to the naked human eye, are now widely present in our surroundings and are contaminating the air, water, and soil. Humans can be exposed to microplastics through skin contact, inhalation, and ingestion (through carrier food and drink sourced from infected sources).

Rampant Exposure: Microplastics have been found in human blood, placenta, and even breast milk, indicating widespread exposure (Elbaz et al., 2024). The long-term health effects of microplastic accumulation in the body are still being studied, but early research suggests potential links to inflammation, oxidative stress, and damage

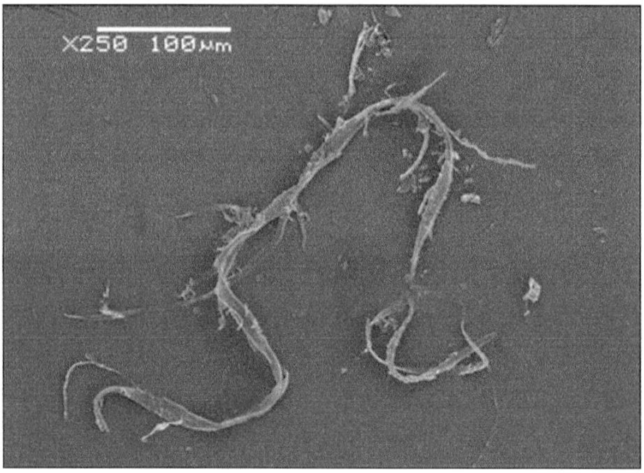

FIGURE 12.4 Microplastics on a microscale.

to cellular function. The FDA knows and acknowledges many studies on microplastics and some early studies on nanoplastics. But while there are many studies on such a topic, there exists a major barrier in the form of the ability to access medical risks.

12.3.1.3 The Vulnerable and the Unseen

It's important to acknowledge that certain populations may be more vulnerable to the health risks associated with plastics. Children, pregnant women, and individuals with pre-existing health conditions may be more susceptible to the effects of chemical leaching and microplastic exposure. Furthermore, communities living near plastic production facilities or waste disposal sites may face widely more intense exposure to harmful chemicals and pollutants.

12.4 ECONOMIC IMPACT

It is evident that the traditional linear economy leads to the generation of plastic waste at unsustainable levels, with disposal posing a serious challenge due to plastics' inherent durability and resistance to degradation. The costs associated with managing this waste can be broadly categorised into several areas. Collection and transportation of plastic waste from dispersed sources like households and businesses require significant logistical coordination, fuel, and labour, making it both complex and expensive. Sorting and processing add further challenges, as plastic waste often needs manual separation, and contamination with other waste streams can lower the value of recyclable materials while increasing processing costs. Landfilling plastic waste involves expenditures related to land use, site operation, and long-term monitoring, with these costs expected to rise as space becomes limited; additionally, environmental impacts such as emissions and soil contamination are often underestimated. Incineration, though it reduces landfill volume, contributes to air pollution and greenhouse gas emissions, creating regulatory and health-related costs. Finally, ocean cleanup represents one of the most daunting aspects of plastic waste management—cleaning up plastic from marine environments is not only technically complex and costly but also essential, as ocean health is critical to global ecological balance and human survival (Patel and Thakrar, 2024).

12.5 CIRCULAR ECONOMY APPROACH TO PLASTICS

Designing for recycling is a critical component of achieving a sustainable future for plastics, involving the selection of recyclable materials, simplification of product design, and reduction of additives to maximise recyclability. This approach also emphasises standardisation and disassembly to enhance recycling efficiency, conserve resources, minimise waste, and promote a circular economy. Key strategies include selecting materials that are abundant, easily procured, environmentally benign, and readily reusable or disposable at the end of a product's life cycle. Products should be designed for disassembly, allowing valuable components to be recovered and repurposed rather than ending up in landfills. Adopting a Cradle to Cradle (C2C) philosophy from the outset marks a shift away from the traditional "cradle to grave" model, fostering closed-loop systems that prioritise sustainability and resource efficiency. Additionally, designing products with shared parts bins can streamline manufacturing, reduce costs, simplify repairs, and extend product lifespans, all of which contribute to waste reduction. These design principles not only support environmental goals but also help minimise the economic burden associated with the accumulation and disposal of plastic waste, making design for recycling a foundational step in building a circular economy for plastics (Vlasopoulos et al., 2023).

12.6 POLICIES AND REGULATIONS FOR PLASTICS

Policies and regulations are essential drivers in the transition to a circular economy for plastics, providing the framework, incentives, and enforcement mechanisms needed to shift away from the traditional linear model towards more sustainable practices. One key policy tool is Extended Producer Responsibility (EPR), which places the responsibility for end-of-life product management on producers, incentivizing them to design for durability, recyclability, and reuse by bearing the costs of collection, sorting, and recycling through mechanisms like deposit refund schemes, take-back programmes, and producer responsibility organisations. Another significant approach involves bans and restrictions on single-use plastics—such as straws, bags, and cutlery—to curb unnecessary plastic consumption and promote reusable alternatives; however, the effectiveness of such bans is often limited by weak enforcement and lack of accountability. Notable policy initiatives include the European Union's Plastics Strategy, which aims to make all plastic packaging recyclable by 2030 through reduced single-use plastic consumption, increased use of recycled materials, and improved recycling infrastructure. On the global scale, international agreements like the Basel Convention regulate the transboundary movement of hazardous and plastic waste, reflecting a growing international commitment to coordinated plastic waste management and environmental protection (Fruergaard et al., 2011).

12.7 RECYCLING LABELS

Recycling labels have been around since the 1970s. The chasing arrows on plastics were the first of their kind; they were made to signify recyclability in general. Presently, there is a push for clearer and more accurate labelling due to the significant strides made in recycling technology. Some regions are moving away from the chasing arrows on plastics entirely and opting for simpler labels that clearly state whether an item is recyclable in that area.

12.7.1 Importance of Recycling Labels

Recycling labels are not merely informative; they are a critical component of a functioning circular economy for plastics. They bridge the gap between consumers, waste management systems, and recycling facilities. By providing clear instructions on material composition and recyclability, labels enable efficient sorting and processing of plastic waste. Clear and accurate recycling labels are essential in the context of a plastics circular economy. In the end, they guarantee that valuable plastic materials are recovered and reused, reducing waste and optimising resource efficiency. They also enable customers to make educated selections and streamline sorting at recycling facilities.

12.7.2 Types of Recycling Labels

12.7.2.1 Resin Identification Codes (RICs)

The "Chasing Arrows" with a Number: This is the most common symbol you see on plastics. It has three chasing arrows in a triangle with a number inside (1–7).

While many people think this means the item is recyclable, it actually only identifies the type of plastic resin.

The Numbers: Each number corresponds to a different type of plastic:

1 = PET (Polyethylene Terephthalate)—often used for bottles
2 = HDPE (High-Density Polyethylene)—often used for milk jugs, detergent bottles
3 = PVC (Polyvinyl Chloride)—less common in packaging, used in pipes, etc.
4 = LDPE (Low-Density Polyethylene)—often used for plastic bags
5 = PP (Polypropylene)—often used for yoghurt containers, bottle caps
6 = PS (Polystyrene)—often used for Styrofoam, disposable cups
7 = Others (a mix of various plastics that are hard to recycle)

12.7.2.2 How2Recycle Label

This label is becoming more common in the USA and is designed to be more straightforward and understandable by the consumers. It provides you with specific instructions on what to do with the product or the product packaging.

Currently, How2Recycle has implemented four major labels:

1. **Widely Recyclable:** Implemented on products that are easily recyclable and are 60% of America's recycling facilities and 50% of Canada's.

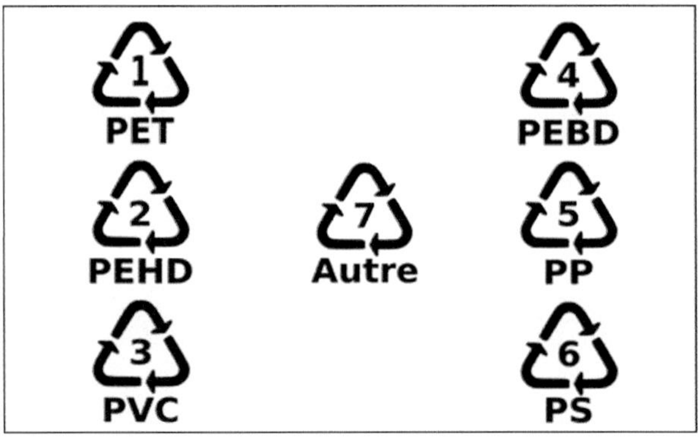

FIGURE 12.5 Chasing arrow labels.

2. **Check Locally:** Recyclability of the product is subject to the recyclable programme of the communities. The How2Recycle organisation is currently working to make the process of checking the recyclability of products with the said labels easier.
3. **Store Drop-offs:** Most curbside or drop-off programmes in the USA do not accept plastic bags, wraps, or films. Products with this label qualify for recycling through store drop-offs.
4. **Not Yet Recyclable:** Packages labelled "Not Yet Recyclable" are unfortunately not recyclable. Sorting, reprocessing, and end markets provide major hurdles, and less than 20% of Americans and Canadians are able to recycle this sort of product.

Includes Information on Different Parts: It may also specify if you need to separate different components (like a lid from a container) **[9]**.

12.7.2.3 Other Symbols

- **Mobius Loop:** This is the classic "chasing arrows" symbol without a number. It generally indicates that a product could be recycled but doesn't guarantee it.
- **Green Dot:** This symbol, found mainly in Europe, indicates that the manufacturer fulfils the obligations of the Packaging Act and takes responsibility for the recycling of the product or its packaging.
- **Compostable Symbols:** These symbols indicate that a product has the ability to breakdown and decompose in a compost bin (either industrial or home compost). It's generally good practice not to put compostable plastics in the regular recycling stream, as they can contaminate it.

One of the biggest challenges to a circular economy for plastics is the current hotch-potch, which is recycling regulations and labelling around the globe. This misunderstanding results in inefficiency, contamination, and eventually increased plastic waste. It is crucial to harmonise various systems and establish a strong standard, which will need coordinated international cooperation to bring about significant change.

12.8 TYPES OF PLASTIC RECYCLING METHODS

Plastic recycling involves several approaches, each contributing differently to a circular economy (Carriere and Horne, 2020). **Mechanical recycling** is the most widely used method, involving the collection, sorting, cleaning, and reprocessing of plastics without altering their chemical structure. While cost-effective and supported by established infrastructure, it faces limitations with mixed or degraded plastics. Innovations like cold fusion and electrostatic separation are being explored to improve its efficiency. **Chemical recycling** offers a promising alternative by breaking plastics down into monomers or chemical feedstocks through methods like depolymerisation, pyrolysis, and catalysis, enabling high-quality material recovery, though challenges remain in cost and scalability (Rathi et al., 2023). **Physical recycling** focuses on reshaping plastic waste into new products via processes like sorting (using NIR sensors, density separation), cleaning, shredding, and melting through techniques such as extrusion and injection moulding. It is effective but limited by material degradation over multiple cycles (Goodship, 2007). **Biological recycling**, still in early development, uses microorganisms and enzymes to degrade plastics into harmless by-products, with recent discoveries highlighting enzymes and insect-based methods for breaking down persistent plastics like polyethylene. Finally, **energy recovery** involves incinerating plastic waste to generate heat or electricity and is considered a last resort when material recovery is not feasible. Each method plays a role in reducing plastic waste and advancing towards a more sustainable, circular plastic economy (Agenda, 2016).

Energy recovery from plastic waste offers several advantages, including reducing the volume of plastic that ends up in landfills or the environment, generating alternative energy sources to lessen dependence on fossil fuels, and enhancing resource efficiency by extracting value from waste, thereby supporting a circular economy (Marczak, 2022). However, this method also presents notable challenges. If not properly managed, processes like incineration can release harmful pollutants, necessitating advanced technologies and strict regulations to mitigate environmental risks. Economic viability is another concern, as setting up energy recovery facilities requires significant investment, with feasibility depending on factors such as scale, technology, and market demand. Additionally, public perception can hinder adoption, as concerns about pollution and health impacts often lead to resistance; therefore, transparent communication and proven safety measures are essential to building public trust and acceptance (Arena et al., 2023).

12.9 CASE STUDY (A MEXICO-BASED START-UP PETGAS)

Petgas, a Mexican enterprise, uses pyrolysis, a thermodynamic process, to convert plastic waste into fuel, aiming to reduce plastic pollution (Zhu, 2023). In this process, plastics are heated in an oxygen-free environment to produce gasoline, diesel, kerosene, paraffin, and coke. Propane initiates the heating, but the pyrolysis-generated gas sustains the process. While fuel combustion releases carbon dioxide, Petgas claims that the overall environmental impact is lower than similar fuels due to a reduced sulfur content, with greenhouse gas emissions being 50% lower. Petgas's facility in Boca del Rio has a weekly capacity of processing 1.5 tons of plastic and producing 356 gallons of fuel, while its Cancún plant processes up to 8,800 pounds of plastic daily, generating over 630 gallons of high-octane gasoline. The company organises collection drives for plastic waste, including from nearby beaches, and encourages local communities to sell raw materials directly to them. Petgas currently supplies fuel to local fire departments and food delivery businesses and aims to contribute to a circular economy by transforming plastic waste into valuable resources. With plans to expand operations, the company seeks to partner with waste management entities and open additional facilities across Mexico, including a fully operational one in California.

12.10 RECYCLING OF PLASTICS—(MARKET SITUATION)

Plastics are in frequent use across a variety of industries, such as packaging, building and construction, and the automotive sector (Al-Salem et al., 2009). Global demand for plastics has been steadily rising over the past few decades. Plastic's distinctive benefits—lightweight quality, durability, chemical and thermal resistance, affordability, and manufacturing diversity—are the cause of this increase in demand. In the context of the escalating environmental problems, this drives government as well as people worldwide to seek sustainable solutions. However, because plastics last for a long time, there is a growing amount of plastic waste that ends up in landfills and the environment, which causes difficulties with waste management and has an effect on the environment. Compound annual growth rates (CAGRs) for the global plastics recycling market are projected to vary from 5.00% to 6.80% between 2018 and 2026, indicating a strong expansion. The market was valued at $34.8 billion in 2017, and some estimate that, at a 7.9% annual growth rate, it will reach $66.9 billion by 2025. This large expansion is likely to benefit all major market categories and resin types.

12.10.1 Global Recycled Plastics by End User

The key markets for recycled plastic are non-food contact packaging, food contact packaging, construction, and the automobile industry. However, packaging continues to be the leading category, accounting for 69% of the market. Population growth and the growing preference for lightweight and sustainable materials are expected to drive up demand in all sectors (Locock, 2017).

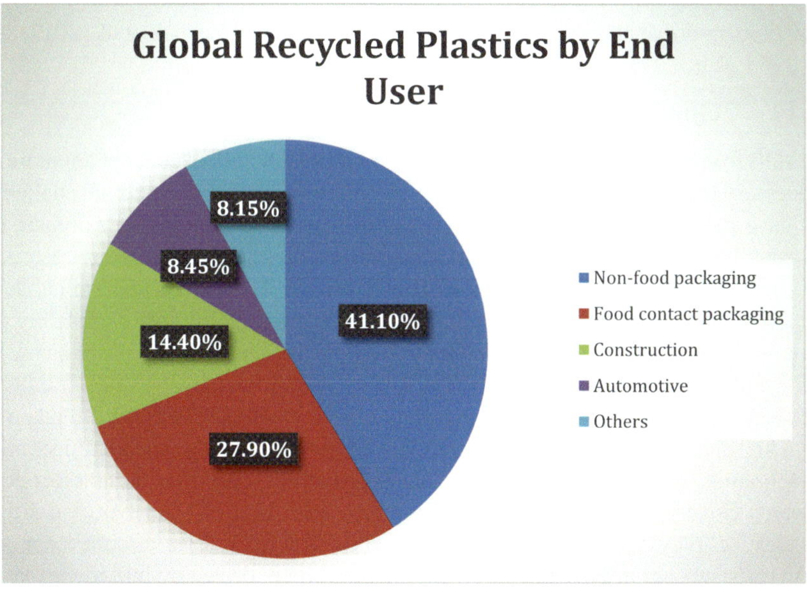

FIGURE 12.6 Global end-user in 2017 [15].

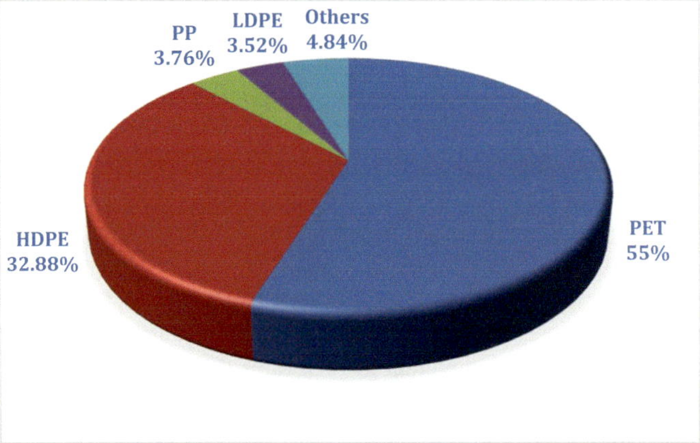

FIGURE 12.7 Global market by type in 2017 (Locock et al., 2017).

12.10.2 Global Recycled Plastics by Type

The global recycled plastics market consists mostly of four resin types: PET, HDPE, PP, and LDPE. Among these, PET has the biggest market share, accounting for 55% in 2017. PET is commonly utilised in the manufacture of plastic bottles and containers for food and consumer goods packaging. Government regulations and resource preservation efforts frequently provide subsidies and incentives to encourage the spread of PET recycling. Demand for recycled PET is predicted to expand in developing countries such as China and India as environmental awareness and the use of PET bottle packaging grows. Similarly, the demand for HDPE is expected to rise as living conditions improve and industries expand in these regions (Locock et al., 2017).

12.11 CONCLUSION

The shift from linear to circular economy is not just a theoretical concept anymore but a necessary transition to address the pressing environmental and economic challenges that have risen due to the overaccumulation of plastics in almost every domain and industry. The archaic linear model, which relies on the extraction of finite resources, mass production, and disposal, has evidently proven itself to be unsustainable. Its much more sustainable counterpart, the circular economy, while not completely perfect in its execution, offers a viable alternative by rethinking how we design, use, and manage plastics, aiming to keep materials in use for as long as possible. Companies like IKEA and Bajaj Motorcycles are a few of many companies that have taken the brave first steps in transitioning to a circular economy-based approach by innovating, collaborating, and pushing customer involvement to drive the circular economy forward. Government policies and regulations, such as EPR and bans on single-use plastics, play a crucial role in creating the framework for a circular economy. Germany's success in recycling and waste management, driven by initiatives like the Green Dot system and the Pfand system, serves as a model for other nations. However, achieving a global circular economy for plastics requires harmonised standards, improved recycling infrastructure, and widespread public awareness. Start-ups like Petgas accumulate waste plastic from beaches and encourage people to dispose of it in a specific location so that it can be easily processed for energy recovery. Since plastic is derived from crude oil, it has a higher calorific value, making it useful for producing heat, steam, and electricity. This is an efficient waste management technique that lessens reliance on landfills. Although there are many obstacles to overcome, action is necessary due to the urgency of the situation. However, a sustainable and effective plastic recycling strategy can be put into place by giving priority to the circular economy idea and incorporating energy recovery where needed. A fully circular system requires cooperation from consumers, legislators, and industries. A circular approach to plastic recycling, when properly implemented, can drastically minimise carbon footprints, reduce plastic pollution, and open the door to a more resource-efficient and sustainable future.

REFERENCES

Agenda, I. (2016, January). The new plastics economy rethinking the future of plastics. In *World economic forum* (Vol. *36*).

Al-Salem, S. M., Lettieri, P., and Baeyens, J. (2009). Recycling and recovery routes of plastic solid waste (PSW): A review. *Waste management*, *29*(10), 2625–2643.

Alzoubi, A., Al-Talib, M., and Al Saad, W. K. (2023). Sustainability as a competitive advantage: IKEA as a case study. In *13th International Conference on Industrial Engineering and Operations Management (IEOM)*, Manila, Philippines, March 7–9, 2023.

An, X., Wang, Y., Adnan, M., Li, W., and Zhang, Y. (2024). Natural factors of microplastics distribution and migration in water: A review. *Water*, *16*(11), 1595.

Arena, U., Parrillo, F., and Ardolino, F. (2023). An LCA answer to the mixed plastics waste dilemma: Energy recovery or chemical recycling?. *Waste Management*, *171*, 662–675.

Carriere, C., and Horne, R. B. (2020). The case for a legislated market in minimum recycled content for plastics. *Environmental Law Reporter*, *50*, 10042.

Chen, Q., Allgeier, A., Yin, D., and Hollert, H. (2019). Leaching of endocrine disrupting chemicals from marine microplastics and mesoplastics under common life stress conditions. *Environment international*, *130*, 104938.

Elbaz, J., Haslam, A., and Prasad, V. (2024). An empirical analysis of overall survival in drug approvals by the US FDA (2006–2023). *Cancer Medicine*, *13*(8), e7190.

Fruergaard, T., and Astrup, T. (2011). Optimal utilization of waste-to-energy in an LCA perspective. *Waste Management*, *31*(3), 572–582.

Goodship, V. (2007). Plastic recycling. *Science Progress*, *90*(4), 245–268.

Kirchherr, J., Reike, D., and Hekkert, M. (2017). Conceptualizing the circular economy: An analysis of 114 definitions. *Resources, Conservation and Recycling*, *127*, 221–232. https://doi.org/10.1016/j.resconrec.2017.09.005

Korhonen, J., Honkasalo, A., and Seppälä, J. (2018). Circular economy: The concept and its limitations. *Ecological Economics*, *143*, 37–46. https://doi.org/10.1016/j.ecolecon.2017.06.041

Locock, K., Deane, J., Kosior, E., Prabaharan, H., Skidmore, M., and Hutt, O. E. (2017). *The recycled plastics market: Global analysis and trends. CSIRO.*

Marczak, H. (2022). Energy inputs on the production of plastic products. *Journal of Ecological Engineering*, *23*(9), 9.

Patel, R. R., and Thakrar, N. (2024). *A comprehensive analysis of corporate social responsibility practices of selected companies of India.*

Ragaert, K., Delva, L., and Van Geem, K. (2017). Mechanical and chemical recycling of solid plastic waste. *Waste Management, 69*, 24–58. https://doi.org/10.1016/j.wasman.2017.07.044

Rathi, B. S., Kumar, P. S., and Rangasamy, G. (2023). A sustainable approach on thermal and catalytic conversion of waste plastics into fuels. *Fuel, 339*, 126977.

Ritchie, H. (2023). *How much plastic waste ends up in the ocean?*. Our World in Data.

Vlasopoulos, A., Malinauskaite, J., Żabnieńska-Góra, A., and Jouhara, H. (2023). Life cycle assessment of plastic waste and energy recovery. *Energy*, *277*, 127576.

Zhu, L., Zhu, J., Zuo, R., Xu, Q., Qian, Y., and Lihui, A. N. (2023). Identification of microplastics in human placenta using laser direct infrared spectroscopy. *Science of the Total Environment*, *856*, 159060.

13 Key Social Factors in Circular Economy

Raajasekar Mathiyalagan and Jayakrishna Kandasamy

13.1 INTRODUCTION

Waste management and the CE are the two interconnected ideas that support sustainable approaches to environmental sustainability are. The former tries to reduce environmental dangers, waste production, and disposal of waste through recycling, reuse, and safe disposal. In the meantime, the latter aims at tackling environmental, economic, and social issues by producing sustainable goods, cutting waste, and improving efficiency. Designing out waste and pollution, reusing resources and products, and recovering natural systems are all fundamental components of a CE. By following these guidelines, we may develop a closed-loop system for recycling and turning junk into new goods, which would lessen our impact on the environment, provide employment, and encourage environmentally responsible businesses. An essential component of the CE is waste management. By reducing trash and turning it into new products, it allows us to build a sustainable system. This in turn promotes the development of eco-friendly industries, creates jobs, and lessens our influence on the environment. The CE's three Rs reduce, reuse, and recycle are essential tenets that direct human behavior. We may successfully lessen our impact on the environment by using energy-efficient equipment and minimum packaging to reduce trash. Energy savings, job development, and lower raw material costs are just a few advantages of recycling. Two key ideas that support environmentally friendly approaches to sustainability are waste management and the CE. We can build a more sustainable future for present and future generations by embracing these ideas.

13.2 SOCIAL FACTORS

Social factors have a significant impact on a CE by influencing consumer behavior, employment markets, community involvement, and resource accessibility. These factors may have positive effects, such as the creation of fresh employment possibilities in repair and recycling, and negative effects, such as the displacement of jobs in some industries and worries about worker health and safety in recycling facilities. This emphasizes the need for careful planning and inclusive implementation approaches to maximize social benefits. The social aspects are listed out in Table 13.1.

DOI: 10.1201/9781003637677-13

TABLE 13.1
List of Social Aspects.

S.No	Social Aspects	S.No	Social Aspects
1.	Employment	15.	Security practices
2.	Labor/management relations	16.	Human rights mechanisms
3.	Occupational health and safety	17.	Social inclusion
4.	Training and education	18.	Social networks
5.	Diversity and equal opportunity	19.	Social cohesion
6.	Fair distribution of income	20.	Participation and local democracy
7.	Quality and well-being	21.	Anticorruption
8.	Investment	22.	Public policy
9.	Nondiscrimination	23.	Compliance
10.	Collection bargaining	24.	Supplier assessment for impacts on society
11.	Cultural traditions	25.	Customer health and safety
12.	Tourism and recreation	26.	Product and service labelling
13.	Local communities	27.	Marketing communications
14.	Child labor	28.	Costumer privacy

13.2.1 Employment

Jobs are generated by the CE in a number of industries, such as waste management, renewable energy, and repair. By generating demand for products and services, these jobs can boost the economy (Mies and Gold, 2021). People with different skill sets might find employment and also be able to access greater employment options by enhancing their skill sets through the CE. In addition to creating strands, observed climatic and technological changes also present some opportunities related to the development of green jobs (Sulich and Sołoducho-Pelc, 2022). Green jobs have a variety of advantages since they improve the environment in both established and emerging industries.

13.2.2 Labor/Management Relations

CE is anticipated to have a more skilled and diverse workforce with an emphasis on repair, refurbishment, remanufacturing, and design for longevity compared to a traditional linear economy. It also places more emphasis on worker well-being and sustainable practices all through the production cycle, which means that workers, businesses, and consumers will work together more to maximize resource usage and minimize waste (Galatti and Baruque-Ramos, 2022). Increasing the economy's circularity could generate jobs for people of all skill levels and have a beneficial net employment impact. The shift to a CE may have a range of implications on labor markets, such as the creation, replacement, loss, and redefinition of jobs (Luthin et al., 2023). While poor labor relations can result in strikes, increased accidents, disruptions in waste collection, and a negative public image, which ultimately affects the total efficacy of waste management services, positive labor relations in waste management can greatly improve operational effectiveness, worker morale, security measures, environmental sustainability, and society perception.

13.2.3 Occupational Health and Safety

The well-being of waste workers is greatly impacted by Occupational Health and Safety (OHS), which lowers the risk of accidents, illnesses, and injuries related to heavy lifting, exposure to hazardous materials, extreme weather, and poor ergonomics. This, in turn, leads to a healthier workforce and increased efficiency within waste management operations. Additionally, OHS protects the environment by ensuring that waste materials are handled and disposed of properly. Since product reuse, recycling, and remanufacturing can frequently involve novel and potentially hazardous processes and materials, OHS is essential to a CE. As such, proactive steps are needed to reduce risks and safeguard worker well-being throughout the product lifecycle (Héry and Malenfer, 2020). Working conditions and procedures are also expected to be changed by the CE.

13.2.4 Training and Education

Training people with the skills and knowledge they need to design, produce, and oversee products using a circular approach can help them to move towards CE. Encouraging a cultural shift toward sustainable consumption practices at all societal levels from consumers to business professionals, training and education are essential to the advancement of a waste management in a CE. Training in supply chain management, creating circular business models, and increasing awareness are important components. The best approach to facilitate the shift from a linear economy to a CE is through education, which should begin in preschool and continue through university (Tiippana-Usvasalo et al., 2023). Additionally, this enhances waste management practices by raising awareness of appropriate disposal techniques, encouraging recycling, and lowering waste generation, which results in a cleaner community and positive environmental impacts. This is primarily accomplished by improving knowledge, altering attitudes, and developing the skills required to manage waste at the individual and community levels. This is demonstrated by improved waste sorting, decreased landfill usage, and boosted resource recovery.

13.2.5 Diversity and Equal Opportunity

To create and promote circular practices, everyone irrespective of background or identity, has the opportunity to engage in the CE (Palm et al., 2024). Diversity and equal opportunity play a critical role in creating and executing circular practices. This leads to more equitable and sustainable solutions through a wider understanding of community needs, diverse perspectives, and creative ideas (Pansera et al., 2024). Community involvement, gender equality, and social effect are a few of the initiatives under diversity. Equal opportunities promote accessibility to recycling and garbage disposal services for underserved populations, foster greater community involvement, and improve knowledge of waste creation patterns.

13.2.6 Fair Distribution of Income

Achieving sustainability requires addressing potential inequalities arising from new jobs, markets, and access to circular products and services. A fair income distribution is crucial to an economy because it ensures that the benefits of resource reuse and recycling are distributed fairly throughout society, prevents vulnerable groups from being marginalized, and fosters social acceptance of the shift to a more sustainable economic model. Sharing of income is a useful incentive for boosting manufacturers' and merchants' earnings when the manufacturer markets products to final consumers. However, when merchants run their own promotions based on various demand function settings, they might not want to split the profits. This needs to change if CE needs to be implemented effectively (Tsao and Lee, 2020). Access to circular opportunities, social safety nets, and job creation and reskilling are key components of equitable income distribution.

13.2.7 Quality and Well-being

Maintaining quality and well-being throughout the life cycle leads to a number of benefits in the CE, including improved public health, a less environmental burden, and financial gains (Henkens et al., 2021). To guarantee that products are made and maintained to last longer, fewer hazardous materials must be used, and prioritize environmental sustainability and human health throughout their lifecycle, functionality and well-being perform a critical role in a CE, which ultimately contributes to a more comprehensive and advantageous economic system for the community as a whole (Padilla-Rivera et al., 2021). By lowering pollution, stopping the spread of disease, and enhancing general quality of life through appropriate waste collection, separation, and disposal methods, waste management quality and well-being have a substantial impact on environmental health. In other words, effective waste management results in cleaner air, water, and soil, all of which have a positive impact on human health and well-being.

13.2.8 Investment

One of the main factors influencing business cycles is investment. Investment allows investors to identify viable solutions promoting a CE that have been tested by companies, as well as best practices in each sector. Investment adds to the stock of capital, and the quantity of capital available to an economy is a crucial determinant of its productivity. Investment thus contributes to economic growth. Over the medium to long term, companies will be encouraged to go beyond marginal adjustments in their business models to implement more radical change (Lehmann et al., 2022). Businesses invest more during economic booms, which boosts growth. On the other hand, investment typically decreases during recessions, making them worse.

13.2.9 Nondiscrimination

For a supply chain to develop a sustainable, all-encompassing, and equitable model to provide systemic solutions, a gender-equal social, economic, and political framework is a prerequisite (Dumitrica et al., 2023). The importance of adopting a gender perspective in public policies that direct and shape the move towards circularity is highlighted by acknowledging the significance of structural determinants and comprehending how they affect and, to some extent, condition the potential outcomes of the change to a CE (Liu, 2024). Businesses can have access to a greater variety of viewpoints and talents, which will result in more creative solutions, by encouraging equal access to occupations, such as recycling facilities, repair services, and design teams (Rossi, 2022). By tackling challenges like poverty and lack of access to resources, involving those who are underrepresented in the CE can open up fresh economic possibilities in communities that might otherwise be left behind. By guaranteeing that managing waste and recycling services are equally available to all communities, nondiscriminatory laws help to avoid marginalized areas being disproportionately affected by environmental constraints.

13.2.10 Collection Bargaining

Through negotiating for improved working conditions, training opportunities related to CE initiatives, and protections against potential job losses due to changes in production processes aimed at reuse and recycling, collective bargaining plays an essential role in promoting waste management by ensuring workers' rights and fair treatment during the transition to more sustainable production practices. In other words, collective bargaining serves as an opportunity to ensure a "just transition" to a CE by addressing worker concerns and fostering collaboration between employers and employees (Johansson, 2021). Unions may advocate for businesses to offer training courses so that workers can learn new skills necessary for an effective waste management, such as how to disassemble, refurbish, or recycle products. In order to guarantee that workers participate in the collection and processing phases, unions may bargain for the inclusion of systems whereby businesses return used goods for recycling or reuse.

13.2.11 Child Labor

Child labor often being exploited in the informal recycling and waste collection sectors within the CE, where low-skilled, cheap labor is needed, perpetuating a cycle of poverty and hindering the true sustainability goals of a CE by undermining ethical practices and human rights (Barford and Ahmad, 2021). Child labor runs counter to the fundamental ideas of a CE, which maximizes resource reuse and minimizes waste in order to achieve sustainable growth. Businesses boost their output to keep up with the rising demand for a product or service. Businesses accomplish this by increasing their hiring and investment. The cycle is restarted with more workers and more money spent in the economy, which raises demand even more. The health effects of child labor in waste management are particularly concerning because children are more susceptible to these risks because of their developing bodies and smaller size. These health effects include respiratory problems, skin diseases, musculoskeletal problems, exposure to toxic chemicals, a higher likelihood of infections, psychological distress, and significant long-term health troubles due to exposure to hazardous waste materials, often without proper protective gear.

13.2.12 Security Practices

The integration of CE principles with cybersecurity measures in modern business practices has emerged as a critical strategy for sustainable development. This approach not only addresses environmental concerns but also ensures the resilience and security of business operations in the digital age (Upadhyay et al., 2021). The integrity and protection of the systems and procedures used in the CE should be guaranteed by cybersecurity. In this regard, companies, businesses, and corporations can improve cooperation, communication and the sharing of new ideas among stakeholders, a change in focus, the creation of digital, secure platforms that facilitate transparent transactions and boost productivity, and most importantly, innovation, with the aid of emerging technologies (Petelin, 2024). In the realm of cybersecurity, the concepts of a CE can also be applied to ensure the protection of important digital assets and data. However, much like physical resources, digital assets too need to be properly managed and maintained to reduce the risk of data theft and cyber-attacks.

13.2.13 Human Rights

A clean, healthy and sustainable environment is essential to human rights in an economic model that focuses on reducing, reusing, recycling, and recovering materials. Right to a healthy environment, right to information, right to participate in policy decisions, right to a just transition and right to decent work are some of the rights entitled to workers in a CE implemented society. People and the environment are at the center of economic decisions in a human rights economy (Setiawan and Wismayanti, 2023). It uses human rights commitments that governments have already made to ground all corporate, investment, budgetary, and economic decisions in respect to human rights.

13.2.14 Social Inclusion

Social inclusion is the process of enhancing the conditions under which people and groups participate in society; enhancing the capacity, opportunity, and dignity of those who are marginalized due to their identity (Souza Piao et al., 2023). Essentially, social inclusion means actively engaging diverse groups to prevent exacerbating existing social inequalities. It is essential because it ensures that everyone, including marginalized communities, has the chance to participate in and benefit from circular practices by creating equal access to jobs, resource recovery, and sustainable consumption. Social inclusion strategies can increase trash diversion rates and advance the concepts of the CE at the local level by actively including underserved populations in waste gathering and recycling initiatives. By considering the needs of a larger consumer base, incorporating input from various social groups during the product design process can result in more inclusive and sustainable features (Liu et al., 2023). When CE projects are planned with social inclusion in mind, they can give underserved areas access to new employment opportunities, fostering social mobility and economic empowerment. When actively participating in decision-making processes, local communities can play a crucial role in putting circular economy practices like community gardens, repair cafes, and sharing platforms into practice.

13.2.15 Social Networks

Social networks facilitate information sharing, raising awareness about sustainable practices, connecting potential collaborators within the CE ecosystem, enabling peer-to-peer sharing initiatives, and empowering consumers to make more informed choices towards circular consumption habits (Pusz et al., 2024). A larger audience can learn about sustainable consumption and waste minimization by using social media platforms to exchange information about CE activities, advantages, and difficulties. People with a shared interest can be connected through online platforms, allowing communities to be formed where members can exchange ideas and best practices and encourage one another to adopt circular behaviors. Social networks can impact consumer decisions and stimulate demand for more sustainable solutions by highlighting beneficial experiences with CE goods and services (Tsironis et al., 2022). Peer-to-peer lending, garment exchanges, and tool sharing programs are examples of sharing platforms that may be created and used with the help of social networks, enabling greater resource use and decreased consumption.

13.2.16 Social Cohesion

Social cohesion helps to form collaboration, community engagement, and shared responsibility across different stakeholders, which is essential for successfully implementing circular practices like recycling, reuse, and repair, ultimately leading to a more sustainable and inclusive economic system. It ensures that the benefits of a CE are distributed fairly and accessible to all members of society. People are inclined to actively engage in projects like repair cafés, sharing platforms, and neighborhood recycling programs when societies are cohesive because they feel a sense of shared responsibility (Schwanholz and Leipold, 2020). People are more likely to prioritize circular behaviors in their daily lives when there is strong social cohesion, which can create common principles around sustainability and preservation of resources (Bolger and Doyon, 2019). Socially cohesive collaborative settings facilitate the sharing of knowledge regarding waste management solutions, which in turn leads to the creation of new and enhanced circular practices. Communities can start and lead waste management initiatives that are suited to their particular requirements when they have strong social linkages, which increases local ownership and involvement.

13.2.17 Local Democracy

Local democracy enables local governments to implement policies that directly impact waste management, urban planning, public procurement, and community engagement, essentially creating the necessary infrastructure and incentives to facilitate the transition towards a circular system at the grassroots level. Key aspects are urban planning, waste management, and public procurement. Municipalities can put CE principles into practice by enacting waste management, planning for urban areas, and public procurement laws that encourage repair services, encourage reuse packaging, and require durable design (Bauwens et al., 2020). Through public discussions and awareness campaigns, local governments may inform citizens about the CE and promote involvement in neighborhood projects like product reuse and waste sorting (Bolger and Doyon, 2019). Local governments can stimulate demand for sustainable solutions by emphasizing circular options in public procurement and using their purchasing power to build markets for circular goods and services. In order to develop new CE solutions that are suited to local needs, local governments should encourage innovation by setting up forums for cooperation between companies, academic institutions, and community organizations.

13.2.18 Anticorruption

Anticorruption ensures fair and transparent practices in waste management, resource recovery, and recycling processes, preventing corrupt actors from manipulating regulations or contracts to exploit environmental vulnerabilities, thereby promoting sustainable practices and maximizing the benefits of a closed-loop system (Zhou and Li, 2021). Inappropriate waste disposal methods, such as unlawful dumping, are frequently brought on by corruption and impede recycling and resource recovery initiatives. A CE can be facilitated by anti-corruption measures that guarantee appropriate waste gathering and management procedures are in place. Businesses may be able to circumvent environmental laws pertaining to recycling and garbage disposal by bribing officials in areas where corruption is common. Robust anticorruption measures can guarantee adherence to these rules, encouraging environmentally friendly behavior (Liu and Lyu, 2024). An efficient waste management requires accurate material tracking throughout its lifecycle, which is made possible by a transparent market. Increased market transparency brought about by anticorruption initiatives can improve resource flow and recycling monitoring.

13.2.19 Public Policy

Public policy creates a regulatory framework that incentivizes businesses and consumers to adopt circular practices, including designing products for reuse and recycling, implementing extended producer responsibility, promoting green procurement, and facilitating innovation through financial incentives and collaboration initiatives (de Melo et al., 2022). Thus, ultimately driving the transition towards a more sustainable economic model. To promote adoption, governments can offer tax breaks to companies who use recycled materials or invest in CE technologies. Thus, creating a market for sustainable products by giving CE items priority in government procurement (Hartley et al., 2020). Strict waste disposal regulations and recycling rate goals should be put in place to encourage businesses to adopt circular practices. Standards can be upheld by establishing rules for product design that encourage durability, reusability, and simplicity of repair. Clear product labelling that identifies recycled content and makes material tracking easier over the course of a product's lifecycle can also be regulated (Niang et al., 2024). CE policies can help reduce pollution and greenhouse gas emissions by encouraging resource reuse and reducing waste production, generating new business prospects in product-as-a-service, recycling, and remanufacturing models that could result in the creation of jobs.

13.2.20 Compliance

Compliance ensures that companies adhere to environmental regulations and standards, thereby facilitating the sustainable use of resources, minimizing waste, and promoting practices like recycling, reuse, and repair (Bovea and Pérez-Belis, 2018). This ultimately drives the transition towards a more circular economic model. Companies are compelled to redesign products to satisfy requirements that require specific recycling or reuse rates, encouraging circularity from the very beginning of the design process (Woodard, 2021). Businesses must create new materials, methods, and strategies for recycling, refurbishing, and reusing products in order to abide by rules pertaining to the CE. This requires innovation in a variety of industries. By showcasing their dedication to sustainability and drawing in eco-aware customers, businesses who aggressively adhere to CE standards can obtain a competitive advantage. Companies may be compelled by compliance to assess their whole supply chain, pinpointing opportunities for waste reduction and resource efficiency enhancement. Labeling requirements are frequently included in regulations, which can inform customers about waste management principles and motivate them to make more environmentally friendly decisions.

13.2.21 Supplier Assessment for Impacts on Society

Supplier assessment evaluates a company's social impact by assessing their practices related to labor standards, waste management, resource usage, and community engagement, ensuring that the entire supply chain aligns with CE principles (Haleem et al., 2021). The assessment also minimizes negative societal impacts and promotes positive social outcomes. Ethical treatment of employees is ensured by evaluating working conditions, pay, hours, and adherence to labor rules. Assessing the supplier's community engagement activities, such as community partnerships, infrastructure upgrades, and skill development programs will give their role in CE (Handfield et al., 2002). In order to reduce their negative effects on the environment and source materials responsibly, suppliers might be evaluated based on how they manage the extraction and use of resources (Luthin et al., 2023). The environmental impact of the supplier's waste reduction plans, recycling procedures, and attempts to integrate recycled materials into their manufacturing processes can be assessed. The CE will be enhanced by evaluating whether suppliers place a high priority on product design that permits reuse, refurbishing, and recycling, prolonging product lifecycles, defining precise standards and measurements to regularly assess social impacts across various supplier categories and industries.

13.2.22 Cultural Traditions

Cultural traditions promote a CE by fostering practices like repairing, reusing, and sharing resources, which are core principles of a circular model, often due to historical values that prioritize resource conservation and sustainability within communities (Beaurain et al., 2023). CE techniques like product repair and refurbishment are easily adopted by cultures with a strong heritage of mending clothing or recycling old furniture. The design tenets of a circular economy, which place an emphasis on product longevity and durability, are more likely to be supported by cultures that place a high value on long-lasting goods and products. Concepts of the waste management, which seek to reduce waste and restore natural systems, may be more acceptable in cultures that have a close relationship with nature and place a high value on sustainability (Gonzalez et al., 2022). Implementing circular behaviors like recycling and reuse may be difficult in cultures where things are easily thrown away after a single usage.

13.2.23 Tourism and Recreation

The World Tourism Organization (UNWTO) issued a warning that tourism contributes significantly to global warming because of its high consumption, high pollution, and low utilization, as well as its large-scale tourist flows and activities (Xu et al., 2022). Given that visitors produce twice nearly as much waste as locals, tourism has a big influence on waste output. By reducing trash, energy use, and carbon emissions, tourism can lessen its environmental impact. Supporting local companies and reinvesting tourism earnings back into the community and combining tourism with CE fosters local economic prosperity. Tourism also improves the well-being of the community by protecting cultural assets and generating job opportunities. The CE can be enhanced by promoting the sale of locally produced trinkets manufactured from recycled materials. Tourism and recreation in CE provides an opportunity to minimize waste, reuse resources, and support local economies through practices like sustainable transportation, locally sourced food, and the refurbishment of existing infrastructure, ultimately contributing to a more environmentally friendly and socially responsible tourism model (Strippoli et al., 2024). It allows tourism businesses to operate in a "closed loop" where resources are continuously recycled and reused throughout the value chain.

13.2.24 Local Communities

Local communities can be advised to actively participate in practices like reuse, repair, recycling, and sharing, which helps minimize waste, conserve resources, and promote sustainable local economies by creating jobs and supporting small businesses within their communities (Bradley and Persson, 2022). They can also influence policy decisions at the local level to facilitate CE transition. Local initiatives for reuse and repair have the potential to create new jobs in the community, especially for small enterprises that specialize in maintenance and renovation. Residents can save money on goods and services by implementing community-driven circular practices, such as sharing systems. Engaging in local CE initiatives can promote social interaction and teamwork by fostering a sense of pride in the community and ownership (Ghinoi et al., 2020). Various community members can strengthen their bonds and address common issues by cooperating on sustainable projects. Wider adoption of the waste management practices can result from local communities exchanging best practices and information. The quantity of waste that ends up in landfills can be greatly decreased by community recycling and reuse initiatives.

13.2.25 Customer Health and Safety

Achieving a completely sustainable approach to product use and reuse, fostering consumer participation, and establishing confidence in a CE all depend on putting the health and safety of customers first (van Boerdonk et al., 2021). Customer health and safety ensures that products designed for reuse, repair, and recycling are safe for consumers throughout their lifecycle, including during the collection, processing, and re-manufacturing stages, thereby promoting trust and encouraging participation in circular practices. Product design, material selection and information transparency. When products are recycled or reused, nontoxic, eco-friendly materials that do not endanger human health must be employed. Product design must include safety elements that take into account potential risks during use, maintenance, and disassembly. Transparency and labeling will be enhanced by giving customers clear information about the materials used, possible safety issues, and how to take care of products (Ellsworth-Krebs et al., 2022). Product life cycle management will be enhanced by monitoring product history to spot possible safety concerns and guarantee appropriate upkeep and repairs over the course of the product's life.

13.2.26 Product and Service Labelling

Product and service labelling informs consumers about a product's environmental impact, durability, reparability, and recyclability, thereby encouraging them to make more sustainable purchasing decisions (Boyer et al., 2021) and making manufacturers to design products with a longer lifespan and greater potential for reuse and recycling, ultimately contributing to a closed-loop system. Labels enable consumers to choose products that support their sustainability objectives by clearly stating a product's recyclability, durability, reparability, and material composition (Gåvertsson et al., 2020). This increases demand for circular products. By emphasizing a product's anticipated lifespan, labels can persuade buyers to keep using it longer, which lowers waste production from premature disposal. Recycling processes can be made more efficient by clearly labeling appropriate recycling techniques, which can greatly lower contamination in recycling streams. Labels can increase consumer confidence in manufacturers' sustainability claims by offering verifiable details on a product's circularity attributes. Manufacturers are encouraged to create more environmentally friendly designs as well as supplies to satisfy consumer demand as consumers place a higher value on circular products, which spurs market innovation.

13.2.27 Marketing Communications

Marketing communications help in educating consumers about circular practices, promoting sustainable products and services, encouraging reuse and recycling. This leads to transition towards a more circular model by highlighting the environmental and economic benefits of circular consumption patterns. Marketing communications act as a bridge between companies and consumers to facilitate the adoption of CE principles throughout a product's lifecycle (Carbonell-Alcocer et al., 2025). Marketing can inform customers about the value of circular practices by outlining the fundamentals of a CE, including its advantages for the environment and the economy (Romero-Luis et al., 2022). Customers may be persuaded to select circular items over conventional ones by emphasizing their design attributes, such as robustness, reusability, and reparability (Aguilar-Morales et al., 2023). Consumer involvement in recycling and reuse initiatives can be increased by effective communication regarding take-back activities, including incentives and convenient drop-off locations. Alternative consumption models that support the waste management, such as sharing platforms, and subscription services, can be promoted through marketing. Open communication about the difficulties and advancements in waste management can foster consumer trust and motivate them to take an active role in the shift to a CE.

13.2.28 Costumer Privacy

Customer privacy fosters trust and encourages participation in circular practices, as it allows companies to collect and utilize data responsibly to optimize product lifecycles, enable efficient reuse and repair programs (Okazaki et al., 2020). Having assured privacy over their data also incentivize consumers to engage in sustainable choices while protecting their personal information. Customers are more inclined to engage in circular behaviors, such as product take-back programs, reselling used goods, and choosing product-as-a-service models, which depend on data collecting for effective management, when they believe their data is secure. Building consumer trust requires openness about data collection and use, which enables businesses to clearly explain the advantages of exchanging data in the framework of a CE (Saura et al., 2021). Based on usage trends and comments, organizations may optimize circular processes and create more environmentally friendly product designs using well-managed consumer data.

13.3 THEMATIC AREAS WITHIN CE

Thematic areas within CE includes labor practices, work place, society, product responsibility and green purchasing. The social aspects of CE are grouped together related to their thematic areas within CE for better understanding and purpose. This is essential in creating a new CE business model to manage the waste effectively.

13.3.1 Labor Practices

By creating high-quality jobs, addressing potential workforce inequality, and fostering social sustainability in addition to environmental benefits, good labor practices and "decent work" guarantees a just transition. By taking into account the human element involved in the process, they ultimately encourage broader adoption of CE principles. Waste collection, remanufacturing, product redesign and recycling are just a few of the industries where a CE may provide a sizable number of new jobs—but only if these positions are planned with respectable work standards in mind. Providing upskilling and reskilling training for employees is crucial to ensuring they can adjust to the new demands of the waste management, which calls for new knowledge and abilities.

13.3.2 Work Place

Implementing circular practices, reducing internal waste, optimizing resource reuse, and encouraging sustainable product design all take place in the workplace, which eventually helps create a business model that is more resource-efficient and ecologically friendly. Waste collection, recycling, remanufacturing, and product redesign are just a few of the industries where a CE may provide a sizable number of new jobs but only if these positions are planned with respectable work standards in mind. Maintaining respectable ethical standards helps to reduce the hazards of poor labor practices in recycling or waste management operations, which can have a detrimental effect on the environment.

13.3.3 Society

Community recycling facilities, repair cafés, and sharing platforms are examples of local projects that can encourage circular practices and a sense of shared accountability in communities. By supporting laws that encourage sustainable product design, encourage circular business models, and enhance waste management systems, citizens can have an impact on governmental decisions. Making the shift to a CE can lead to the creation of new jobs in sectors such as waste management, recycling, repair services, and product redesign. By giving underserved communities access to resources and employment possibilities, CE programs can strengthen them.

13.3.4 Product Responsibility

Manufacturers must be held responsible for how their products are managed at the end of their useful lives, which should motivate them to create recycling and take-back schemes. examining a product's environmental effects at every stage of its lifecycle, from the extraction of raw materials to disposal, in order to pinpoint areas that might use improvement. Innovation in business strategies, manufacturing techniques, and materials can be sparked by the demand for more environmentally friendly product design. Product responsibility reduces pollution and climate change by reducing waste production and resource usage.

13.3.5 Green Purchasing

Green purchasing encourages energy-efficient and ecologically conscious suppliers and gives preference to recycled or recyclable goods over conventional ones (Wijayasundara et al., 2022). By extending the life of materials to a second or possibly third life, this kind of purchase lowers waste. Furthermore, this approach not only supports the CE but also builds partnerships and business between organizations or entities that share similar environmental commitment standards. Buying green goods and services, especially those with more recycled content and better recyclability, could support the CE by increasing market scale and facilitating resource cycling in the closed-loop system.

13.4 SHARING ECONOMY AND SUBSCRIPTION MODELS

An economic model known as a "Sharing Economy" (SE) is one in which resources and things are cooperatively shared by people and groups, turning tangible assets into services. People and organizations can profit from unused resources thanks to the SE. By encouraging reduced consumption through resource sharing, the SE can have a positive impact on waste management by reducing trash formation because people need to buy less products and can make better use of what they already have, which helps create a more CE. Access to things through sharing platforms rather than private ownership reduces the need to produce new goods, which reduces production and disposal waste. The SE is a subset of the CE, and the two concepts are closely related. The sharing economy is based on sharing underutilized resources, while the CE focuses on maximizing the value of products and materials (Henry et al., 2021). The sharing economy is also focused on the use phase of a product's journey. However, unlike in a linear economy, products in a sharing economy are constantly in use to ensure maximum utilization of resources within a CE. It's about the sharing of resources by multiple people to allow existing resources to be used more fully, rather than letting them sit idly. In order for the sharing economy to be successful, peer-to-peer or community interaction needs to be accessible and products need to be designed for easy repair, upgrade, and disassembly. An example for SE is represented through a diagram in Figure 13.1, where the wastes are separated at the wastes collection bin rather than in a different place. A lot of labor work and resources are saved by the consumers and this also helps the circularity of the products.

13.4.1 Uber and the Sharing Economy

In just 10 years, Uber has grown to become the biggest taxi business in the world without possessing a single car or hiring a single driver. By connecting drivers directly

FIGURE 13.1 Shared waste management system.

with two individuals (the independent driver and the customer) via a digital platform, this services network for independent contractors has completely transformed the transportation industry. The economic phenomena known as "uberization," which has grown significantly in recent years, involves doing away with middlemen and digitizing services and administrative convenience.

13.4.2 Airbnb and the Sharing Economy

Airbnb is a cooperative platform that offers lodging from a person after they have made a reservation. The website gives its users the chance to learn about other cultures and share experiences in addition to facilitating financial transactions. In addition to upending the travel industry since its founding in 2008, Airbnb has also caused a stir in several cities' real estate markets.

13.5 THEORETICAL APPROACHES WITHIN CE

Theoretical approaches offer a framework for comprehending the intricate dynamics of CE practices by elucidating "why" certain CE practices occur and how to optimize them in various contexts. This framework helps researchers and practitioners identify important factors determining implementation, design policies, analyze possible effects, and guide the development of new circular solutions. It provides an in-depth comprehension of the mechanisms driving circularity. Considering elements

TABLE 13.2
Theories Related to Social Aspects in CE

Theory	The Overall Goal of Theory and Its Impact on Measuring Social Performance
Stakeholder theory	Stakeholder theory can be viewed as a tool for the CE application, extent, and caliber of social performance measurement
Socio-technical transition theory	The socio-technical component of CE case studies may be significantly altered by this theory
Network theory	Understanding how a network's location affects inter-organizational decision-making and social performance monitoring is made easier with the use of network theory
Gidden's structuration theory	Physical resource management can be enhanced by comprehending this relationship and the feedback that exists among social structures and human behavior. Therefore, clearly connecting the institutional forces behind material change with material elements would enable the CE projects to become more effective
Social economy theory	The social economy serves as a useful and informative model for the CE, boosting labor-intensive operations while improving the caliber and variety of human labor engaged in recycling and remanufacturing
Social integration and capital theory	By measuring the existence of these aspects and their relationship to CE, a combination of both theories may be utilized to comprehend and compare the importance of social traits in CE activities worldwide
Institutional theory	Understanding how various institutional contexts influence the social performance is measured and how it aids organizations in evaluating their adherence to institutional regulations is made easier by institutional theory
Resource-based theory	The resource-based view aids in the analysis of how businesses might use their skills and talents in social performance evaluation to gain a sustained competitive advantage

like financial incentives, environmental consciousness, social norms, and legal frameworks, theories can assist in explaining why people, organizations, or communities decide to implement particular circular practices. New circular economy models that consider the interdependencies among various stakeholders and system-level effects can be developed using theories as a basis. A list of theories related to social aspects in CE have been listed out in Table 13.2.

13.6 CONCLUSION

Higher employment opportunities, better local economies, better community engagement, improved resource utilization, and an eventual move towards more sustainable consumption are just a few of the positive social effects of a CE that can improve people's quality of life. However, potential issues like skills gaps and unequal access to CE initiatives must be carefully considered. A CE can boost local economies by encouraging recycling, repair, and renovation projects. This can lead to the creation of new jobs in a variety of industries. The feeling of ownership and involvement can be promoted via CE and waste management projects that require community involvement, such as neighborhood recycling programs or repair cafes. A CE may lessen disparities in access to necessities by encouraging the reuse and repair of long-lasting and reasonably priced products. A CE can lessen the strain on natural resources by lowering dependency on virgin materials, which will assist sustainability in the long run. Putting CE principles into effect could persuade customers to take part in product return programs, purchase durable goods, and repair rather than throw away items. Significant financial and environmental advantages come from effective waste management techniques, such as increased public health, reduced pollution, cost savings, resource conservation, and job creation. Governments and people may contribute to the implementation and upkeep of effective waste management systems, including individuals and corporations. Societies can shift to a better environmentally friendly waste management strategy by enacting laws that support it, increasing public awareness, and encouraging cooperation and collaborations. In addition to being essential for a healthier and cleaner environment, effective waste management is also a means of achieving a more sustainable and prosperous future.

REFERENCES

Aguilar-Morales, S. Y., Negrete-Cardoso, M., Rosano-Ortega, G., Sánchez-Ruíz, F. J., Sánchez-Baltasar, L. B., Vega-Lebrún, C. A., and Schabes-Retchkiman, P. S. (2023). Marketing strategies for waste recycling: A bibliometric analysis towards the circular economy. *Environmental Science and Pollution Research*, *30*(25), 67565–67581. https://doi.org/10.1007/s11356-023-27040-y

Barford, A., and Ahmad, S. R. (2021). A call for a socially restorative circular economy: Waste pickers in the recycled plastics supply chain. *Circular Economy and Sustainability*, *1*(2), 761–782. https://doi.org/10.1007/s43615-021-00056-7

Bauwens, T., Hekkert, M., and Kirchherr, J. (2020). Circular futures: What will they look like? *Ecological Economics*, *175*, 106703. https://doi.org/10.1016/j.ecolecon.2020.106703

Beaurain, C., Chembessi, C., and Rajaonson, J. (2023). Investigating the cultural dimension of circular economy: A pragmatist perspective. *Journal of Cleaner Production*, *417*, 138012. https://doi.org/10.1016/j.jclepro.2023.138012

Bolger, K., and Doyon, A. (2019). Circular cities: Exploring local government strategies to facilitate a circular economy. *European Planning Studies*, *27*(11), 2184–2205. https://doi.org/10.1080/09654313.2019.1642854

Bovea, M. D., and Pérez-Belis, V. (2018). Identifying design guidelines to meet the circular economy principles: A case study on electric and electronic equipment. *Journal of Environmental Management*, *228*, 483–494. https://doi.org/10.1016/j.jenvman.2018.08.014

Boyer, R. H. W., Hunka, A. D., Linder, M., Whalen, K. A., and Habibi, S. (2021). Product labels for the circular economy: Are customers willing to pay for circular? *Sustainable Production and Consumption*, *27*, 61–71. https://doi.org/10.1016/j.spc.2020.10.010

Bradley, K., and Persson, O. (2022). Community repair in the circular economy – Fixing more than stuff. *Local Environment*, *27*(10–11), 1321–1337. https://doi.org/10.1080/13549839.2022.2041580

Carbonell-Alcocer, A., Romero-Luis, J., and Gertrudix, M. (2025). The role of communication in environmental awareness according to circular economy stakeholders. *Journal of Environmental Management*, *374*, 124112. https://doi.org/10.1016/j.jenvman.2025.124112

de Melo, T. A. C., de Oliveira, M. A., de Sousa, S. R. G., Vieira, R. K., and Amaral, T. S. (2022). Circular economy public policies: A systematic literature review. *Procedia Computer Science*, *204*, 652–662. https://doi.org/10.1016/j.procs.2022.08.079

Dumitrica, C.-D., Grigorescu, A., and Davidescu, A. A. (2023). Circular economy a push and pull mechanism – Recycling starts before purchasing. *Journal of Cleaner Production*, *430*, 139363. https://doi.org/10.1016/j.jclepro.2023.139363

Ellsworth-Krebs, K., Rampen, C., Rogers, E., Dudley, L., and Wishart, L. (2022). Circular economy infrastructure: Why we need track and trace for reusable packaging. *Sustainable Production and Consumption*, *29*, 249–258. https://doi.org/10.1016/j.spc.2021.10.007

Galatti, L. G., and Baruque-Ramos, J. (2022). Circular economy indicators for measuring social innovation in the Brazilian textile and fashion industry. *Journal of Cleaner Production*, *363*, 132485. https://doi.org/10.1016/j.jclepro.2022.132485

Gåvertsson, I., Milios, L., and Dalhammar, C. (2020). Quality labelling for re-used ICT equipment to support consumer choice in the circular economy. *Journal of Consumer Policy*, *43*(2), 353–377. https://doi.org/10.1007/s10603-018-9397-9

Ghinoi, S., Silvestri, F., and Steiner, B. (2020). The role of local stakeholders in disseminating knowledge for supporting the circular economy: A network analysis approach. *Ecological Economics*, *169*, 106446. https://doi.org/10.1016/j.ecolecon.2019.106446

Gonzalez, C., Agrawal, V., Johansen, D., and Hooker, R. (2022). Green supply chain practices: The role of institutional pressure, market orientation, and managerial commitment. *Cleaner Logistics and Supply Chain*, *5*(July). https://doi.org/10.1016/j.clscn.2022.100067

Haleem, A., Khan, S., Luthra, S., Varshney, H., Alam, M., and Khan, M. I. (2021). Supplier evaluation in the context of circular economy: A forward step for resilient business and environment concern. *Business Strategy and the Environment*, *30*(4), 2119–2146. https://doi.org/10.1002/bse.2736

Handfield, R., Walton, S. V., Sroufe, R., and Melnyk, S. A. (2002). Applying environmental criteria to supplier assessment: A study in the application of the analytical hierarchy process. *European Journal of Operational Research*, *141*(1), 70–87. https://doi.org/10.1016/S0377-2217(01)00261-2

Hartley, K., van Santen, R., and Kirchherr, J. (2020). Policies for transitioning towards a circular economy: Expectations from the European Union (EU). *Resources, Conservation and Recycling*, *155*, 104634. https://doi.org/10.1016/j.resconrec.2019.104634

Henkens, B., Verleye, K., and Larivière, B. (2021). The smarter, the better?! Customer well-being, engagement, and perceptions in smart service systems. *International Journal of Research in Marketing*, *38*(2), 425–447. https://doi.org/10.1016/j.ijresmar.2020.09.006

Henry, M., Schraven, D., Bocken, N., Frenken, K., Hekkert, M., and Kirchherr, J. (2021). The battle of the buzzwords: A comparative review of the circular economy and the sharing economy concepts. *Environmental Innovation and Societal Transitions*, *38*, 1–21. https://doi.org/10.1016/j.eist.2020.10.008

Héry, M., and Malenfer, M. (2020). Development of a circular economy and evolution of working conditions and occupational risks—A strategic foresight study. *European Journal of Futures Research*, *8*(1), 8. https://doi.org/10.1186/s40309-020-00168-7

Johansson, N. (2021). Circular agreements—Exploring the role of agreements and deals as a political tool for a circular economy. *Circular Economy and Sustainability*, *1*(2), 499–505. https://doi.org/10.1007/s43615-021-00004-5

Lehmann, C., Cruz-Jesus, F., Oliveira, T., and Damásio, B. (2022). Leveraging the circular economy: Investment and innovation as drivers. *Journal of Cleaner Production*, *360*, 132146. https://doi.org/10.1016/j.jclepro.2022.132146

Liu, B., and Lyu, Y. (2024). Economic corruption, green recovery, and mineral trade relationships in emerging economies. *Resources Policy*, *90*, 104725. https://doi.org/10.1016/j.resourpol.2024.104725

Liu, K. (2024). Circular economy and the separated yet inseparable social dimension: Views from European circular city experts. *Sustainable Production and Consumption*, *51*, 474–483. https://doi.org/10.1016/j.spc.2024.09.016

Liu, Z., Schraven, D., de Jong, M., and Hertogh, M. (2023). The societal strength of transition: A critical review of the circular economy through the lens of inclusion. *International Journal of Sustainable Development & World Ecology*, 30(7), 826–849. https://doi.org/10.1080/13504509.2023.2208547

Luthin, A., Traverso, M., and Crawford, R. H. (2023). Assessing the social life cycle impacts of circular economy. *Journal of Cleaner Production*, *386*, 135725. https://doi.org/10.1016/j.jclepro.2022.135725

Mies, A., and Gold, S. (2021). Mapping the social dimension of the circular economy. *Journal of Cleaner Production*, *321*, 128960. https://doi.org/10.1016/j.jclepro.2021.128960

Niang, A., Bourdin, S., and Torre, A. (2024). The geography of circular economy: Job creation, territorial embeddedness and local public policies. *Journal of Environmental Planning and Management*, *67*(12), 2939–2954. https://doi.org/10.1080/09640568.2023.2210749

Okazaki, S., Eisend, M., Plangger, K., de Ruyter, K., and Grewal, D. (2020). Understanding the strategic consequences of customer privacy concerns: A meta-analytic review. *Journal of Retailing*, *96*(4), 458–473. https://doi.org/10.1016/j.jretai.2020.05.007

Padilla-Rivera, A., do Carmo, B. B. T., Arcese, G., and Merveille, N. (2021). Social circular economy indicators: Selection through fuzzy delphi method. *Sustainable Production and Consumption*, *26*, 101–110. https://doi.org/10.1016/j.spc.2020.09.015

Palm, J., Lazoroska, D., Valencia, M., Bocken, N., and Södergren, K. (2024). A gender perspective on the circular economy: A literature review and research agenda. *Journal of Industrial Ecology*, *28*(6), 1670–1683. https://doi.org/10.1111/jiec.13554

Pansera, M., Barca, S., Martinez Alvarez, B., Leonardi, E., D'Alisa, G., Meira, T., and Guillibert, P. (2024). Toward a just circular economy: Conceptualizing environmental labor and gender justice in circularity studies. *Sustainability: Science, Practice and Policy*, *20*(1). https://doi.org/10.1080/15487733.2024.2338592

Petelin, E. (2024). Security priorities in circular economy: A conceptual review. *Sustainable Production and Consumption*, *47*, 655–669. https://doi.org/10.1016/j.spc.2024.05.004

Pusz, M., Jonas, A. E. G., and Deutz, P. (2024). Knitting circular ties: Empowering networks for the social enterprise-led local development of an integrative circular economy. *Circular Economy and Sustainability*, *4*(1), 201–232. https://doi.org/10.1007/s43615-023-00271-4

Romero-Luis, J., Carbonell-Alcocer, A., Gertrudix, M., Gertrudis Casado, M. D. C., Giardullo, P., and Wuebben, D. (2022). Recommendations to improve communication effectiveness in social marketing campaigns: Boosting behavior change to foster a circular economy. *Cogent Social Sciences*, *8*(1). https://doi.org/10.1080/23311886.2022.2147265

Rossi, M. A. (2022). The advent of 5G and the non-discrimination principle. *Telecommunications Policy*, *46*(4), 102279. https://doi.org/10.1016/j.telpol.2021.102279

Saura, J. R., Ribeiro-Soriano, D., and Palacios-Marqués, D. (2021). From user-generated data to data-driven innovation: A research agenda to understand user privacy in digital markets. *International Journal of Information Management*, *60*, 102331. https://doi.org/10.1016/j.ijinfomgt.2021.102331

Schwanholz, J., and Leipold, S. (2020). Sharing for a circular economy? An analysis of digital sharing platforms' principles and business models. *Journal of Cleaner Production*, *269*, 122327. https://doi.org/10.1016/j.jclepro.2020.122327

Setiawan, H. H., and Wismayanti, Y. F. (2023). The green economy to support women's empowerment: Social work approach for climate change adaptation toward sustainability development. In Uday Chatterjee, *Climate Change, Community Response and Resilience* (pp. 225–240). Elsevier. https://doi.org/10.1016/B978-0-443-18707-0.00012-6

Souza Piao, R., de Vincenzi, T. B., da Silva, A. L. F., de Oliveira, M. C. C., Vazquez-Brust, D., and Monteiro Carvalho, M. (2023). How is the circular economy embracing social inclusion? *Journal of Cleaner Production*, *411*, 137340. https://doi.org/10.1016/j.jclepro.2023.137340

Strippoli, R., Gallucci, T., and Ingrao, C. (2024). Circular economy and sustainable development in the tourism sector – An overview of the truly-effective strategies and related benefits. *Heliyon*, *10*(17), e36801. https://doi.org/10.1016/j.heliyon.2024.e36801

Sulich, A., and Sołoducho-Pelc, L. (2022). The circular economy and the Green Jobs creation. *Environmental Science and Pollution Research*, *29*(10), 14231–14247. https://doi.org/10.1007/s11356-021-16562-y

Tiippana-Usvasalo, M., Pajunen, N., and Maria, H. (2023). The role of education in promoting circular economy. *International Journal of Sustainable Engineering*, *16*(1), 92–103. https://doi.org/10.1080/19397038.2023.2210592

Tsao, Y.-C., and Lee, P.-L. (2020). Employing revenue sharing strategies when confronted with uncertain and promotion-sensitive demand. *Computers & Industrial Engineering*, *139*, 106200. https://doi.org/10.1016/j.cie.2019.106200

Tsironis, G., Daglis, T., and Tsagarakis, K. P. (2022). Social media and EU companies' engagement in circular economy: A LinkedIn approach. *Sustainable Production and Consumption*, *32*, 802–816. https://doi.org/10.1016/j.spc.2022.06.006

Upadhyay, A., Mukhuty, S., Kumar, V., and Kazancoglu, Y. (2021). Blockchain technology and the circular economy: Implications for sustainability and social responsibility. *Journal of Cleaner Production*, *293*, 126130. https://doi.org/10.1016/j.jclepro.2021.126130

van Boerdonk, P. J. M., Krikke, H. R., and Lambrechts, W. (2021). New business models in circular economy: A multiple case study into touch points creating customer values in health care. *Journal of Cleaner Production*, *282*, 125375. https://doi.org/10.1016/j.jclepro.2020.125375

Wijayasundara, M., Polonsky, M., Noel, W., and Vocino, A. (2022). Green procurement for a circular economy: What influences purchasing of products with recycled material and recovered content by public sector organisations? *Journal of Cleaner Production*, *377*, 133917. https://doi.org/10.1016/j.jclepro.2022.133917

Woodard, R. (2021). Waste management in small and medium enterprises (SMEs): Compliance with duty of care and implications for the circular economy. *Journal of Cleaner Production*, *278*, 123770. https://doi.org/10.1016/j.jclepro.2020.123770

Xu, A., Wang, C., Tang, D., and Ye, W. (2022). Tourism circular economy: Identification and measurement of tourism industry ecologization. *Ecological Indicators*, *144*, 109476. https://doi.org/10.1016/j.ecolind.2022.109476

Zhou, A., and Li, J. (2021). Impact of anti-corruption and environmental regulation on the green development of China's manufacturing industry. *Sustainable Production and Consumption*, *27*, 1944–1960. https://doi.org/10.1016/j.spc.2021.04.031

14 Use of I4.0 Tools and Techniques in Handling Waste

Mahajan Gargi Ravindra, Rajyalakshmi G, Jayakrishna Kandasamy, and Jishnu Priyan G

14.1 INTRODUCTION TO I4.0 TOOL

The way garbage is gathered, processed, and recycled has significantly changed with the introduction of Industry 4.0 (I4.0) technologies. Conventional waste management techniques frequently have negative environmental effects, high operating costs, and inefficiency. Nonetheless, waste management may become a more data-driven, sustainable, and effective process with the use of cutting-edge technologies. Real-time waste level monitoring is made possible by IoT devices, such as sensors placed in trash cans. Waste management businesses may minimize needless collections and related fuel expenses by using this expertise to optimize collection schedules and routes. Organizations may save money and lessen their environmental impact by switching from a fixed collection schedule to a responsive system based on real garbage amounts (Sanghavi et al., 2019).

Big data analytics, in addition to IoT, is essential for enhancing waste management. Analysis of the enormous amounts of data produced by IoT devices can reveal trends in waste production. Municipalities may more efficiently allocate resources and plan for periods of high waste thanks to this data-driven strategy. Waste management tactics may be improved to increase operational effectiveness and better adapt to shifting waste trends by employing analytics. Recycling facilities are using AI more and more to improve and automate sorting procedures (Fatimah et al., 2020). The efficiency of recycling may be increased by using machine learning algorithms to precisely detect and classify different waste items. Reducing contamination and human error in recycling streams is crucial to increasing the number of recyclables that are kept out of landfills and reducing waste in general. Because cloud computing makes it possible to store and process waste data centrally, it also makes better waste management possible. Real-time information is accessible to stakeholders, such as recycling centres and garbage collectors, improving cooperation and communication. Those involved in waste management can make better decisions and work together more effectively because of this consolidated access. To improve operational efficiency, automated mobile robots, or AMRs, are being implemented in trash processing

DOI: 10.1201/9781003637677-14

settings (Das et al., 2019). AMRs lessen the need for physical labour in trash processing plants by moving items on their own, which lowers labour costs and enhances worker safety. Their capacity to independently traverse intricate surroundings enables more efficient operations and increased output.

Additionally, blockchain technology is becoming a useful tool for waste management since it offers an unchangeable, transparent record that follows garbage from production to disposal. By improving accountability and traceability, this technology guarantees adherence to environmental laws (Strazzullo et al., 2024). Blockchain promotes appropriate waste management practices and builds stakeholder confidence by recording each stage of the waste treatment method. There are several benefits to using I4.0 technologies in trash management. First off, the integration of IoT and data analytics significantly improves operational efficiency, resulting in fewer collections, more efficient trash collection procedures, and lower expenses. Automated sorting systems that enhance recycling procedures and guarantee that more materials are kept out of landfills lead to higher recycling rates (Abdulnour et al., 2022). Additionally, by streamlining routes and collection schedules, firms can achieve considerable cost savings.

Another important advantage is sustainability, as these technologies support recycling and reduce the amount of garbage sent to landfills, which is in line with international environmental objectives. Better resource management is also made possible by real-time data on trash creation, which enables businesses to divert items from disposal to recycling. Even with all of the advantages, there are several obstacles to overcome while using I4.0 technologies for trash management. Adoption of innovative technology by waste management businesses or smaller towns is hindered by high initial costs. As more networked systems are created, data privacy and security issues surface, calling for rigorous evaluation of data protection strategies (Marinelli et al., 2021).

The adoption process may also be made more difficult by the integration of new I4.0 technologies with old waste management systems. Organizations must implement training and development initiatives to close the skills gap caused by the need for competent workers with expertise in data analysis and technology (Mohamed et al., 2024). Lastly, because workers may be used to conventional waste management techniques, cultural opposition may potentially impede the implementation of new technologies. Waste management might undergo a substantial transformation with the application of Industry 4.0 technologies, which provide a route to increased operational effectiveness, sustainability, and efficiency (Figure 14.1).

Trash management operations can adapt to the problems presented by rising trash creation by integrating IoT, big data analytics, artificial intelligence, and automation (Javaid et al., 2022). The potential advantages of these technologies much exceed the difficulties, even though there are some barriers to overcome in their implementation. The more ecologically conscious and sustainable approach to trash management will be made possible by the effective application of I4.0 technologies in garbage handling, putting businesses in a better position to support global sustainability objectives and better serve their communities (Toth-Peter et al., 2023).

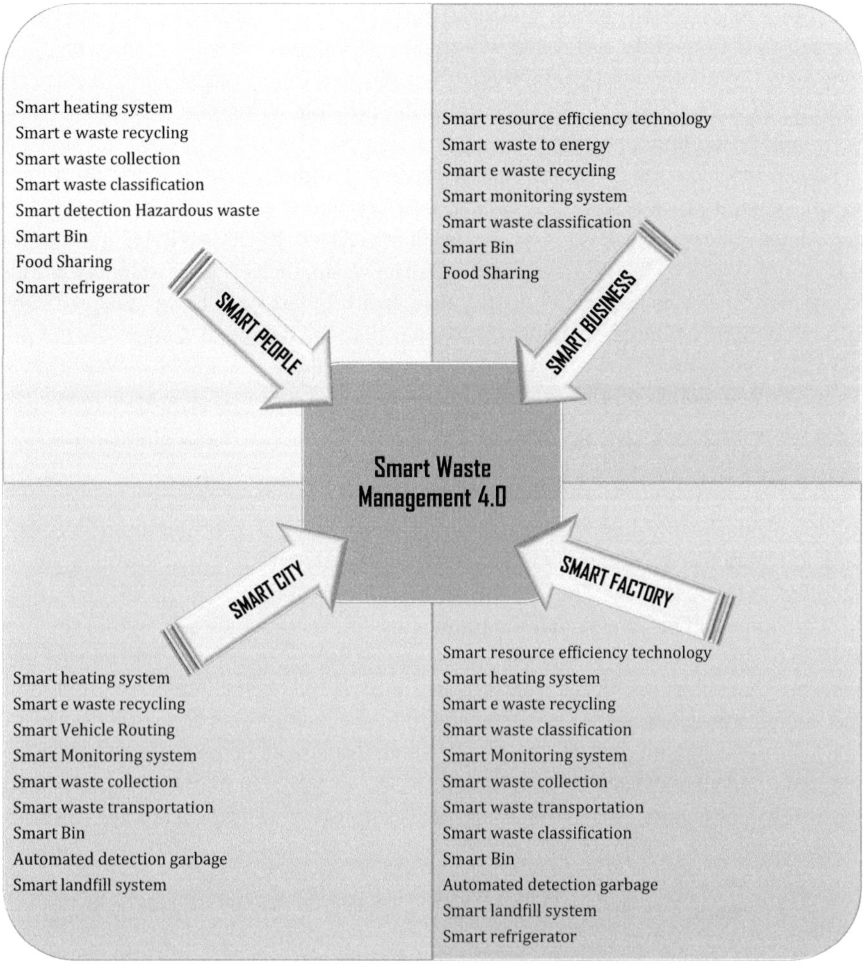

FIGURE 14.1 Smart Waste management 4.0.

14.2 IMPORTANT INDUSTRIAL SOLID WASTE

A typical result of contemporary industry, manufacturing, and other industrial operations is industrial waste. It is made up of a variety of materials, ranging from commonplace to dangerous, and handling it presents serious problems for both the environment and human health. To reduce, treat, and dispose of industrial waste effectively and eventually contribute to a more sustainable and healthier planet, it is essential to comprehend its nature, origins, and possible effects (da Rocha et al., 2022). Any undesired material—solid, liquid, or gaseous—produced by industrial processes is referred to as industrial waste. Since this broad term covers a wide range of chemicals, it is crucial to comprehend the many categories and the unique difficulties that each one poses. Industrial garbage is distinguished by its varied composition and frequently necessitates intricate procedures for safe management, in contrast to municipal waste, which mostly originates from homes. These wastes are often governed by certain laws because of the possible harm they may cause to the environment and human health.

Numerous industries, such as manufacturing, building, mining, agriculture, and energy generation, produce industrial waste. Industrial waste may or may not be hazardous, whereas hazardous waste is characterized by its toxicity, ignitability, corrosivity, or reactivity. While certain industrial waste, such as wastewater containing heavy metals, can be inert and not very dangerous, others need to be managed carefully. Usually, the source of the waste—rather than the degree of danger—is the primary differentiator.

14.2.1 Types of Solid Industrial Waste

14.2.1.1 Construction and Demolition Debris

Bricks, plasterboard, concrete, metal scraps (such as steel and aluminum), asphalt from streets and pavements, wooden beams, plastics, and roofing materials are all included in this category. The construction of new buildings, the demolition of old infrastructure and buildings, and the renovation of existing ones all result in the production of this waste. No matter the size of the project, it might generate a lot of garbage, which contributes to the overall waste stream. Inadequate handling of building and demolition debris might result in a significant landfill overflow. Soil and water can also get contaminated by the incorrect management of potentially hazardous materials, such as asbestos in older structures. A circular economy may be advanced, and this effect significantly reduced by recycling concrete and metals, for instance (Monferdini et al., 2024).

14.2.1.2 Manufacturing Byproducts

This type of waste comprises wasted packaging, plastic shavings, scrap metal, residual rubber, excess raw materials, and defective products. The manufacturing of consumer goods, electronics, textiles, and vehicles are just a few of the sectors that produce manufacturing byproducts. Every stage of the production process, including product assembling, shape, and cutting, can result in waste. If not managed appropriately, these byproducts might damage the environment and result in hazardous discharge (Asteriaa et al., 2018). Two advantages of introducing recycling processes in production include lowering the need for virgin resources and keeping a significant quantity of waste out of landfills.

14.2.1.3 Mining Waste

This includes overburden rock (material removed to reach ore), tailings (fine particles left behind after the extraction of valuable minerals), and sludge that may contain harmful compounds. The primary source of solid waste is mining for minerals, metals, and fossil fuels. The process of extracting resources usually results in large volumes of waste rock, which can disturb ecosystems. One of the primary ways that mining waste may pollute land and water is through acid mine drainage, which occurs when sulfide minerals are exposed to air and water. Appropriate waste management practices, such as controlling tailings and repairing impacted regions, are essential to lowering these worries (Asteriaa et al., 2018).

14.2.1.4 Ash and Slag

Slag, a byproduct of smelting metal ores, bottom ash, a heavy residue that sinks to the bottom of furnaces, and fly ash, which are light particles removed during combustion, are all included in this group. Ash and slag are mostly produced by burning fossil fuels in power plants, waste incinerators, and metal production operations. These leftovers may include heavy metals and other dangerous elements that, if handled poorly, might leak into the environment. Adding fly ash to concrete is a common recycling method that benefits the construction industry and environmental sustainability (Asteriaa et al., 2018).

14.2.1.5 Agricultural Trash

This category includes organic elements such agricultural leftovers, animal dung, leftover pesticides, food product plastic packaging, and trash from food processing.

Farming operations such as planting, harvesting, raising animals, and food processing generate this waste. For instance, substantial amounts of waste are produced by animal faces and leftover plant material during harvest. Soil erosion, water contamination from runoff (particularly from fertilizers and manure), and elevated greenhouse gas emissions (especially methane) can result from agricultural waste that is not properly handled. Composting and other similar practices can help turn organic waste into useful soil nutrients (Singh, 2023).

14.2.1.6 Commercial Waste

Paper and cardboard, plastic packaging, food waste, metal containers, and general office garbage—which may include electronics and furniture—make up this category of waste. This garbage, which is produced by offices, retail stores, restaurants, and enterprises during regular operations, might differ greatly depending on the type of business. The amount of garbage generated overall is greatly increased by commercial waste. Implementing waste minimization techniques can lower landfill contributions and encourage sustainable behaviours, since many components are recyclable (Chandrappa and Das, 2024).

14.2.1.7 E-Waste

This includes circuit boards, batteries, and other metal and plastic parts from electronics as well as abandoned electronic gadgets. This trash is produced when businesses and people replace outdated or malfunctioning electronics, such as computers, televisions, cellophanes, and appliances. Hazardous substances including lead, mercury, and cadmium are frequently found in e-waste; improper recycling can endanger human health and the environment. For the safe processing and recovery of important metals, specialized recycling systems are essential (Quan et al., 2024).

14.2.1.8 Hazardous Solid Waste

Solid wastes that contain hazardous compounds, such as chemical residues, heavy metal-laden items (like batteries), polluted soils, and specific industrial byproducts, fall under this category (Table 14.1). Numerous activities, such as the production of chemicals, industrial operations, building projects, and pharmaceutical or medical procedures, produce hazardous solid waste. The safety of the environment and human health are seriously threatened by this kind of trash. To reduce these dangers and safeguard ecosystems and populations, certain handling, storage, and disposal techniques are necessary.

14.3 GENERAL WASTE MANAGEMENT APPROACHES

Effective waste management is crucial in today's industrial setting for sustainability, regulatory compliance, and environmental preservation (Figure 14.2). Industrial solid waste encompasses a broad range of materials generated by construction, processing, and manufacturing activities. Businesses are focusing increasingly on implementing effective waste management programs that not only satisfy their trash disposal needs but also lessen their environmental impact as environmental issues get more attention.

TABLE 14.1
Annual Waste Generation

S. No	Name	Quantity (million tonnes per annum)	Source/Origin
1.	Steel and blast furnace	35.0	Conversion of pig iron to steel and manufacture of iron
2.	Brine mud	0.02	Caustic soda industry
3.	Copper slag	0.0164	By product from smelting of copper
4.	Fly ash	70.0	Coal-based thermal power plants
5.	Kiln dust	1.6	Cement plants
6.	Lime sludge	3.0	Sugar, paper, fertilizer tanneries, soda ash, calcium carbide industries
7.	Mica scraper waste	0.005	Mica mining areas
8.	Phosphogypsum	4.5	Phosphoric acid plant, ammonium phosphate
9.	Red mud/bauxite	3.0	Mining and extraction of alumina from Bauxite
10.	Coal washery dust	3.0	Coal mines
11.	Iron tailing	11.25	Iron ore
12.	Lime stone wastes	50.0	Lime stone quarry

Source: National Waste Management Council- Ministry of Environment & Forests-1990/1999.

FIGURE 14.2 Waste management cycle.

Numerous industrial solid waste management (SWM) solutions are examined in this comprehensive overview, along with their methodology, best practices, and significance.

Reducing the quantity and toxicity of trash generated at its source is the cornerstone of effective waste management. In order to use this proactive approach, industrial processes need to be thoroughly analysed and improved. Industrial process modifications can lead to significant waste savings. For instance, switching to continuous processing from batch processing might improve material efficiency and decrease waste. Manufacturers should regularly evaluate their production methods to identify areas for waste reduction and the integration of more effective processes and technologies. When industries replace hazardous and waste-intensive items with safer, more environmentally friendly ones, garbage production may be significantly reduced. To reduce waste and toxicity, for example, use water-based paints instead of solvent-based ones (Samreen et al., 2023).

Performing materials flow analysis can help organizations identify operational inefficiencies. Long-term waste reduction necessitates a culture of continuous improvement, which training initiatives that empower employees to conserve raw materials, optimize processes, and apply best practices will foster. When businesses use life cycle assessment (LCA), they may evaluate how their products affect the environment from birth to death. Businesses may reduce waste and advance sustainability by making educated decisions based on a breakdown of each step's inputs and outputs.

14.3.1 Reuse and Recycling

For businesses to reduce their environmental effect and derive value from waste products, reuse and recycling are essential components of sustainable industrial processes (Figure 14.3). Facilities can create systems to recycle waste products from their activities. To close the material loop, metal scrap from production, for example, can be melted and recast into new goods. Campaigns to raise awareness and provide training to employees support workplace recycling programs.

By forming alliances with recycling businesses, industry may redirect garbage to specialized processing facilities. Industries make sure that recyclables are sorted and processed effectively by working with regional recycling organizations. To monitor recycling rates, find new recycling prospects, and maximize these collaborations, regular audits of recycling streams are helpful. Industries can see their waste as a possible resource for other organizations thanks to the idea of byproduct synergy. Waste products may be recycled into raw resources by working with local companies, which reduces the amount of waste produced overall. To encourage inter-industry cooperation, waste heat from one industrial process, for instance, can be diverted to supply energy for nearby facilities. Industries may systematically sort and treat mixed waste streams by investing in MRFs. In order to maximize recycling rates and overall efficiency, these facilities use sophisticated sorting systems to collect valuable materials.

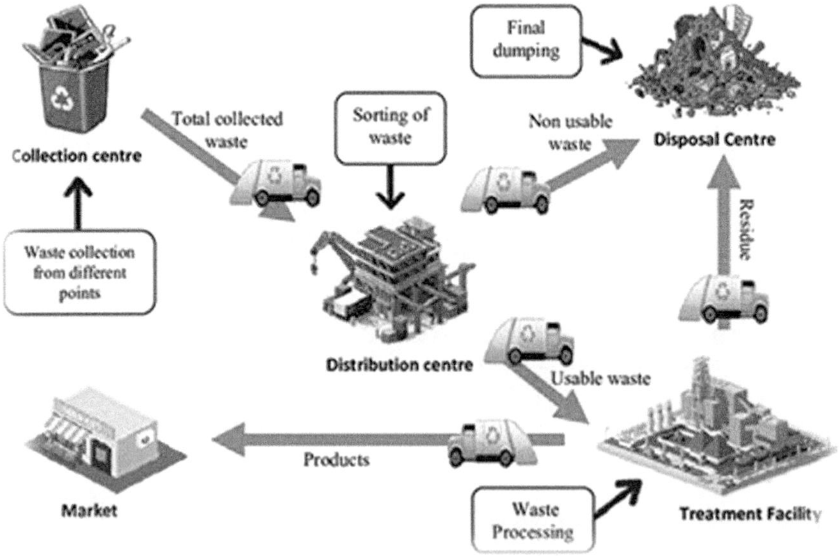

FIGURE 14.3 Solid waste management and transportation.

14.3.2 Composting

For organic wastes produced by industrial processes, especially in the food processing and agricultural industries, composting offers an efficient waste management solution (Figure 14.4). Natural microorganisms are used in the composting process to convert organic matter into compost that is high in nutrients. Providing composting systems with the ideal aeration, moisture, and temperature conditions to promote microbial activity can speed up this process. By using composting techniques, businesses may remove organic waste from landfills, lowering methane emissions and offering a more ecologically responsible way to dispose of garbage. To guarantee that food scraps, plant materials, and other organic waste are effectively collected for composting, industrial enterprises should set up extensive organic waste collecting procedures (Agarwal et al., 2015).

The compost that is produced may be used to agriculture, horticulture, and landscaping, improving soil quality and lowering the demand for artificial fertilizers. Businesses may establish a sustainable cycle that helps local agriculture and their waste management initiatives by integrating compost into their operations. Nowadays, a lot of businesses are spending money on on-site composting systems, which enable them to treat organic waste effectively and save money on transportation expenses related to off-site disposal. Employees may gain a greater grasp of sustainability practices by using an on-site composting system as a useful teaching tool.

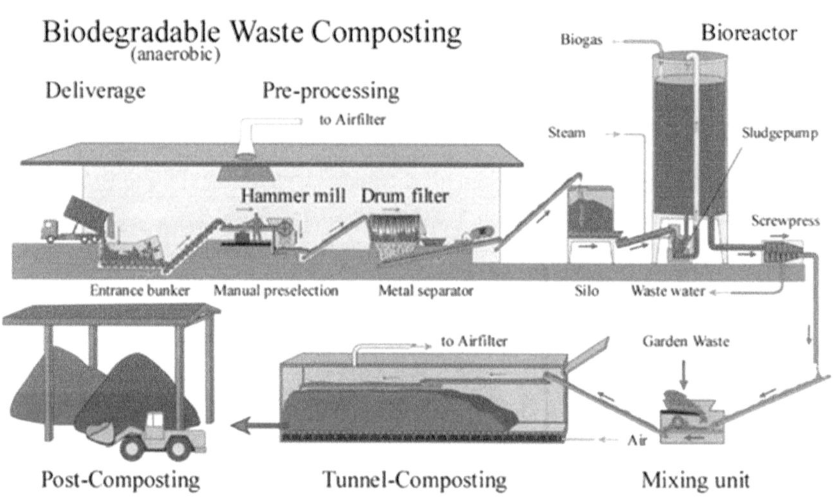

FIGURE 14.4 Biodegradable waste composition.

14.3.3 Landfilling

Although landfilling is frequently viewed as a last choice, it is nonetheless a crucial component of waste management for materials that cannot be recycled, reused, or treated. Selecting appropriate disposal sites is essential. Before a landfill is built, a number of aspects must be carefully considered, including as geological stability, groundwater accessibility, and ecological sensitivity. Additionally, to ensure the preservation of local resources, ongoing monitoring and control of water quality are required. Industries are required to abide with local, state, and federal rules governing landfill operations. By guaranteeing safe and responsible waste disposal, compliance lessens the risks landfilling brings to the environment and public health. Mechanisms for managing landfill gases—especially methane—and leachate—the liquid that drains or leaches from a landfill—must be in place.

Adequate containment and treatment technologies can lessen contamination of nearby water sources, even though biogas recovery systems can capture methane emissions for electricity generation, supporting energy recovery and sustainability projects. Regular monitoring of the disposal site is necessary to ensure environmental safety and compliance. In addition to conducting routine inspections to identify any issues that might jeopardize the environment, this should entail monitoring gas emissions, leachate outflows, and waste composition.

14.3.4 Incineration and Energy Recovery

As a waste management technique, incineration involves burning waste items to produce energy and reduce their volume. Advanced incinerator methods, which enable effective combustion and energy recovery, should be used by industries. High-efficiency incinerators maximize energy production while reducing hazardous emissions. One especially efficient technique for turning waste into useable energy is combined heat and power (CHP) systems. To trap and neutralize dangerous pollutants before they are discharged into the environment, modern incinerators are outfitted with several emission control equipment, including as scrubbers and electrostatic precipitators. For these systems to meet air quality requirements, regular maintenance and updates are essential (Joshi and Ahmed, 2016).

Businesses may lessen their dependency on conventional energy sources and promote sustainability by using incineration to recover energy from garbage. By using this energy for on-site operations, total energy costs may be decreased, and corporate sustainability initiatives can be supported.

Industries should interact with local communities to inform them of the advantages of contemporary incineration technology, such as its role in waste reduction and energy recovery, considering any public concerns about incineration (Marshall and Farahbakhsh, 2013).

14.3.5 Hazardous Waste Management

Because of the possible threats to the environment and human health, hazardous industrial waste management is vital. Hazardous waste must be correctly identified

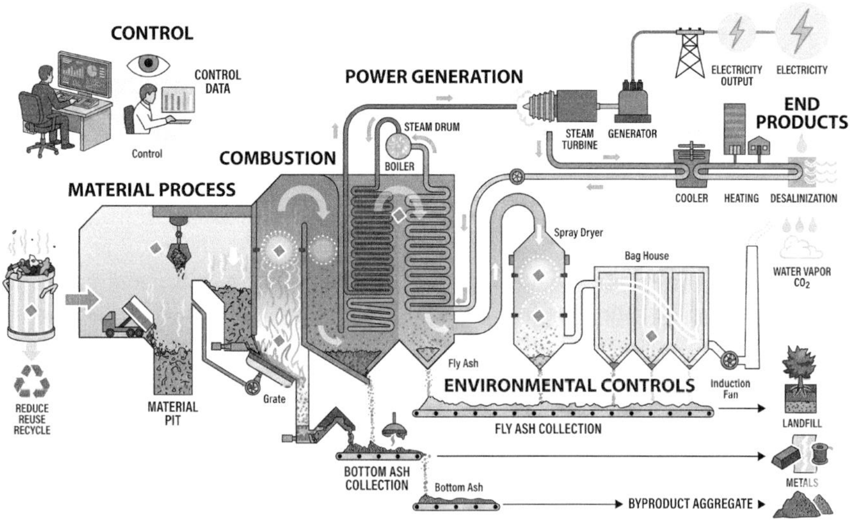

FIGURE 14.5 The rapid growth of energy creation through waste.

and categorized by industrial operators in accordance with legal requirements. By putting in place a strong waste monitoring system, hazardous items are handled and disposed of properly, lowering the risks of poor management. Hazardous waste needs to be handled in ecologically safe ways, which requires the use of specialized treatment and disposal facilities. To neutralize hazardous qualities, these facilities should employ techniques like heat treatment and stabilization (Demirbas, 2011).

Emergency response plans must be created and updated on a regular basis by enterprises in order to handle hazardous waste situations efficiently. Workers are more prepared to deal with spills, leaks, and explosions when they regularly participate in training and exercises. Respecting local and federal laws is essential when it comes to handling hazardous trash. Regular inspections and audits by regulatory bodies can assist guarantee adherence to these rules and pinpoint areas where hazardous waste management procedures need to be enhanced.

14.3.6 Regulatory Compliance

Promoting sustainable waste management techniques requires adherence to environmental standards. Businesses need to be aware of how waste creation, treatment, recycling, and disposal standards are changing. Enhancing comprehension and compliance can be achieved by funding employee training or expert consultation. Industries must correctly monitor and report their waste management procedures. Keeping accurate records shows an organization's dedication to sustainable practices and aids in regulatory compliance. Working proactively with regulatory agencies to advance and carry out sustainability programs builds cooperation and goodwill while coordinating waste management procedures with governmental and community goals (Shahbazi et al., 2013).

14.3.7 Employee Training and Involvement

For waste management initiatives to be successful, personnel at all levels must be involved. Regular training sessions should be held by organizations to teach staff members about sustainable practices, environmental effects, and appropriate trash disposal methods. A sense of individual accountability for trash development and management is fostered by such initiatives. Organizations might create incentive programs that recognize departments or people for creative ideas and successful waste reduction projects to increase involvement in waste reduction activities. Creating unambiguous avenues for staff members to offer input on waste management procedures aids in fostering a continuous improvement culture. Stronger dedication to sustainability goals and creative solutions might result from employee insights.

14.3.8 Partnerships and Collaboration

Working together with stakeholders and outside organizations is frequently necessary for effective waste management. Companies may exchange resources, best practices, and work together on multi-sector waste management projects by joining industry trade associations. Cooperation in advancing recycling programs, community education, and projects that strengthen SWM efforts results from interacting with local governments. Innovation in waste management techniques and technology may be promoted by collaborating with educational institutions and research groups. New materials, procedures, and technologies that improve sustainability can be found through collaborative research efforts (Deshpande et al., 2022).

14.3.9 Continuous Improvement and Monitoring

Sustaining efficient waste management procedures requires the adoption of a culture of continual improvement. Organizations may track the success of their waste management initiatives and pinpoint areas for development by establishing key performance indicators, or KPIs. Recycling rates, trash diversion rates, and regulatory compliance are examples of common indicators. Regular waste audits and evaluations help identify inefficiencies and facilitate the implementation of corrective actions. Employee participation in waste reduction initiatives, trash creation, and disposal techniques should all be assessed in these assessments. Because waste management is a dynamic field, firms must continue to be flexible. Companies may innovate their operations and stay in compliance with changing standards by keeping up with industry trends, regulatory changes, and emerging technology (Rajakumaran et al., 2023).

A successful industrial SWM plan uses a variety of techniques, all of which are designed to maximize waste reduction, advance sustainability, and satisfy legal requirements. Industries must employ a variety of strategies that complement one another in order to reduce their environmental impact, from proactive waste minimization and creative recycling initiatives to conscientious landfilling and hazardous waste management. By involving staff members at all levels, building solid relationships with external stakeholders, and making a commitment to ongoing development, organizations may cultivate an internal culture of sustainability (Zailan

TABLE 14.2
Waste Management Approaches (Narayanswamy et al., 2022)

Method	Description	Advantages	Disadvantages
Segregation and recycling	Segregation and recycling, separating recyclable materials from total waste	It lowers the fraction of waste being landfilled and conserves resources	It requires good training and established systems to perform
Source reduction	The prevention of waste by minimization at the source	Lowers volume of wastes and costs	Initial investment required for materials
Composting	Conversion of organic waste to fertilizer	Reduces landfill amount and nurtures soil	Limited application to organic types of waste only
Plasma gasification	Waste is heated to produce energy	It gives a zero-waste solution and produces usable by-products	High installation cost with the complexity of technology
Landfills	Burying nonrecyclable wastes in some designated areas	Simple, inexpensive to dispose of	Environmental concern, limited spaces available

et al., 2022). By adhering strictly to rules and prioritizing cutting-edge technology, enterprises may improve their waste management procedures while also supporting more general environmental sustainability objectives. In the end, including efficient waste management into business strategy helps sectors function more ethically, cut expenses, and advance a sustainable future for all. Industries may set the standard for a cleaner, greener world by promoting an inclusive and comprehensive approach to waste management (Table 14.2).

14.4 MACHINE LEARNING MODELS FOR WASTE SEGREGATION

Machine learning models for trash segregation are innovative approaches that use automation and sophisticated data analysis to maximize the waste management process. By analyzing photos taken from various trash locations, these models primarily use deep learning, especially convolutional neural networks (CNNs), to reliably categorize waste elements. When garbage is introduced into the system, the CNN analyzes the visual input and uses patterns it has learnt from large training datasets to detect and classify things as either biodegradable or nonbiodegradable (Munir et al., 2023). In addition to improving accuracy, this automated method drastically lowers the need for manual sorting, reducing the health risks connected to human

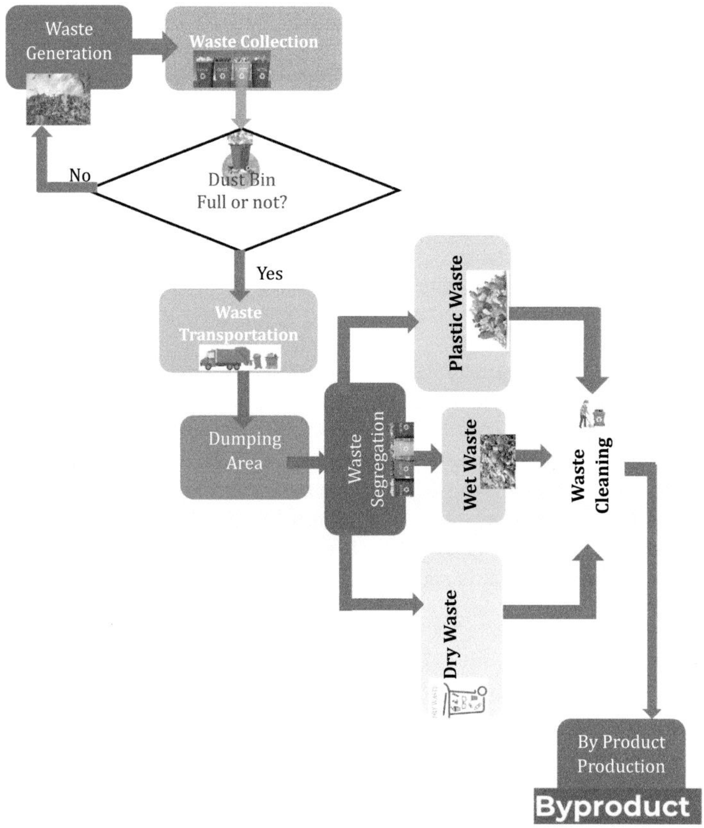

FIGURE 14.6 Flowchart for smart waste management.

involvement in waste segregation. Utilizing Python libraries like Matplotlib for result visualization, NumPy for effective matrix operations, and TensorFlow for model training, these systems are able to manage massive amounts of data and make choices in real time (Bobulski and Kubanek, 2021). A key component of contemporary urban waste management methods, machine learning-driven trash segregation improves recycling rates and reduces the environmental effect of garbage disposal, therefore promoting sustainability (Xia et al., 2022).

14.4.1 Artificial Neural Networks (ANN)

Interconnected nodes or neurons arranged in layers (input, hidden, output) make up artificial neural networks, or ANNs. As learning takes place, the weight assigned to each relationship changes (Namoun et al., 2022). ANNs are appropriate for classification jobs because of their exceptional capacity to identify patterns in big, complicated datasets. To reduce prediction errors, ANNs use a technique known as backpropagation to learn from training data by modifying weights. They can efficiently extract characteristics at various degrees of abstraction by processing data across many layers. With the use of historical records, ANNs may be taught to recognize and categorize different kinds of garbage according to characteristics such as size, texture, and material composition (Pillai et al., 2023).

A research study conducted in Bengaluru, India, involved the development of an ANN for waste classification. Images of several trash kinds, including plastic, paper, biological, and metal, were used to train the model. The ANN achieved an impressive accuracy rate of 92% in classifying waste correctly. This enabled waste management companies to streamline sorting processes and increase efficiency in recycling operations (Abu-Qdais et al., 2023).

14.4.2 Geographic Information Systems (GIS)

Understanding links in geographic settings is made easier with the use of GIS, a collection of tools for organizing, analyzing, and displaying geographical data. In order to capture, manage, and analyze all types of spatially linked data, it combines technology, software, and data. Geographical demographics, trends affecting trash management, and waste generation patterns may all be evaluated with the use of GIS. Maps and other visual representations that draw attention to links and patterns that conventional data analysis might overlook can be made by users. Waste management companies may create customized collection routes by using GIS to analyse waste generating hotspots, which lowers operating expenses and fuel usage. GIS was used by the local government to map the amount of rubbish generated in various neighborhoods. Planners were able to determine the regions with the most trash output by superimposing socioeconomic data and population density. As a result, collection schedules were modified, increasing efficiency by 30% and lowering greenhouse gas emissions related to garbage transportation (Khan et al., 2021).

14.4.3 Adaptive Neuro-Fuzzy Inference System (ANFIS)

By creatively combining the powers of neural networks with fuzzy logic, the Adaptive Neuro-Fuzzy Inference System (ANFIS) establishes itself as a very successful model for enhancing waste sorting procedures. In this system, fuzzy logic is essential since it makes it possible to analyze imprecise and ambiguous data pertaining to a variety of waste kinds. This feature enables ANFIS to offer a thorough and nuanced understanding of the diverse characteristics linked to distinct waste types, such as recyclable plastics and biodegradable materials.

This improved interpretive capacity in waste management results in improved decision-making since ANFIS skillfully captures the nuances and complexity of waste classification. Additionally, the capacity of ANFIS to dynamically modify its settings in response to incoming real-time data greatly improves its accuracy and dependability in waste classification. This adaptable characteristic makes the system especially useful in settings where trash generation patterns are always changing since it guarantees that it will continue to be sensitive to variations in waste composition over time.

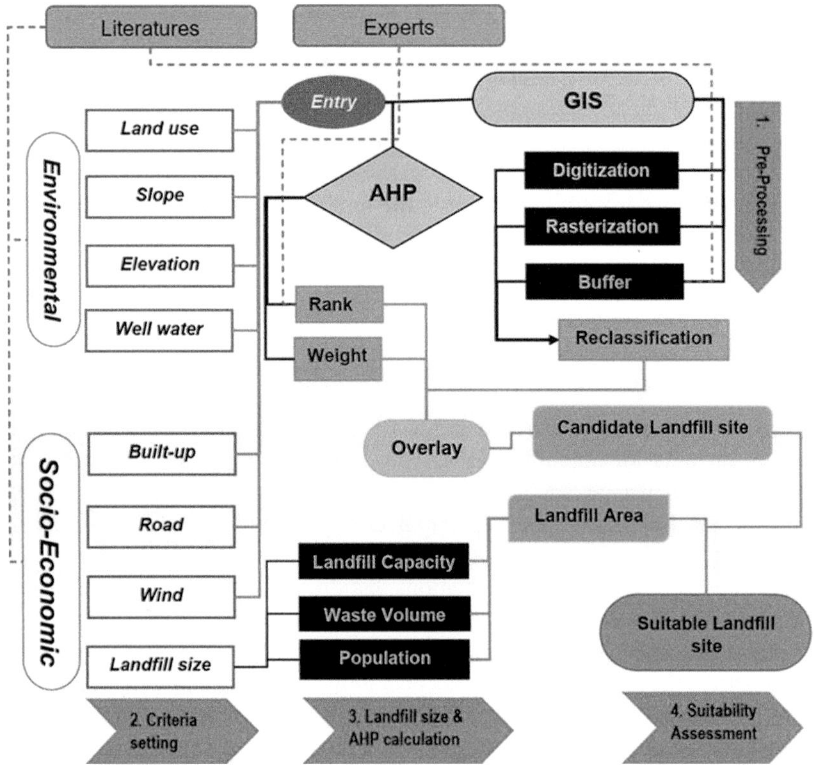

FIGURE 14.7 Application of geographic information system in solid waste management.

ANFIS's useful features facilitate the automation of trash sorting procedures, resulting in more efficient operations that can greatly raise recycling rates. Through the provision of an advanced framework for trash analysis and classification, ANFIS helps to improve waste management tactics, which in turn leads to improved environmental results and the advancement of sustainability in urban areas (Manikandan et al., 2023).

14.4.4 Solid Waste Management Strategies

The effectiveness of waste segregation procedures may be greatly increased by machine learning models, which are a revolutionary development in SWM techniques. These models can effectively categorize different forms of trash, such as organic materials, recyclables, and non-recyclables, by using complex algorithms to examine a wide range of data, including waste composition, generation patterns, and collection logistics.

In addition to streamlining the sorting process, this improved classification capabilities enhances the waste management workflow, enabling more effective material processing and disposal. Furthermore, by using previous data to forecast future trash

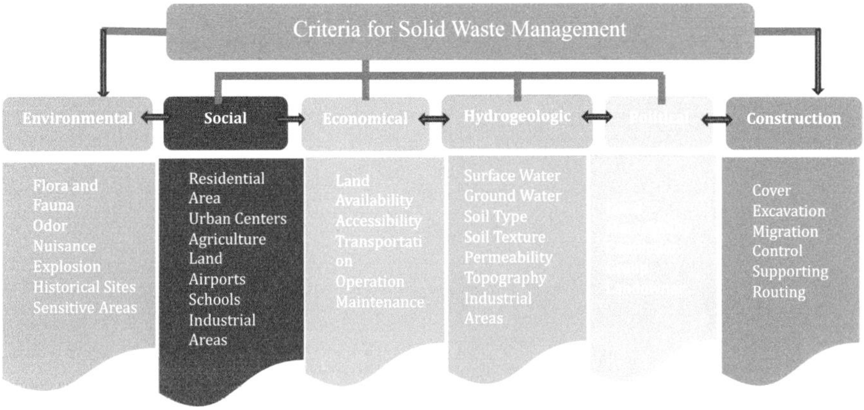

FIGURE 14.8 Criteria used for decision making in SWM.

generation trends, machine learning helps waste management authorities make proactive strategy adjustments that improve resource allocation and save operating costs (Cha et al., 2024).

Municipalities may further encourage efficient recycling programs and reduce landfill usage by incorporating this cutting-edge technology into SWM procedures, which will eventually increase environmental sustainability. A cleaner and more sustainable urban environment is achieved via the effective identification and processing of recyclable materials made possible using machine learning. In conclusion, the application of machine learning to SWM improves the environmental stewardship and operational effectiveness of waste management plans, opening the door to more intelligent and sustainable urban life.

14.4.5 Global System for Mobile Communications (GSM)

Through efficient data transfer and monitoring, machine learning models that include the global system for mobile communications (GSM) architecture may greatly improve waste segregation procedures. Data analysis and decision-making may be done in real time thanks to the reliable platform that GSM technology offers for sending information obtained from waste management systems. GSM makes it possible for garbage sorting facilities to gather and send data in real time to central management systems. This feature improves reaction to problems as they emerge by enabling real-time monitoring of trash segregation activities.

Garbage management officials may remotely monitor garbage containers and segregation systems using GSM technology. This eliminates the need for continual onsite supervision and guarantees that sorting procedures are operating at their best while facilitating timely collection (Sami et al., 2020).

To find patterns and trends in garbage creation and segregation, machine learning algorithms may be used to evaluate data sent over GSM. These kinds of information help maximize operational efficiency and guide improved management tactics.

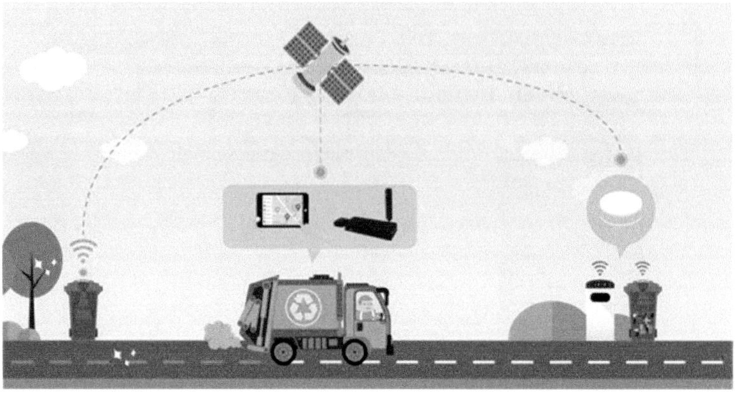

FIGURE 14.9 Real time tracking of garbage collection vehicles and bins.

Geographic tracking of garbage collection vehicles and bins is made possible by GSM technology, which improves waste collection route optimization and lowers operating expenses. This geographic data may be analyzed by machine learning algorithms to enhance the logistics of waste management. Trash management systems may interact with consumers via mobile applications by utilizing GSM, raising knowledge of appropriate trash disposal practices and enticing neighborhood involvement in recycling initiatives. For trash segregation, combining GSM technology with machine learning models offers a complete solution for data analysis, real-time monitoring, and increased operational effectiveness. Better decision-making is facilitated by this synergy, which also helps to improve waste management techniques (Bristol et al., 2024).

14.4.6 You Only Look Once (YOLO)

YOLO, an acronym for "You Only Look Once," is a well-known real-time object recognition system designed to recognize and categorize items in pictures or video frames with speed and accuracy. Yolo handles object recognition as a single regression issue, going straight from image pixels to bounding box coordinates and class probabilities, in contrast to conventional object detection methods that could apply a classifier to various areas of an image several times. A solitary CNN is utilized by YOLO to partition the input image into grid data. Bounding box and class probability predictions are made for objects whose center is inside each grid cell. In addition to predicting a certain number of bounding boxes, each grid cell also provides a confidence score that shows the probability that the box contains an item and the prediction's accuracy.

Each grid cell forecasts the probability distribution for the identified items over the selected classes in addition to bounding boxes. YOLO's quickness is one of its main benefits. It is appropriate for uses like autonomous driving and spying since it can process photos at rates close to real-time. YOLO gathers contextual information by considering the entire image while making predictions, which aids in its

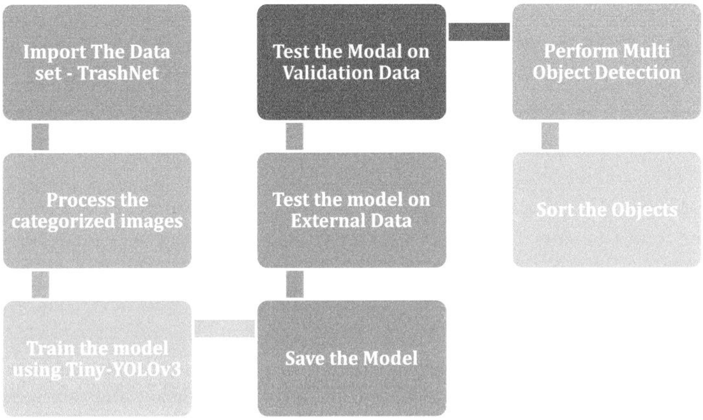

FIGURE 14.10 Steps for the multi-object detection model.

comprehension of the relationships between various things in a scene. Because of its end-to-end training methodology, YOLO frequently generates less false positives than other object detection algorithms (Zailan et al., 2022).

The initial iteration that established the foundation for the YOLO methodology introduced multi-scale detection, which improved the handling of numerous sizes and the ability to recognize tiny objects improved the model's accuracy by linking characteristics from various network sizes using a better backbone (Darknet-53). Using a variety of strategies, including data augmentation, enhanced attention mechanisms, and the use of CSPDarknet as the backbone, it is further optimized for speed and accuracy. This project, which was not officially created by the original YOLO authors, used PyTorch to implement YOLO and provided a user-friendly API and several pre-trained models. It is renowned for being useful in practical settings.

Security camera monitoring in real time to identify and follow people or cars. Using object detection, one may recognize other cars, pedestrians, and traffic signs, giving robots the ability to comprehend and navigate their surroundings and identification of irregularities in medical pictures. Frameworks like TensorFlow or PyTorch may be used to build YOLO; pre-trained models are frequently used to improve performance with minimal training time. Using transfer learning techniques, the model may be trained on unique datasets, enabling it to adjust to certain item categories or applications (Shah et al., 2022).

14.4.7 Cyber-Physical Systems (CPS)

Cyber-physical systems (CPS) waste segregation using machine learning models is a novel way to improve waste management procedures. These models use sophisticated machine learning algorithms to examine information gathered from cameras and sensors that continuously monitor garbage. The algorithms can accurately discriminate between biodegradable and nonbiodegradable materials by training on extensive datasets that include pictures and details about different trash types. Sensors take pictures of incoming waste items on conveyor belts in a standard CPS configuration for trash sorting. CNNs, which are excellent in image identification tasks, are among the methods used by the machine learning model to process these photos. The automatic sorting system is guided by the model's predictions, which classify waste items according to learning patterns and route each item to the proper recycling or disposal receptacle (Kannan et al., 2024).

The sorting process is made more efficient and automated by the incorporation of machine learning into CPS. These devices can greatly increase trash management facilities' throughput by lowering the requirement for manual sorting and human error. Additionally, by learning from fresh information and feedback, they may dynamically adjust to a variety of waste characteristics, improving their effectiveness over time. In metropolitan settings, where waste composition can vary greatly, this feature is very beneficial. No matter how waste streams alter, the segregation process will continue to work since machine learning models can swiftly adjust to these fluctuations. This flexibility maximizes resource recovery and fosters sustainability by reducing contamination rates in recycling procedures.

Additionally, using CPS for waste segregation supports more general environmental objectives. These systems may assist enterprises and governments in optimizing their waste management plans, minimizing the usage of landfills, and promoting circular economy efforts by using real-time data and clever algorithms.

14.5 MUNICIPAL SOLID WASTE MANAGEMENT (MSWM)

In municipal solid trash management (MSWM), machine learning models for trash segregation are ground-breaking instruments that greatly increase the effectiveness and precision of waste material sorting. These models use sophisticated algorithms to examine large datasets, such as pictures and details of different waste materials. Machine learning algorithms may categorize garbage into several groups, such as recyclable, biodegradable, and nonbiodegradable, by using this data as training data (Monferdini et al., 2024).

These models usually use methods like CNNs, which are particularly good at picture recognition. Cameras using cutting-edge imaging equipment take real-time pictures of waste materials moving around conveyor belts in a trash processing plant. When these photos are processed by the machine learning model, waste categories are accurately identified and categorized. By automating the sorting process, less manual work is required, and the possibility of human categorization mistakes is greatly decreased (Zaman, 2022).

These models can increase their categorization accuracy when they meet a greater range of waste items since they are always learning from fresh data. In metropolitan environments, where waste composition can change dramatically daily, this is especially crucial. The quality and quantity of recyclables are eventually improved by machine learning algorithms that adapt to these variances and further lower contamination rates in recycling streams.

Moreover, machine learning's use into MSWM supports more general sustainability objectives. Initiatives to reduce landfill usage and promote resource recovery are supported by these models, which enhance waste segregation and enable better

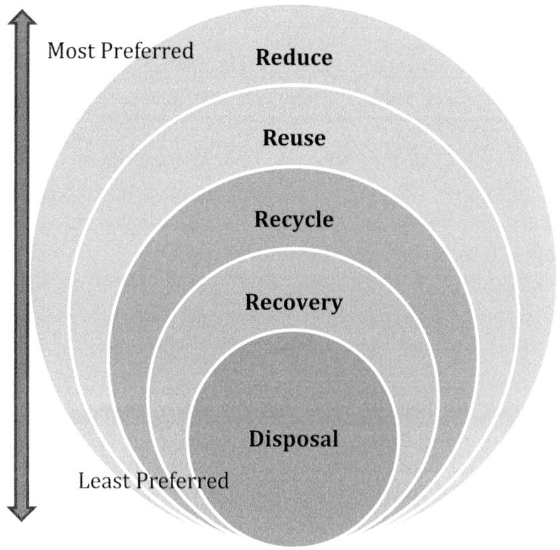

FIGURE 14.11 Waste management hierarchy.

recycling rates. By guaranteeing that valuable resources are recycled and repurposed instead being thrown away, they support a circular economy.

In municipal solid trash management, machine learning models for trash segregation provide an advanced method for addressing the intricacies of garbage sorting. Their capacity to increase accuracy, automate procedures, and adjust to shifting waste patterns promotes more effective and ecologically friendly waste management techniques, which are essential for promoting sustainable urban living and lessening the environmental impact of waste disposal.

14.6 CONCLUSION

- The implementation of Industry 4.0 technologies has significantly transformed waste management, enhancing its efficiency and reliance on data.
- IoT devices facilitate immediate monitoring of waste levels, enabling optimized collection schedules and lowering operational expenses.
- AI advancements improve the sorting of recyclable materials, reducing contamination and increasing recycling effectiveness.
- The use of AMRs enhances operational efficiency and ensures worker safety by reducing the need for manual labor in waste processing.
- Blockchain technology provides a clear and unalterable record of waste management processes, improving accountability and adherence to regulations.
- While there are many advantages, challenges such as high initial costs, concerns over data privacy, and difficulties in system integration can hinder the adoption of these technologies.
- The incorporation of Industry 4.0 technologies presents a substantial opportunity to revolutionize waste management practices and address the challenges associated with increasing waste generation.

REFERENCES

Abdulnour, S., Baril, C., Abdulnour, G., and Gamache, S. (2022). Implementation of industry 4.0 principles and tools: Simulation and case study in a manufacturing SME. *Sustainability*, *14*(10), 6336.

Abu-Qdais, H., Shatnawi, N., and Esra'a, A. A. (2023). Intelligent solid waste classification system using combination of image processing and machine learning models.

Agarwal, R., Chaudhary, M., and Singh, J. (2015). Waste management initiatives in India for human well being. *European Scientific Journal*.

Asteriaa, D., Herdiansyahb, H., and Haryantoc, D. T. (2018). Women's enviromental literacy in development of human capital for the management of household waste. In *Urbanization, human capital, and regional development: The Indonesian experiences* (pp. 81–108). Gramedia Pustaka Utama.

Bobulski, J., and Kubanek, M. (2021). Deep learning for plastic waste classification system. *Applied Computational Intelligence and Soft Computing*, *2021*(1), 6626948.

Bristol, D. M. N., Gue, I. H. V., and Ubando, A. T. (2024). A state-of-the-art review on machine learning based municipal waste to energy system. *Cleaner Energy Systems*, 100143.

Cha, G. W., Park, C. W., and Kim, Y. C. (2024). Optimal machine learning model to predict demolition waste generation for a circular economy. *Sustainability*, *16*(16), 7064.

Chandrappa, R., and Das, D. B. (2024). *Solid waste management: Principles and practice*. Springer Nature.

da Rocha, A. B. T., de Oliveira, K. B., Espuny, M., da Motta Reis, J. S., and Oliveira, O. J. (2022). Business transformation through sustainability based on Industry 4.0. *Heliyon*, *8*(8).

Das, S., Lee, S. H., Kumar, P., Kim, K. H., Lee, S. S., and Bhattacharya, S. S. (2019). Solid waste management: Scope and the challenge of sustainability. *Journal of Cleaner Production*, *228*, 658–678.

Demirbas, A. (2011). Waste management, waste resource facilities and waste conversion processes. *Energy Conversion and Management*, *52*(2), 1280–1287.

Deshpande, A., Birla, I., Deshpande, S., Haribhakta, Y., and Niture, D. (2022, August). Smart and lucrative waste segregation. In *2022 International Conference on Signal and Information Processing (IConSIP)* (pp. 1–5). IEEE.

Fatimah, Y. A., Govindan, K., Murniningsih, R., and Setiawan, A. (2020). Industry 4.0 based sustainable circular economy approach for smart waste management system to achieve sustainable development goals: A case study of Indonesia. *Journal of Cleaner Production*, *269*, 122263.

Javaid, M., Haleem, A., Singh, R. P., Suman, R., and Gonzalez, E. S. (2022). Understanding the adoption of Industry 4.0 technologies in improving environmental sustainability. *Sustainable Operations and Computers*, *3*, 203–217.

Joshi, R., and Ahmed, S. (2016). Status and challenges of municipal solid waste management in India: A review. *Cogent Environmental Science*, *2*(1), 1139434.

Kannan, D., Khademolqorani, S., Janatyan, N., and Alavi, S. (2024). Smart waste management 4.0: The transition from a systematic review to an integrated framework. *Waste Management*, *174*, 1–14.

Khan, R., Kumar, S., Srivastava, A. K., Dhingra, N., Gupta, M., Bhati, N., and Kumari, P. (2021). [Retracted] machine learning and IoT-based waste management model. *Computational Intelligence and Neuroscience*, *2021*(1), 5942574.

Manikandan, V., Arjun, S., Visweswaran, A., and Sruthi, N. (2023). Automated waste segregation using machine learning. *AIJR Proceedings*, 264–268.

Marinelli, M., Deshmukh, A. A., Janardhanan, M., and Nielsen, I. (2021). Lean manufacturing and Industry 4.0 combinative application: Practices and perceived benefits. *IFAC-PapersOnLine*, *54*(1), 288–293.

Marshall, R. E., and Farahbakhsh, K. (2013). Systems approaches to integrated solid waste management in developing countries. *Waste Management*, *33*(4), 988–1003.

Mohamed, N. H., Khan, S., and Jagtap, S. (2024). Waste 4.0: Transforming medical waste management through digitalization and automated segregation. *Discover Sustainability*, *5*(1), 353.

Monferdini, L., Pini, B., Bigliardi, B., and Eleonora, B. (2024). Implementing Industry 4.0 technologies for enhanced material flow and handling management: A case study in logistics. *Procedia Computer Science*, *232*, 626–634.

Munir, M. T., Li, B., and Naqvi, M. (2023). Revolutionizing municipal solid waste management (MSWM) with machine learning as a clean resource: Opportunities, challenges and solutions. *Fuel*, *348*, 128548.

Namoun, A., Tufail, A., Khan, M. Y., Alrehaili, A., Syed, T. A., and BenRhouma, O. (2022). Solid waste generation and disposal using machine learning approaches: A survey of solutions and challenges. *Sustainability*, *14*(20), 13578.

Narayanswamy, N., Rajak, A., and Hasan, S. (2022). Development of computer vision algorithms for multi-class waste segregation and their analysis. *Emerging Science Journal*, *6*(3), 631–646.

Pillai, K. S., Sneha, M. L., Aiswarya, S., Anand, A. B., and Prasad, G. (2023). Municipal solid waste management: A review of machine learning applications. In *E3S Web of Conferences* (Vol. *455*, p. 02018). EDP Sciences.

Quan, Z., Xu, X., Wang, W., Jiang, J., and Gao, S. (2024). Do industrial solid waste recycling and technological innovation promote low-carbon development in China? New insights from NARDL approach. *Science of the Total Environment*, *916*, 170446.

Rajakumaran, G., Usharani, S., Vincent, C., and Sujatha, M. (2023, April). Smart waste management: Waste segregation using machine learning. In *Journal of Physics: Conference Series* (Vol. *2471*, No. 1, p. 012030). IOP Publishing.

Sami, K. N., Amin, Z. M. A., and Hassan, R. (2020). Waste management using machine learning and deep learning algorithms. *International Journal on Perceptive and Cognitive Computing*, *6*(2), 97–106.

Samreen, S., Agarwal, M. K., Gautam, A. K., and Khajanchi, Y. (2023). Comprehensive review of solid waste management strategies and disposal approaches. *Multidisciplinary Reviews*, *6*.

Sanghavi, D., Parikh, S., and Raj, S. A. (2019). Industry 4.0: Tools and implementation. *Management and Production Engineering Review*.

Shah, J., and Kamat, S. (2022, April). A method for waste segregation using convolutional neural networks. In *2022 Second International Conference on Advances in Electrical, Computing, Communication and Sustainable Technologies (ICAECT)* (pp. 1–5). IEEE.

Shahbazi, S., Kurdve, M., Bjelkemyr, M., Jönsson, C., and Wiktorsson, M. (2013). Industrial waste management within manufacturing: A comparative study of tools, policies, visions and concepts. In *The 11th International Conference on Manufacturing Research (ICMR2013),* Cranfield UK, 19 September–20 September 2013 (pp. 637–642).

Singh, D. (2023). Advances in industrial waste management. In Pardeep Singh, *Waste management and resource recycling in the developing world* (pp. 385–416). Elsevier.

Strazzullo, S., Cricelli, L., Troise, C., and Camilleri, M. A. (2024). Leveraging Industry 4.0 technologies for sustainable value chains: Raising awareness on digital transformation and responsible operations management. *Sustainable Development*.

Toth-Peter, A., de Oliveira, R. T., Mathews, S., Barner, L., and Figueira, S. (2023). Industry 4.0 as an enabler in transitioning to circular business models: A systematic literature review. *Journal of Cleaner Production*, *393*, 136284.

Xia, W., Jiang, Y., Chen, X., and Zhao, R. (2022). Application of machine learning algorithms in municipal solid waste management: A mini review. *Waste Management and Research*, *40*(6), 609–624.

Zailan, N. A., Azizan, M. M., Hasikin, K., Mohd Khairuddin, A. S., and Khairuddin, U. (2022). An automated solid waste detection using the optimized YOLO model for riverine management. *Frontiers in Public Health*, *10*, 907280.

Zaman, A. (2022). Waste management 4.0: An application of a machine learning model to identify and measure household waste contamination—A case study in Australia. *Sustainability*, *14*(5), 3061.

15 Upcycling Waste into Wealth

J C Nikhil Rao, Surriya S, Kota Suraj, and Koushik Manchala

15.1 INTRODUCTION

Over the years, particularly over the last few decades, waste management has emerged as an important issue on the global landscape. The phenomenon of industrialization, along with rapid urbanization and increased consumerism, has raised the levels of waste production in society to unprecedented levels. Landfilling and incineration, the most common waste disposal techniques, have long ceased to be viable owing to their social and environmental impacts. A key area of concern to add value to waste materials is where upcycling works tremendously. Often called creative reuse, upcycling is the process of transforming waste materials, by-products, or unwanted products into new materials or products of higher value (Cooper and Kettley, 2021).

Upcycling does not demolish waste into raw materials as recycling does. Instead, it takes waste in its original form and enhances its value through craftsmanship and innovation. This activity is in line with the goals of the circular economy since it enhances resource productivity and minimizes the negative environmental impacts that result from conventional waste management techniques. The process of upcycling is not novel. At every point in time, societies have devised creative ways of waste management. The Romans and Egyptians were some well-known ancient civilizations who reused and reprocessed materials because resources were scant. Upcycling emerged as a practice during dire periods like the Great Depression and World War II. People started extending the use of available resources as much as possible. Today, upcycling has emerged as a well-planned activity used for environmental sustainability instead of just a survival tactic during upcycles strategy (McDonagh, 2021).

Additionally, there is an innovative strategy, "upcycling," has emerged which is different from the previous process of recycling because waste can now be transformed into fuels, chemicals, and even valuable carbon nanomaterials. The process of upcycling waste into fuels requires an input of heat energy, but the upcycled product can be utilized in other ways. Upcycling into carbon fuels is an innovative waste management technique. The valuable carbons nanomaterials include adsorbents, water purifiers, and devices for energy storage. While these techniques solve a part of the problem, methods like landfilling, or incineration, or even recycling, tend to be more insufficient. For this reason, a lot of studies have been done to transform waste into valuable products. This chapter explores various forms of waste and the methods for

DOI: 10.1201/9781003637677-15

recycling and upcycling them. The advantages of upcycling go beyond just protecting the environment; it significantly contributes to economic growth by encouraging the emergence of new industries, generating employment, and aiding local craftspeople and small enterprises. Additionally, upcycling enhances social responsibility by increasing awareness of sustainable consumption practices and decreasing reliance on new resources (Terzi, 2020).

This chapter investigates the importance of upcycling in managing waste, focusing on its effects on the environment, economy, and society. It investigates the core concepts of upcycling, various methods for putting it into practice, and successful examples that showcase its possibilities. Furthermore, it addresses the obstacles to broader acceptance and suggests ways to improve its impact. By recognizing and adopting upcycling, individuals, companies, and policymakers can help create a more sustainable future, turning waste into valuable resources while alleviating pressure on the Earth.

15.2 WASTE AS RESOURCES AND WEALTH

The approach toward waste as a discard of human production processes has transformed as scientists now recognize waste materials as a lucrative source with dual economic and ecological applications. Environmental issues including resource shortages have made the circular economy model popular because it uses waste materials to revitalize production systems instead of disposing of them. The change in how we perceive waste has motivated people to develop innovative waste management systems that transform waste materials into profitable economic sources for materials and energy and sustainable development opportunities (Van Woerden, 2020).

The transition toward sustainable waste management depends on diverse physical, chemical, and biological along with thermal waste treatment processes which reduce environmental problems while boosting resource utilization rates. The essential elements of sustainability to minimize landfills include recycling together with composting and waste-to-energy conversion alongside proper waste disposal. Manufacturers in various industries have started to develop waste utilization plans using unwanted materials as manufacturing ingredients while building structures and producing energy. Waste recovery effectiveness improves due to implemented waste management policies together with technological breakthroughs and thus secures both resource efficiency and sustainability in future operations.

15.2.1 Extracting Value from Waste Streams

Many waste streams contain valuable raw materials that can be recovered and repurposed. Metals, for instance, require significant energy for extraction and refining. However, metals such as aluminum, copper, and steel, which are commonly found in discarded products, can be recovered with minimal processing. For example, aluminum cans present in waste streams can reach nearly 100% purity, making them ideal for reuse. Producing new aluminum products from recycled cans requires 95% less energy and 97% less water than processing aluminum from raw bauxite ore.

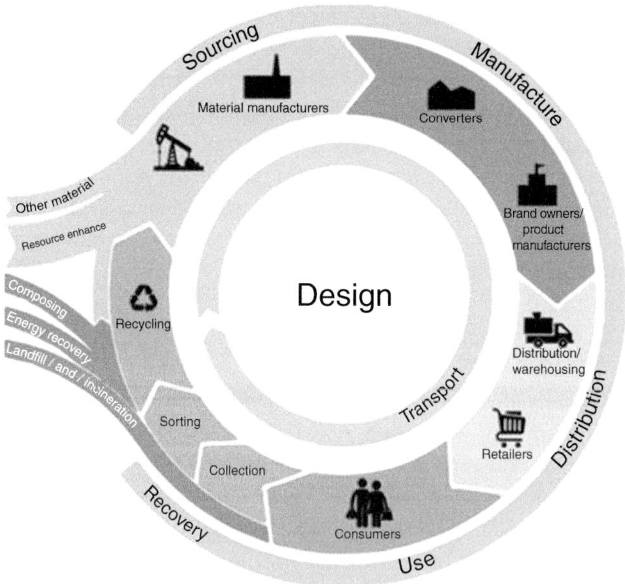

FIGURE 15.1 Circular economy (Van Woerden, 2020).

Organic waste including food waste and agricultural byproducts presents extensive prospects to accomplish sustainable farming practices while generating energy at the same time. Sugarland farmers should grow their soil quality through composting processes that develop sustainable natural fertilizers for minimizing chemical fertilizer usage while enhancing farming conditions. The process of anaerobic digestion takes organic waste to produce biogas that functions as electricity fuel for heating systems and transportation power to decrease fossil fuel consumption. Industrial and power generation facilities can recover energy from plastic rubber leather and wood materials due to their high calorific value. The use of waste-derived energy sources enables industries to enhance their total energy efficiency along with reduced environmental pollution and decreased usage of nonrenewable fossil fuels which results in a sustainable and resource-efficient upcoming era (Bhada-Tata, 2022)

15.2.2 Waste as a Resource in Manufacturing and Industry

Manufacturing plants need enormous quantities of raw materials comprising metals and wood together with plastics and leather and glass in their production. The extraction method together with processing techniques and transportation routes leads to environmental damage and diminishes original resources. Waste materials operated as substitute raw materials to decrease industries' demand for natural resources (Fujita, 2020).)

The wastewater treatment plants provide a clear illustration of waste material exploitation that functions as an effective resource. A variety of valuable uses exists for wastewater after proper treatment since it can become suitable for agricultural irrigation alongside functioning as a source of energy and producing fertilizer. The process of mining nutrient and mineral values from wastewater leads to diminished requirements for new raw materials which reduces both manufacturing costs and environmental expenses. Firms in various industries have started executing closed-loop procedures through which process waste transforms into useful input materials for other operations. Recycling and recovering raw materials after use allows re-engineering to increase production efficiency and discover new revenue streams through new industrial supply networks.

15.2.3 The Economic and Social Benefits of Waste Utilization

Numerous economic advantages emerge because waste utilization replaces waste disposal methods to turn unwanted materials into new valuable products. New waste management industries have developed their focus on waste repurposing which produces both environmental and profit gains. The businesses operating in waste recycling alongside upcycling and energy recovery programs build a circular economy system which cuts down raw material usage and creates minimal environmental harm. Private organizations with government support invest in waste management technology which has stimulated the market for new solution development. The use of waste streams results in manufacturing cost reductions since companies depend less on original materials thus maximizing overall economic performance (Zaman, 2020).

The applications of waste utilization led to social benefits that establish environmental stewardship coupled with healthcare improvement. The waste management and recycling industry expansion adds jobs to the workforce because it requires personnel in waste collection and processing as well as value-added production stages. The removal of environmental pollution from waste utilization minimizes health dangers caused by landfills and improper waste disposal methods. People join community recycling projects and waste-to-energy initiatives through public engagement while learning sustainable practices because these initiatives provide civil participation opportunities. The implementation of waste utilization practices creates dual benefits for the economy while establishing a society that is cleaner, healthier, and socially responsible at the same time.

15.2.3.1 Waste-to-Energy Plants

Air pollution enhances municipal solid waste conversion to electricity and heat at waste-to-energy plants thus decreasing fossil fuel use. Different waste-to-energy technologies at Waste-to-Energy (WTE) plants use incineration together with anaerobic digestion and gasification methods to extract energy from waste materials.

Reduction in Landfill Waste: waste-to-energy facilities achieve waste reduction by sending minimal waste to landfills thus preventing environmental pollution while suppressing methane emissions.

Renewable energy production: The biogas generated from organic waste consists mainly of methane that leads to power generation while also providing heating functions and becoming a fuel source for vehicles. The sustainable power generated from WTE plants supports two environmental improvements: it cuts down greenhouse gas amounts while lessening reliance on traditional fossil fuel sources to power our world.

Economic Savings: WTE facilities reduce municipal waste management expenses by creating two-fold financial benefits through their ability to generate power for sale. The nations of Sweden together with Japan and Germany have proven that their introduction of WTE plants results in successful implementation for their energy systems (Elgizawy et al., 2021).

15.2.3.2 Recycling and Upcycling Enterprises

By transforming waste into fresh consumer items and textiles and construction materials driving green innovation businesses both establish employment positions and introduce fresh market potentials. The global recycling sector maintains hundreds of billion-dollar worthwhile experiencing increasing market demands for materials such as recycled plastics and metals as well as paper (Prajapati, 2021).

Economic Growth: The collection processing and sorting sectors generate new job opportunities in which economic growth supports small businesses and creates dependable employment. The increasing market demand for recycled materials creates new business opportunities that draw investment capital to advance industrial growth. The economic growth of sustainable livelihoods counts on governmental incentives while organizations drive waste management initiatives that advance economic advancement.

Product Innovation: The practice of making new products from surplus materials represents both product development and value addition because it turns waste elements into premium goods for the market. The creation of sustainable alternatives through this process leads to technological advancements which result in alternatives that decrease virgin resource usage. Superior green consumers fuel the market for recycled products which forces companies to switch to environmentally friendly operations and support sustainable fiscal growth.

Resource Conservation: Resource conservation benefits substantially from recycling because it eliminates the requirement for raw materials thus preserving natural resources and stopping industrial carbon emissions. Chemical-based plastic recycling and automated material sorters have been developed by advanced recycling technology which improves both efficiency and large-scale operation.

15.2.3.3 Biofuel Production

Sustainable energy promotion includes transforming agricultural waste and organics into biofuels that serve as green energy alternatives. The renewable energy sources of ethanol biodiesel and biogas serve as environmentally friendly options to petroleum-derived fuel products (Yang, 2022). Currently, biofuel manufacturing provides several main benefits to the sector.

Reduced Carbon Footprint: The creation of biofuels results in lower carbon emission levels than fossil fuel products which helps control climate change by decreasing carbon dioxide and particulate matter pollution levels. Biofuel manufacturing through plant oils along with algae and waste biomass enables us to decrease our use of petroleum as a fuel source. Biofuels generate cleaner air quality through their use, and they enable worldwide carbon neutrality targets.

Utilization of Agricultural Residues: Renewable energy production from agricultural waste materials through anaerobic digestion and bioethanol methods helps decrease waste management challenges in the environment. Farmer income increases through organic waste conversion into biogas and biofertilizers while simultaneously blocking damaging methane formation from decomposed organic materials. The adoption of agricultural residue use drives the circular economy model to create sustainable agricultural systems.

Energy Independence: Extracting energy from biofuels makes nations less dependent on imported fossil fuel supplies which improves their energy security position. Several nations worldwide expand biofuel consumption while Brazil together with the United States currently leads ethanol manufacturing through sugarcane and corn resources.

15.2.3.4 Spare Parts and Equipment Reuse

Industrial waste reduction requires products to last longer as scrap equipment and electronic pieces, and auto elements become vital resources for reuse. Organizations that specialize in remanufacturing take end-of-life equipment components to regenerate and reintroduce them as market products. Spare parts reuse delivers these main advantages:

Cost Savings: The price tag on modified spare parts reaches much lower levels than fresh products thus enabling consumers and businesses to maintain their equipment integrity while managing costs effectively. The total expense to operate equipment becomes more affordable thus benefiting especially the automotive electronics and manufacturing sectors. The use of refurbished components in repair work decreases both the need for raw material extraction and the depletion of natural resources and eventually reduces production expenses (Liu et al., 2021).

Reduced E-Waste: Market demand for sparse metals declines due to the electronics industry recovering them from junked electronics systems which in turn stops new mining operations thereby protecting both mining areas and natural resources. These materials become available after recycling thus keeping them out of landfills while blocking harmful pollutants from getting into soil and water. Promotions of e-waste recycling establish new positions for workers who collect devices and perform dismantled tasks and material recovery activities thus generating employment while building a circular economic framework.

Support Recovery in a Circular Economy: Recovered components and parts contribute to a sustainable life cycle of manufacturing by extending a merchandise lifespan, decreasing industrial waste, and lowering the demand for uncooked substances. The burgeoning e-waste recycling industry, for example, aims to reclaim precious metals such as gold, silver, copper, and rare earth metals from electronics waste, reducing the need for environmentally damaging mining operations. These preserves limited natural reserves and reduces carbon emissions from material extraction and processing. Additionally, transitioning towards a circular economy paves the way for economic growth through innovation in product development, refurbishment, and resource recovery, resulting in a more sustainable and enduring economic framework.

15.2.3.5 Environmental Conservation and Social Impact Beyond Economic Advantages

15.2.3.5.1 Waste Utilization Helps the Environment: Waste utilization significantly helps the environment by eliminating pollution, conserving natural resources, and decreasing greenhouse gas emissions. Sustainable waste management reduces landfill accumulation, prevents hazardous substances from leaching into soil and water bodies, and mitigates air pollution caused by uncontrolled burning of wastes. Waste wrapping also demonstrates social responsibility by encouraging sustainable consumption behaviors in people and raising public awareness of responsible waste disposal and community recycling programs (Bocken et al., 2020).

15.2.3.5.2 Community-Led Recycling Programs: In many cities, there are initiatives to involve the citizens in recycling and waste segregation, thereby enhancing community engagement and fostering an environment of environmental responsibility. These initiatives raise awareness around appropriate waste disposal and engage the community in reducing landfill contribution. Local governments and nongovernmental organizations (NGOs) also play a crucial role by offering incentives, building infrastructure, and promoting educational programs that complement these initiatives, making them even more effective.

15.2.3.5.3 Creating Jobs for the Underserved: Structured waste management not only employs residents discussed above but also provides informal waste collectors and recyclers, an important group, especially in developing countries, with job opportunities in this field. These initiatives provide training, fair pay, and safer working environments, decreasing exploitation and health hazards from unregulated waste picking. Thus, inclusion of informal labor in formal recycling systems allows for greater efficiency and a collaborative process in developing a more inclusive and sustainable waste economy.

15.2.3.5.4 Health Benefits: Reaching all the way to the hospital, slowing landfill waste lessens air and water pollution and thereby reduces the rates of respiratory diseases and waterborne infections. The integrated advantages of waste valorization will pave the way for a more sustainable and resilient future as industries and societies continue to adopt waste valorization practices. The central challenge lies in transforming waste into a resource, a source of wealth, through high-efficiency waste treatment technologies, sustainable waste management policies, and a culture of responsible consumption.

15.2.3.6 Waste Utilization in Construction and Infrastructure Development

Construction and infrastructure development activities generate one of the largest, and certainly one of the bulkiest, waste streams. This waste includes all materials, including, but not limited to, concrete, bricks, gypsum, wood, glass, metals and excavated soil, and many of these materials can be recycled or reused. As construction and infrastructure waste accounts for a sizeable proportion of total waste generation in urbanized regions, proper management becomes high in importance due to its effects in resource depletion and environmental harm. Waste materials can potentially be used in recycling for instance as aggregates in road construction, drainage systems, and other infrastructure projects. Such materials help conserve the natural resources like sand and gravel which are obtained by mining processes that threaten the environment. The recycling of waste from construction and infrastructure development can present a means to reduce carbon emissions in the manufacturing of new materials thereby helping with the combat against climate change (Pacheco-Torgal and Jalali, 2020).).

Innovations in Waste Management for Construction and Infrastructure Technology: The infrastructure for the segregation, recovery, and treatment of construction and infrastructure waste is well defined and economically sound. Advanced sorting techniques with automatic material recovery premises enable efficient separation of materials such as concrete, metals, and plastics. Mechanical treatment solutions require a systematic blending of solid waste enormously into small-sized particles and a wide dimensional-crushed or grounded solid waste into reusable aggregates. Hereafter, thermal and chemical treatments further enhance the durability and certain mechanical properties of some recycled materials.

Intelligent demolition made possible via technologies like digital modelling and AI enables the maximum recovery of materials. Techniques such as the building information modelling (BIM) allow those involved in the planning to identify ahead of time all materials providable for reuse before the commencement of any demolition work, thus reducing waste. Essentially, deconstruction, as opposed to standard demolition, refers to taking apart the building in such a way that valuable materials would be suitable for redoing.

Applications of Recycled Waste in Construction and Infrastructure: Recycled construction and infrastructure materials have widespread applications across various industries. Some key applications include:

Road Construction: Recycled concrete and asphalt can be used as base layers for roads, reducing reliance on virgin aggregates.

Building Materials: Recycled bricks, wood, and metal can be repurposed for new construction projects, lowering production costs and minimizing environmental impact.

Landscaping and Urban Development: Crushed concrete and recycled wood are commonly used in landscaping projects, parks, and urban beautification initiatives.

Energy Production: Certain waste materials, such as wood and biomass, can be processed into biofuels and used for renewable energy generation.

15.3 WASTE TREATMENT

Waste treatment is one among fundamental processes inside modern waste management, that minimizes environmental contamination, preserves natural resources and enhances material recovery. Waste generation has risen to unprecedented levels with the immense pace of urbanization, industrial take off, and population growth. Waste, if not treated properly can lead to major environmental issues such as land, water, and air pollution and dangerous greenhouse gases like methane, carbon dioxide etc. Different waste treatment approaches, based on physical, chemical, and biological processes and some thermal are important to tackling these problems (Huang, 2021).

The execution of sustainable waste treatment methodologies enables recycling processes alongside composting techniques and waste-to-energy transformations while ensuring responsible disposal practices which together establish a circular economy where materials experience perpetual repurposing instead of being discarded. Physical waste treatment processes involve segregation shredding and compaction while chemical treatments include neutralization and incineration. The administration of organic waste treatment depends on biological processes like composting and anaerobic digestion while thermal methods such as pyrolysis and plasma gasification provide energy recovery solutions. To address global waste challenges and ensure a sustainable future, the combined implementation of effective waste management policies and technological innovations emerges as an essential requirement.

15.3.1 Types of Waste Treatment Methods

15.3.1.1 Physical Waste Treatment

Segregation, shredding, dewatering, and compaction are adopted in this method for efficient waste management. Segregation separates recyclable as well organic materials from general waste to enhance the effectiveness of resource recovery. Shredding and crashing: breaks down materials of a plastic/metal/con-construction debris for easier handling/rerouting items. Dewatering and drying solve the industrial waste as well sludge waste management problem by reducing moisture content, hence disposal will be easier. Compaction reduces the total waste volume for more convenience in storage and transport (Rajendran and Singh, 2022).

15.3.1.2 Chemical Waste Treatment

An essential step for one of secretive and challenging waste materials. Incineration high temperature combustion of waste results in significant volume reduction and, with modern air pollution controls, little emissions, though the waste must still be disposed of somewhere. Chemical neutralization is essential for the treatment of industrial acids and bases, hazardous waste so as to dispose that safely. In this case, other chemical processes (e.g., oxidation and reduction break down pollutants in wastewater) while precipitation and coagulation aid in the removal of heavy metals as well as other contaminants so that a safer waste management at sites as well environmentally protection can be achieved.

15.3.1.3 Eco-friendly Way

Biological waste treatment purposely meant to be eco-friendly and sustainable, biological treatment works best on the organic matter. Composting enables biodegradable materials like food waste or any other organic residues to break down normally and produce nutrient-rich compost for agriculture and gardening. Anaerobic digestion is a process where organic waste is digested in the absence of oxygen to make biogas, which is used as an alternative renewable source for energies. Bioremediation utilizes microorganisms in contaminated soil and water to breakdown, detoxify hazardous pollutants, thus, can be said to have played the most important role in ecological restoration as well as pollutants control (Chaturvedi and Pandey, 2022).

15.3.1.4 Wastewater Treatment

Given the challenges of water pollution from municipal, industrial and agricultural sources, wastewater treatment is mandatory. It is commonly divided into three phases. It covers the primary treatment which consists of the capture of large solids and sediments in wastewater. Secondary treatment uses biological processes to degrade organic pollutants, diminishing the environmental effect when discharging wastewater. Tertiary treatments are filtered and disinfected with advanced chemicals so that the water can be reused or released into natural bodies of water. With increasing worries of water scarcity, wastewater recycling and reusing have turned into a necessary for proper water resource management.

15.3.1.5 Innovation for Technological Advancements

A progress in waste management leads to more effective ways of sustainability. Plasma gasification, a new technology that's taking off where it is used for gasification of waste to produce syngas (useful to produce energy) Lastly, pyrolysis is an even more advanced process that effectively decomposes the organic Waste into its raw materials (bio-oil gas and char) at high temperatures without oxygen; both could be alternate fuels replacement. Electrochemical treatment for wastewater is investigated to control electrolysis in the removal of heavy metals and other contaminants. But to that end, modern recycling companies are embracing 3D printing to turn plastic and metal waste into new goods so that they start relying less on virgin materials. Efficient treatment of waste reduces greenhouse gases, escapes land and water pollution, and

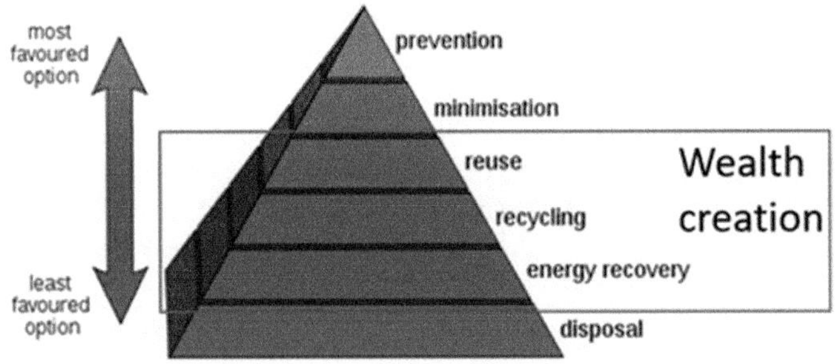

FIGURE 15.2 Waste hierarchy pyramid (Rios and Vassilev, 2020).

TABLE 15.1
Hierarchy Level

Hierarchy level	Description
Prevention	Avoiding waste generation through sustainable production and consumption practices
Minimization	Reducing waste by optimizing resource use in manufacturing and consumption
Reuse	Extending the lifecycle of products by repurposing or repairing them
Recycling	Processing waste materials to create new products, reducing the need for virgin materials
Recovery	Extracting energy or valuable materials from waste that cannot be recycled
Disposal	Safely discarding waste through landfilling or incineration when no other options are viable

a circular economy balances the use of materials all around creating new industries with jobs plus sustainable models of business. Still there are obstacles such as high-scale deployment costs and insufficient waste management infrastructure/enforcement amongst others which prevent wider adoption (Yuan and Zhang, 2021).

15.3.2 The Waste Hierarchy and Its Role in Waste Treatment

Organized in the waste hierarchy, strategies are organized according to their priority from the environment harm prevention. Ranking prevention, minimization, reuse, recycling, and recovery finally disposal are our hierarchical steps:

15.3.3 Regulatory Frameworks and Policies for Waste Treatment

Waste treatment strategies are influenced by government policies that create a legal base, guidelines, and regulatory systems. Several international agreements and national statutes emphasize the need for waste reduction, resource recovery, and sustainable waste management. They reduce waste, drive recycling, and disposal out of business ensuring sustainability and preventing illegal dumping (Woerden, 2018).

15.3.3.1 International and Regional Waste Management Regulations

15.3.3.1.1 European Union Waste Framework Directive

Comprehensive on waste management in member states, EU Waste Framework Directive (2008/98/EC) The Circular Economy Package has waste prevention, recycling, and resource efficiency at the core, and this directive sets out priorities that contribute to these principles. The core elements of the directive consist of:

Waste hierarchy: Waste prevention, reuse, recycling and recovery and energy recovery from incineration permit disposal as a last resort.

Recycling targets: Directs member states to have a minimum of 55% municipal waste recycling by 2025, 60% by 2030 and ratio must reach its target-recycled mass at least 65% by 2035.

Hazardous Waste Management: Defines very stringent requirements to protect the health of humans and the environment, during the entire process of handling and disposal.

Extended Producer Responsibility (EPR): Requires producers to manage the end of life of their products, from postconsumer waste through to disposal.

Permeability of landfill restrictions: Landfill permits limited biodegradable waste input to mitigate methane emissions, environmental assaults.

15.3.3.1.2 Basel Convention

Basel Convention on the Control of Transboundary Movements of Hazardous wastes and their Disposal 1989, it is an international agreement to stop unwanted dumping of hazardous waste, especially in poor countries. This encompasses essential provisions like:

PIC (Prior informed consent): Requires exporter countries in this regard to seek approval from importing countries before hazardous waste shipment.

Ban on export to non-OECD countries: It bans sending hazardous waste to countries not meeting minimum standards of waste management infrastructure.

Waste tracking: Mandates documentation and tracking of hazardous waste movement to ensure compliance.

Facilities: Assists developing countries in enhancing their capacities for waste management.

15.3.3.1.3 Resource Conservation and Recovery Act (RCRA) USA:

Enacted in 1976, the RCRA is the primary U.S. law governing solid and hazardous waste disposal. It establishes a "cradle-to-grave" system for waste management, ensuring that waste is properly handled from generation to final disposal. Key aspects include:

Subtitle C (Hazardous Waste Management): Regulates hazardous waste generators, transporters, and treatment, storage, and disposal facilities (TSDFs).

Subtitle D (Solid Waste Management): Encourages state and local governments to develop environmentally sound solid waste disposal plans.

Corrective actions: Requires facilities handling hazardous waste to implement cleanup procedures if contamination occurs.

Promoting recycling and waste minimization: Provides incentives for businesses to reduce waste generation and adopt sustainable practices.

15.3.3.1.4 India's Solid Waste Management Rules (2016)

India's solid waste management (SWM) Rules, 2016, introduced comprehensive waste management regulations aimed at urban and rural sustainability. The rules include:

Mandatory waste segregation: Households, institutions, and businesses must separate waste into biodegradable, nonbiodegradable, and domestic hazardous waste categories.
Composting and bio-methanation: Encourages composting organic waste at the source to reduce landfill dependency.
Extended Producer Responsibility (EPR): Requires manufacturers and brand owners to collect and recycle packaging waste, particularly plastic waste.
Strict landfill norms: Limits the disposal of nonrecyclable and nonbiodegradable waste to landfills, emphasizing waste-to-energy solutions.
Decentralized waste management: Encourages local bodies to establish recycling and waste treatment facilities within municipal limits.

15.3.3.1.5 China's National Sword Policy (2018)

China was historically the world's largest importer of recyclable waste, but the National Sword Policy imposed strict bans on low-quality and contaminated recyclables. Key aspects of the policy include the following:

Import restrictions: Limits the import of 24 categories of solid waste, including mixed paper and plastic waste.
Higher contamination standards: Reduces allowable contamination levels in imported recyclables to 0.5% or lower.
Encouraging domestic recycling: Promotes investments in domestic recycling infrastructure and circular economy initiatives.
Impact on global recycling: Forces exporting nations (like the USA and EU countries) to improve waste sorting and processing capabilities.

15.3.4 Global Waste Treatment Strategies and Best Practices

Different countries have adopted innovative approaches to waste treatment:

Sweden's Waste-to-Energy Program: Converts 99% of household waste into energy, significantly reducing landfill dependency.
Japan's 3R Strategy (Reduce, Reuse, Recycle): Emphasizes resource efficiency through strict recycling laws and public participation.
Germany's Circular Economy Laws: Enforces high recycling rates and strong incentives for businesses to adopt sustainable waste management practices.
Singapore's Zero Waste Masterplan: Aims to reduce landfill waste by 30% through food waste recycling, extended producer responsibility, and innovative sorting technologies.

15.3.4.1 Waste-to-Energy Technologies

Energy demands are increasing coupled with lack of space in the landfill, and so waste-to-energy (WTE) is a mainstream. As an example of some of the most exciting WTE technologies:

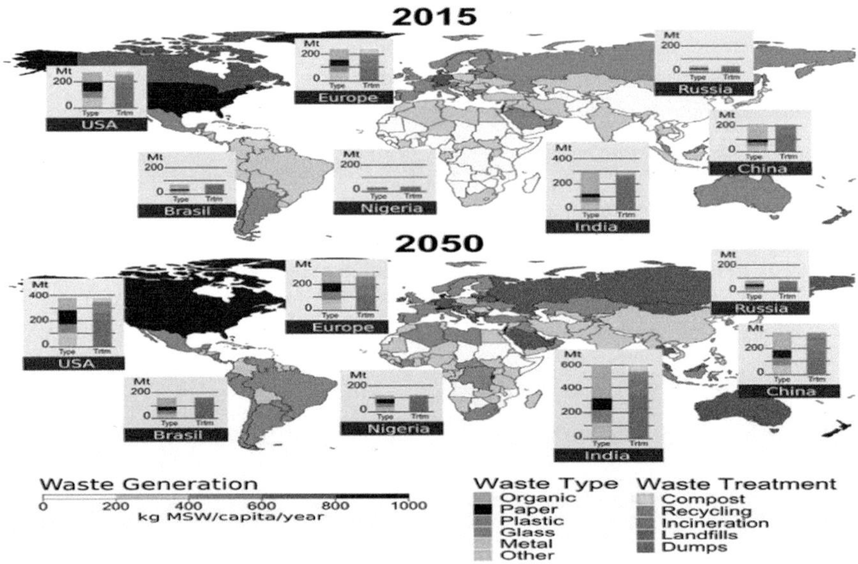

FIGURE 15.3 Waste generation (Singh and Anil, 2023).

Incineration with Energy Recovery: Burns municipal solid waste to generate electricity and heat.

Anaerobic Digestion: Converts organic waste into biogas for energy use.

Plasma Gasification: Uses high temperatures to break down waste into synthetic gas (syngas), which can be used for fuel or electricity generation (Oleskowicz-Popiel et al., 2009).

Pyrolysis: Thermally decomposes organic materials in the absence of oxygen to produce biofuels.

Treatment of waste is an important element of waste management and solves issues of environmental damage reduction, material recovery, and energy usage. This will ultimately require integrated waste management strategies in order to implement a sustainable transition towards a more circular economy driven by technological innovations. Physical, chemical, thermal, and biological treatment ways can be integrated in a holistic manner by the societies in waste management to deliver environmental protection as well as economic development.

15.4 CREATING VALUE TO THE WASTE

Recognition of waste as a resource has made stakeholders to come up with novel approaches for economic, social, and environment value creation. Businesses and governments can simultaneously work towards sustainability as well create new economic value by turning waste into something useful and unique products. This post looks at some approaches to extracting more value from waste, among them industrial symbiosis, models of circular economy, and valorization of waste.

15.4.1 Industrial Symbiosis: Enhancing Resource Efficiency

Industrial symbiosis refers to the collaboration between different industries, where waste from one industry serves as a raw material for another. This process connects waste generators with organizations that can utilize the waste as input, transforming disposal into a productive resource, benefiting both businesses while minimizing waste (Chertow and Ehrenfeld, 2018).

Successful examples of industrial symbiosis include:

- The Kalundborg Eco-Industrial Park (Denmark): One of the earliest models where multiple industries exchange waste heat and byproducts to reduce environmental impact and operational costs.
- Zhangjiagang Industrial Symbiosis Network (China): A large-scale initiative where steel, chemical, and energy enterprises collaboratively reuse industrial waste in an environmentally friendly manner.

By adopting industrial symbiosis strategies, industries can enhance sustainability, reduce costs, and mitigate environmental damage.

15.4.2 Circular Economy and Waste Valorization

A circular economy is the production of goods and taking out waste by designing for longevity, repair, remanufacture, and recycling of products. Circular economy in contrast to conventional linear model of "take, make and dispose" assumes a closed-loop system founded on material recycling and recovery. It reduces the consumption of natural resources and hence brings along ecofriendly as well as economical resiliency.

15.4.2.1 Key Components of a Circular Economy

Product life extension focuses on designing products that can be easily dismantled, modified, and maintained to prolong their lifespan. Modular designs enable the swap-out of parts, enhancing environmental sustainability. Additionally, leasing or take-back programs from manufacturers facilitate the refurbishment and resale of used products, reducing waste and promoting long-term resource efficiency.

Efficient material recovery and recycling systems ensure the collection of postconsumer materials such as plastics, metals, and textiles. Advanced chemical recycling technologies improve material purity and enable closed-loop recycling. extended producer responsibility (EPR) laws hold manufacturers accountable for the end-of-life disposal of their products, prioritizing recycling across various materials.

This approach involves creating high-value products from waste materials. Examples include biodegradable packaging made from agricultural residues and fashion items crafted from upcycled textiles. Industrial symbiosis, where waste from one industry serves as raw material for another, further reduces waste generation. Research into eco-efficient sustainable materials, such as bio-based plastics and composites, continues to drive innovation in this space.

15.4.2.2 Waste Valorization in the Circular Economy

Waste valorization is the first pillar of circular economy, meaning turning waste into value instead of throwing it out. Landfill diversion, resource conservation and support for sustainable industrial processes that is approach reduces the single use pathway.

In recent decades, the biorefinery concept has gained prominence, emphasizing the conversion of organic waste into biofuels, value-added chemicals, and fertilizers. This approach supports sustainable resource utilization by transforming waste into economically and environmentally beneficial products, reducing dependence on fossil fuels and minimizing waste disposal issues (Rajeev and Shukla, 2021).

Through anaerobic digestion processes, biogas is created from food waste and agricultural residuals; by-products derived from biomass drive key pharmaceuticals/professional cosmetic products.

E-Waste Recycling: Precious metals such as gold, silver, and palladium are extracted from discarded electronics and repurposed for new devices. Rare earth elements recovered from batteries and circuit boards contribute to the sustainable production of electronic goods. Urban mining techniques tap into valuable resources stored in end-of-life devices, reducing reliance on virgin resource extraction.

Construction Waste Repurpose: Recycling materials like concrete, glass, and metal from demolished buildings into new construction resources is known as construction waste repurposing. Demineralized concrete serves as an aggregate for road construction or new concrete production, while reclaimed wood and bricks support eco-friendly architectural designs with minimal environmental impact.

Utilization of Agricultural Waste: Agricultural waste is transformed into biochar for soil improvement and carbon sequestration through the pyrolysis of crop residues. By-products are processed into natural dyes, edible additives, and animal feed. Additionally, manure and organic waste are composted, while the resulting biogas is utilized to produce fertilizers and generate renewable energy, promoting soil health and sustainability.

Repurposing Textile Waste: Old clothes and textiles are recycled into fibers for new garments or insulation materials. Closed-loop recycling systems break down discarded fabrics using chemical processes and reform them into high-quality fibers. Meanwhile, circular fashion models encourage renting, reselling, and redesigning clothes, reducing fashion waste and fostering sustainable consumption.

15.4.2.3 Benefits of Circular Economy and Waste Valorization

Environmental Advantages: Some of the environmental advantages that a circular economy might offer are waste being minimized and pollution reduced out of landfill. Resource extraction and production process are reduced to a lower carbon footprint as a result of less using the materials. In addition, biodiversity is maintained through less land use conversion from habitat destruction of raw material extraction.

Economic Impacts: Proven new business models for the circular economy in recycling, remanufacturing and waste-flow based product innovation. Production costs are reduced with reclaimed materials rather than virgin resources. Moreover, creation of green jobs and eco-friendly supply chains are promoted (Rajeev and Shukla, 2021).

Social Benefits: Social benefits are aimed at the promotion of sustainable consumption through information and education. Establishing value-added from local waste streams widens the social benefits of local economies. Solution: Community involvement in Circular initiatives (repair cafes, share platforms, and zero waste).

By following the tenets of a circular economy and waste-to-value industries contribute not only towards more sustainable ways but also make sure that resources are used wisely, and environmental impacts are kept low-end.

15.4.3 Economic and Social Benefits of Waste-Based Enterprises

Waste to wealth: creating jobs, spurring innovation, and environmental stewardship transforming a sector the waste management and recycling industry into a big global industry with vast amounts of capital flowing in the race for sustainable business models. Redirecting waste to useful, environmentally friendly and often economically beneficial enterprises also serves to minimize loss in terms cost redevelopment and social enhancement.

15.4.3.1 Job Creation and Economic Growth

Among the greatest benefits of waste-based enterprises is employment potential at different skill levels one of them. From recycling and upcycling to waste management jobs are created in the areas of collection, sorting, processing, and manufacturing that contribute to both formal and informal employment markets. In the developing countries, a significant share of waste collection and recycling is typically undertaken by an informal sector, supplying livelihoods to millions of waste pickers and recyclers. Legalizing these jobs under some kind of policy framework will lead to higher wages, better working conditions and social security benefits. In developed countries advanced recycling technologies and circular economy actions ensure high-tech job opportunities for engineers, researchers, product developers. Beyond indirectly helping to keep the doors open for more waste-based jobs, it also helps local economies by unlocking entrepreneurship and innovation (Haas and Wijkman, 2021).

15.4.3.2 Cost Savings and Financial Efficiency

Waste-based businesses save businesses and consumers money through resource efficiency. Businesses that embed both waste recovery and recycling aspects into their operations save on the raw material costs with recycling using recycled rather than virgin inputs. This is a big help for industries like construction, manufacturing, and packaging where materials make up a large fraction of production expenses. Residents can also save money buying upcycled or refurbished products, which are frequently cheaper than brand new alternatives. Local governments save money on waste management and landfill disposal as well, so their public purse strings are not being stretched for wasteful services like parks or cleaning. Through waste-based investment, communities can implement a cost-effective model that sacrifices little to noneconomic burden and turbocharges resource productivity.

15.4.3.3 Innovation and Technological Advancements

As sustainability is placed more and more in the public spotlight, there is a proliferation in waste management technologies product innovation. Upcycling and value-added waste processing have enabled companies to innovate effective ways of recycling waste into new products, by transforming plastic [waste] to construction materials or creating biodegradables packaging from agricultural residues. Waste-to-energy methods (biogas production/plastic pyrolysis) convert the otherwise wasted material into usable sources of energy, reducing fossil dependency and increasing energy security. By implementing smart waste management systems using both AI and tracking via IoT, the waste collection routes became efficient and higher in sorting efficiency hence minimizing waste going to landfills. The wider the adoption of these new technologies also stimulates economic benefits and drives towards better performing more sustainable waste infrastructure.

15.4.3.4 Environmental Impact and Sustainability

Waste-based enterprises reduce pollution and save natural resources through the reuse of waste. When we divert waste from landfills the methane that is released in the landfill can be decreased, a very strong GHG contributor to climate change. Recycling metals, plastics and paper saves the earth from mining, deforestation, and water scarcity needed to make new materials from scratch, reducing natural ecosystems. It helps in the betterment of public health by preventing the soil and water bodies from hazardous materials causing an effective proper waste disposal. Also, nonlandfill reliance is to a great degree less contaminating considerably less air and water contamination. The incorporation of sustainable waste-based enterprises not only contributes towards the economic growth but also an ecosystem that is more advantageous for generation mostly (Lund, 2020).

15.4.3.5 Social Development and Community Well-Being

Waste-based enterprises have a major impact for social and community development providing livelihoods and improve life. Numerous waste collection and recycling programs give poor people reliable employment, entrepreneurialism opportunities as well as skill development. Enhanced waste management systems decrease the health risks associated with the unregulated dumping of waste, which contributes better sanitation and general well-being of public. Educational programs on waste segregation and recycling generate a culture of care within communities thus producing green habits in the society. Hence increases the prospects of a fair and socially responsible society, besides economic benefit to waste-based business (Tanikawa and Furumai, 2019).

15.4.3.6 Government Policies and Support for Sustainable Growth

Supportive polices and incentives are needed to drive waste-based business outputs into realty from governments end. Subsidies, tax benefits for recycling and waste that is thermo-chemical derived drive industry development. EPR mandates help in preventing waste by making manufacturers responsible for the disposal of their products, promoting a more sustainable design and practices with respect to circular economy. Large-scale jobs and authority may come via public–private partnerships (PPP) in waste management infrastructure by using the government resources with private sector capabilities. Governments that prioritize policies supporting waste-based businesses aiming at long-term economic sustainability while solving urgent environmental problems can be the enablers.

At the intersection of economic and environmental challenges is a powerful solution: waste-based enterprises. It leads the way for businesses and governments to prosper together while solving global challenges in waste management using novel waste valorization strategies. Well with more investment and supportive policy along with awareness of the public, waste can be made into wealth for human as well as environment.

15.5 SUCCESSFUL IMPLEMENTATION

Waste Upcycling and Proper Management: it takes a technology, regulatory back-up, collaborative industries, and community involvement both at scale. The use of more advanced technologies like AI sorting, bioconversion, and material recovery facilities (MRFs) to improve segregation and resource recovery advanced technologies in turn. Governments have a critical role to play including steering policy like the EU circular economy action plan and pushing EPR provisions through on a national level, incentivizing businesses to design for take back. Public–private partnerships, as the case of Adidas in their collaboration of ocean plastic footwear or TerraCycle upcycling program, are sole indicators of what circular economies may look like. Community engagement with campaigns, promotional recycling models, and community-delivered decentralized waste collection are also equally imperative. New materials science breakthroughs, proper blockchain waste trace in the markets and enduring public–private collaborations will need to overcome hurdles like high upfront costs of implementing solutions for upcycled products, lack of a market and enforcement fossil fuel policies.

15.5.1 Government Policies and Regulations

Effective legislation is foundational for encouraging waste segregation, recycling and upcycling the world over. Governments have put in place policies to shape responsible waste handling. EPR policies mean manufacturers are responsible for the waste their products generate so that it is disposed of correctly and recycled properly. Landfill levies and bans on single-use plastics combat reliance on waste reduction initiatives for landfill closure. So, for example, Germany and South Korea have very good waste segregation laws that make recycling and upcycling a lot more efficient.

15.5.2 Public–Private Partnerships (PPP)

Large-scale waste recovery projects by governments, industries, and nonprofits that cooperate with each other. Public–private partnerships (PPPs) are programs that will drive private sector investment in recycling infrastructure, through subsidies or other incentives to develop waste processing facilities by the governments. Offering tax incentives and funding from the government make it easier for companies to invest in upcycling projects in terms of sustainable way of manufacturing. Local municipal waste management solutions or working together between governments and corporates on building effective solutions for waste collection segregation and processing.

15.5.3 Technology and Innovation

Technological development is very important in making waste upcycle and management more efficient. AI Recycling Plants: These recycling plants are AI and machine learning powered in their approach, therefore ramping up waste sortation to the max while enabling proper recycling. With 3D printing on recycled plastic, it can be manufactured into different pieces and components. The transformation of waste to bricks, tiles, or concrete aggregate also leads to new markets in sustainable infrastructure.

15.5.4 Consumer Awareness and Behavioral Change

For the continuation of waste management projects, education of communities and their responsible use/consumption is a need. Some winning strategies are the educational ad campaigns about waste segregation, the relevance of upcycling and green living. The presence of incentives on waste reduction (i.e. rewards for consumers which opt for upcycled products and sustainable behaviors) make consumers to follow the path down a sustainable lane. Promoting consumers to make sustainable purchasing decisions initiates in a new mindset of circular economy.

15.5.5 Case Studies of Successful Implementation

Multiple countries and institutions have demonstrated what successful waste upcycling and management looks like, creating examples to be followed by future. For example, Sweden has turned over 99% of its waste into energy through waste-to-energy program, this way they use the rate lower and reduce landfill rate. Only 80% of the waste in Japan is circulated globally, and the country's Zero Waste Town Kamikatsu keeps a fair number way cutting edge 45-category sorting system to make this happen, but they are the world to say (Hotta and Nagata, 2023).

In the private sector, with innovative waste minimization and substitution trends companies are starting to look at new ways. Adidas collaborating with environmental organizations to develop shoes out of ocean plastic illustrates the clear trend into sustainable business models from a circular economy approach. Terra Cycle has also created pioneering upcycling initiatives for hard to recycle items, turning what would normally go to landfill into new products.

Both governments and NGOs are critical to this process as well. India has its EPR for manufacturers to manage postconsumer waste, for example, whereas countless organizations strive daily to remove plastic from our oceans and recreate it into viable products.

Ongoing challenges are, however, high upfront costs and logistical complexity that are combined with insufficient consumer awareness. The case studies nonetheless underpin the shift that can be catalyzed by combining technology, policy, and industry collaboration with community input aimed at waste management system sustainability and resource efficiency.

15.5.5.1 Sweden's Waste-to-Energy Model

Over 99% of waste in Sweden is either recycled or converted into energy, through the competent waste-to-energy (WTE) system in place across the country. More than 30 WTEs in operation, burning nonrecyclables to produce electricity and district heating for over 1 MSE Utah residences the highly efficient processing of waste in Sweden causes the country to import around 1.5 million tons waste from EU (especially the UK and Norway as well as Italy) every year allowing for the importation of less waste to be landfilled outside its borders but still satisfies energy demand needs. Rigorous environmental regulations and state-of-the-art filtration mechanisms guarantee that emissions from incineration are very far from EU. Sweden has drastically cut dependency on landfill, with less than 1% of its waste being sent to landfill.

15.5.5.2 The Netherlands' Circular Economy Plan

The Dutch government is aiming to move fully to a circular economy by 2050, with a 2030 interim target of reducing raw material extraction by 50%. Circular Economy Action Plan focuses on waste prevention, material recovery, and sustainable product design. The key sectors that Amsterdam Circular Innovation Program is targeting are construction, textiles, and organic waste valorization. As an example, in construction, the city requires the use of prefabricated components so that buildings are built namely with the intent of being re-used. Furthermore, the Netherlands supports textile recycling via programs such as "ReBlend" turning discarded clothes into quality yarn for new products. Moreover, places like Rotterdam incentivize the trend with biorefineries processing food waste into biofuels and bioplastics among other things (van der Horst–Verschelling, 2020).

15.5.5.3 Eco Bricks Initiative in the Philippines

Eco Bricks tackling plastic pollution by converting nonbiodegradable plastic waste into long-lasting low-cost building materials. Plastic bottles are densely mixed with other plastic waste, to produce robust bricks for schools, community centers, and houses in the slums. The initiative became a big hit at national level as NGOs, schools, and corporates are collaborating for plastic collection drives. While also creating clean energy, Eco Bricks empowers local communities with jobs primarily those of waste collectors and micro producers. It complies with the national level of the Philippines waste management policy such as Ecological Solid Waste Management Act in solving plastic waste due to creativity (Singh, 2017).

15.5.5.4 India's Waste Management Startups

India has been experiencing the rising waste crisis; more than 62 m tonnes of municipal solid waste per year is generated, and only around 30% is managed effectively. Startups Recircle and Saahas Zero Waste are entering this space to solve how waste can be managed end to end for enterprises, residential wings as well as municipalities. Recircle is a digital waste marketplace that matches the demand of waste generators to recyclers, to ensure efficient material recovery. Get waste management directly from the folk, that is decentralized waste management services: set up processing units for composting organic waste and recycling plastics both locally Saahas Zero Waste the Indian government is bolstering such programs through the Plastic Waste Management Rules, 2016 wherein EPR will be operating mandatorily manufacturers have to put recycled plastics in their supply chains. Such initiatives are slowly steaming India from a linear to circular economy by minimizing landfill waste and encouraging to consume responsibly (Gupta and Jain, 2019).

15.5.5.5 Japan's Zero Waste Town Kamikatsu

In Japan Tokushima Prefecture, a small town called Kamikatsu is practicing the zero-waste model declaring landfill waste to be zero and aiming for full continuous flow. They are very serious about waste sorting, with 45 identified categories of everything from plastics to metals, metals (not the kind you recycle), paper, and toothbrushes among others to maximize material recovery as residents. The town is running a "Kuru-Kuru", waste to reuse center so people can return used items rather than throwing. Composting of organic waste for use in the local farms further reduces adverse impacts on environment. This has allowed Kamikatsu to maintain an astounding 80% recycling rate of the waste, which is much higher than average in Japan (about 20%). The project also encouraged mass participation by allowing residents to be more engaged on keeping this system. This has led to other towns all over the world following suit who now use the same waste separation and reuse strategy as this town (Hotta and Nagata, 2023).

15.5.6 Challenges and Future Prospects

While there have been successes in both waste upcycling and strategies, there are many barriers to wider uptake. Some of which are—funding issues, lack of infrastructure, cynicism from consumers, and enforcement loopholes as well as technological gaps. These include financial barriers, lack of infrastructure, market uncertainty (a big problem since consumers do not direct resources), and a need for technology improvements. Solutions to these will come from efforts of government bodies, industries, and communities to generate a systemic waste management model that has sustainability and efficiency in its core.

High Capital Requirement: Most upcycling and recycling technologies are costly to start because they need considerable investment in research, infrastructure, or operational set-ups waste-to-energy plants. Advanced sorting systems and MRFs are very expensive making them out-of-the question for small businesses and many developing economies; Besides it, shifting to manufacturing models of sustainability will also entail there—invention of entire products and supply chains for which financial requirements are far larger. Governments and financial institutions need to intervene by parting with subsidies, low-interest loans, and tax credits for businesses that upcycle. PPPs likewise can be leveraged to share the financial risk and costs so more companies are incentivized to take a stab at waste upcycling.

Absence of Infrastructure: One of the biggest problems, especially in emerging countries, is that they do not have robust enough waste collection segregation and processing facilities. Open dumping and uncontrolled landfills continue to be the method of choice in many regions causing great environmental destruction and harming human health. If not for proper waste transportations and recycling facilities, recyclable materials often wind up in landfills or waters. Acknowledging this will entail substantial spending on new recycling plants, material recovery facilities, and composting plants. Decentralized systems—having processing units located nearer to the collection point to reduce transportation cost and emission should be the priority of governments. In addition, technology-enabled solutions like AI-based sorting, IoT for tracking waste, and blockchain-based waste traceability can operationalize presently leaky waste management systems.

Consumer Perception and Market Demand: Market demand is one of the biggest hurdles in the widespread popularity of upcycled products. As this continues, many people still are not advocating purchasing upcycled products due to concerns over the quality, durability, and aesthetics. It is pronounced in sectors like fashion where upcycled textiles are typically considered only equivalent virgin materials. Businesses need to build product design, brand, and certification programs for customers to have faith in upcycled products to solve this problem. Marketing should highlight the environmental gains, economic viability, and artisanal nature of upcycled items. States can add to this by implementing eco-labelling initiatives, mandating the procurement of sustainable products, or launching public awareness campaigns to encourage consumers towards circular economy routes.

Market and Policy Regulatory issues: There is a high degree of policy support for waste minimization and upcycling in many countries but the implementation and compliance; weak regulatory frameworks continue to be the biggest hurdles here. In a few corners of the world, the lack of strict penalties allows corporations to continue unsustainable waste disposal. The lack of uniform waste classification, recycling, and material reuse protocols also contributes to the inefficiency of cross border waste management. Here, governments need to ramp up EPR laws to enforce mandatory recycling targets and encourage companies producing sustainable production. It is also going to require larger global cooperation to fix the sewage of illegal waste exports and create a uniform sustainability standard throughout global supply chains.

Concluding Remarks: Though these constraints, the trajectory of waste prediction/upcycling and management still looks bright as technology improves, global awareness grows, and business models shift. Innovations in the early stages of development, such as biodegradable plastics, chemical recycling, AI sorting of waste, and upcycled construction materials, are beginning to define an industry. Massive investments in circular economy are being made globally by governments, and due ambitious zero-waste commitment from large corporations and closed-loop manufacturers. We expect the continued involvement of consumers at responsible waste separation to also lead to substantial progress on waste upcycling acceleration, supported by binding regulations and industry-led sustainability aims. If stakeholders choose to then mitigate infrastructure gaps, financial barriers, and consumer concerns, there is the potential for waste upcycling to be that foundation of a global economy built around sustainability.

15.6 CONCLUSION

Upcycling is one of the methodologies that can facilitate conversion of today's waste into valuable resources as an essential to getting society sustainability in modern day societies. This chapter elaborated on how there a move has been away from conventional waste disposal methods to the principles of a circular economy and illustrated the environmental, economic, and social benefits through waste management. Industries can minimize their ecological impact and create economic benefits by shaping waste into fuels, chemicals, and high-value products.

Particularly, successful waste-to-wealth initiatives that include industrial symbiosis waste-to-energy valorization and material recovery processes signified the contributions of sustainable waste management. This chapter presents case studies of different countries (Sweden, Japan, and others) to reveal the effectiveness of policy-driven and technology-based approaches to reduce landfill dependency and enhance resource efficiency. One can only consider these challenges as such high up-front cost and little infrastructure and market hesitancy to buy upcycled products to be though very real regulatory claw-back products, least financial inducement, and consumer education programs need to be put in place, nonetheless.

The upcoming decade sees the evolution of waste management in tandem with innovation, policy enforcement, and behavioral change practiced by government bodies, industries, and communities, as they work together to develop circular economy models. With the help of upcycling and sustainable waste treatment methods, societies not only greatly reduce environmental pollution but also hold the key toward natural resources conservation and long-term economic development in the coming future, thus ultimately leading towards a sustainable and waste free world development.

REFERENCES

Bocken, N. M. P., de Pauw, I., Bakker, C., and van der Grinten, B. (2020). Product design and business model strategies for a circular economy: An overview of the case studies. *Journal of Industrial Ecology,* 24(4), 875-887. https://doi.org/10.1111/jiec.13058

Chaturvedi, A., and Pandey, B. (2022). *Innovative waste upcycling: Circular economy perspectives.* Springer. https://doi.org/10.1007/978-981-16-4879-0

Chertow, M. R., and Ehrenfeld, J. R. (2018). Industrial symbiosis: A promising strategy for enhancing resource efficiency and sustainability. *Journal of Industrial Ecology*, 22(2), 347–359. https://doi.org/10.1111/jiec.12613

Fallon, C., and McDonagh, J. (2021). Creative upcycling: From waste to wealth and well-being in rural communities. *Journal of Rural Studies*, 88, 87–96. https://doi.org/10.1016/j.jrurstud.2021.08.010

Gupta, R., and Jain, M. (2019). Private sector startups and solid waste management in India: Their emerging role in unburdening the burden of formal and informal workforce. In Sadhan Kumar Ghosh, *Waste management as economic industry towards circular economy* (pp. 123–138). Springer

Haas, W., and Wijkman, A. (2021). Circular economy and waste-based businesses: Bridging the gap between waste management and economic opportunity. *Journal of Cleaner Production*, 267, 122018. https://doi.org/10.1016/j.jclepro.2020.122018

Hoornweg, D., and Bhada-Tata, P. (2022). *What a waste 2.0: A global snapshot of solid waste management to 2050 (urban development series).* World Bank Publications. https://openknowledge.worldbank.org/handle/10986/30317

Hotta, Y., and Nagata, Y. (2023). Kamikatsu's zero waste initiative: A model for sustainable waste management. *Journal of Environmental Sustainability*, 15(2), 123–135.

Huang, B., Wang, M., and Xu, Y. (2021). Circular economy and economic development: A new model for upcycling waste into business opportunities. *Journal of Cleaner Production*, 290, 125676. https://doi.org/10.1016/j.jclepro.2021.125676

Liu, G., Wei, W., and Li, Y. (2021). Circular economy practices in spare parts and equipment reuse: Challenges and opportunities in manufacturing industries. *Journal of Cleaner Production*, 280, 124216. https://doi.org/10.1016/j.jclepro.2020.124216

Lund, R. T. (2020). The role of waste-based enterprises in sustainable development: Economic and social impacts. *Sustainable Development*, 28(4), 923–933. https://doi.org/10.1002/sd.1995

Pacheco-Torgal, F., and Jalali, S. (2020). Reuse of industrial waste in construction: A sustainable approach to construction and infrastructure development. *Journal of Cleaner Production*, 274, 123054. https://doi.org/10.1016/j.jclepro.2020.123054

Rajeev, P., and Shukla, A. (2021). Waste valorization for sustainable development: Exploring the pathways of circular economy. *Environmental Science and Pollution Research*, 28(24), 31999–32017. https://doi.org/10.1007/s11356-021-15067-9

Rajendran, S., and Singh, A. (2022). Upcycling of solid waste: A systematic review of approaches, benefits, and challenges. *Waste Management*, 141, 45–56. https://doi.org/10.1016/j.wasman.2022.01.023

Rios, L. M., and Vassilev, S. V. (2020). Waste hierarchy and its role in sustainable waste management: A critical review. *Journal of Environmental Management*, 265, 110489. https://doi.org/10.1016/j.jenvman.2020.110489

Rosa, P., Sassanelli, C., and Terzi, S. (2020). Towards circular business models: A systematic literature review on upcycling in industry. *Sustainability*, 12(6), 2081. https://doi.org/10.1016/j.jclepro.2019.117696

Singh, A., and Anil, A. G. (2023). Carbon nanomaterials from waste: A sustainable approach for energy and environmental applications. *Journal of Environmental Chemical Engineering*, 11(2), 109667. https://doi.org/10.1016/j.jece.2023.109667

Singh, N. (2017). Investigating strength and properties of ecoladrillo: Eco Bricks. *Technology*, 8(7), 134–142.

Tanikawa, H., and Furumai, H. (2019). Economic and social impacts of recycling and waste management practices: A case study of waste-based enterprises. *Waste Management*, 92, 204–211. https://doi.org/10.1016/j.wasman.2019.05.013

van der Horst–Verschelling, J. (2020). A circular economy in 2050.

Yuan, Z., and Zhang, W. (2021). Innovations in waste recycling technologies: Advances in material recovery and energy production. *Renewable and Sustainable Energy Reviews*, 145, 111038. https://doi.org/10.1016/j.rser.2021.111038

Zaman, A. U., and Swapan, M. S. H. (2020). Circular economy and waste management: A review of the benefits and barriers. *Sustainable Production and Consumption*, 26, 232–246. https://doi.org/10.1016/j.spc.2020.10.005

Index